# CONDUCTING POLYMERS

Edited by **Faris Yilmaz**

**Conducting Polymers**
http://dx.doi.org/10.5772/61723
Edited by Faris Yilmaz

**Contributors**

Syed Abdul Moiz, Iqbal Ahmed Khan, Waheed Younis, Khasan Karimov, Rabin Bissessur, Barrit Lane, Alaa Abd-El-Aziz, Wael Alsaedi, Douglas Dahn, Emma McDermott, Andrew Martin, Saiful Izwan Abd Razak, Farah Nuuruljannah Dahli, Noor Fadzliana Ahmad Sharif, Izzati Fatimah Wahab, Nor Syuhada Azmi, Nadirul Hasraf Mat Nayan, Khalida Fakhruddin, Abdul Halim Mohd Yusof, Weina He, Xuetong Zhang, Gheorghe Batrinescu, Andréa Kalendová, Tereza Hájková, Jaroslav Stejskal, Cesar Sierra, Iskandar Saada, Hughes Matthieu, Anca Filimon, Lucas Fugikawa Santos, Jadwiga Soloducho

**Published by InTech**

Janeza Trdine 9, 51000 Rijeka, Croatia



**Publishing Process Manager** Edi Lipovic

**Technical Editor** SPi Global

**Cover** InTech Design team

First published September, 2016

Printed in Croatia

Additional hard copies can be obtained from orders@intechopen.com

Conducting Polymers, Edited by Faris Yilmaz

p. cm.

Print ISBN 978-953-51-2690-4
Online ISBN 978-953-51-2691-1

For EU product safety concerns:
IN TECH d.o.o., Prolaz Marije Krucifikse Kozulić 3, 51000 Rijeka, Croatia,
info@intechopen.com or visit our website at intechopen.com.

# Meet the editor

Dr. Faris Yilmaz was born in Palestine on October 7, 1969. After precollege education in Jenin, West Bank, he was accepted in the Chemical Engineering Department of Middle East Technical University (METU) in Ankara, Turkey. There he obtained his BSc degree in Chemical Engineering and Ms and PhD degrees in Polymer Science and Technology. His research has mainly concentrated on conducting polymers, their solution properties, composites, and nanocomposites. He has publications in different polymer journals. He is married and has two children.

# Contents

# Preface

Conducting polymers—organic conjugated materials are a fast-expanding area of research, part of the growing field of polymer technology. They offer high electrical conductivity through doping by oxidation and a wide range of unique electrochromic and electromechanical characteristics. The properties of conducting polymers depend not only on the properties of their individual states but also on their morphology and intrafacial characteristics.

It might be surprising to those who recognize that our modern lifestyle is dependent, to a large extent, on the use of organic polymers as thermal and electrical insulators and to learn that specific plastics may also be used as conductors of electricity. Besides demonstrating the versatility of polymers, their use as conductors will lead to developments that are not possible with other available materials of construction.

An authentic revolution took place in the area of solid-state chemistry and physics just after World War II. The century of solid state started from the modest beginnings of the transistor at Bell Laboratory. Since then, the area of science and technology has been directed primarily toward the study of alloys, ceramics, and inorganic semiconductors. The size of electronic devices became smaller and smaller, while the dimensionality of materials was also reduced just after the invention of the integrated circuit. It is at this point that the advent of the discovery of quasi one-dimensional conductors has opened up a whole new area of "nonclassical" solid-state chemistry and physics.

In the modern world, plastic and electrical devices are always tightly integrated together. However, it was in 1977 that an electrically conductive, quasi one-dimensional organic polymer, polyacetylene, was discovered. During the past 30 years, a variety of different conducting polymers have been developed. Excitement about these polymeric materials is evidenced by the fact that the field of conducting polymers has attracted scientists from such diverse areas of interest as synthetic chemistry, electrochemistry, solid-state physics, materials science, polymer science, electronics, and electrical engineering.

In the present text, we attempt to do justice to the different topics of conducting polymers and their applications. This text is generally suitable for researchers rather than students. In the first chapter, a discussion is done to illustrate a conductive polymer–based membrane. Conducting polymers in sensor design are illustrated in the second chapter. The third chapter is about conducting polymer aerogels. Also, coating of conducting polymers on natural cellulosic fibers is discussed in the fourth chapter. Space-charge–limited current in conducting polymers; perspectives of conductive polymers toward smart biomaterials for tissue engineering; electrical transport in electronic polymer devices; properties of metal oxide pigments surface modified with polyaniline phosphate and polypyrrole phosphate in corrosion protective organic coatings; exfoliated nanocomposites based on polyaniline and tungsten disulfide; phenylenevinylene systems, the oligomeric approach; and insertion of poly[bis-(methoxyethoxyethoxy) phosphazene] into lithium hectorite are discussed in the fifth, sixth, seventh, eight, ninth, tenth, and eleventh chapters, respectively. Finally, let us point that although many books in the field of conducting polymers appear, none of them are complementary.

**Dr. Faris Yilmaz**
FNSS Savunma Sistemleri A.Ş., Ankara
Turkey

Chapter 1

# Conductive Polymer-Based Membranes

Gheorghe Batrinescu, Lucian Alexandru Constantin,
Adriana Cuciureanu and Mirela Alina Constantin

Additional information is available at the end of the chapter

http://dx.doi.org/10.5772/63560

**Abstract**

This review focuses on an important theme of conductive polymer domain: preparation and applications of advanced materials with permselective properties, such as conductive polymer-based membranes. The most common groups of conductive polymers, their particularities, their use in membranes preparation together with main specific obtaining methods/techniques and conductive polymer-based membrane applications are presented based on a comprehensive documentary study.

**Keywords:** polymers, conductive polymers, membranes, membrane processes, conductive polymers membranes applications

## 1. Conductive polymers general data

### 1.1. Introduction

Polymers represent a larger class of organic compounds, in terms of both diversity and industrial-scale applications. They are used in the majority of industrial branches and have contributed decisively to the last century economic development. Polymers' impact upon the human society progress showed a positive effect, but nowadays, more attention is given to the mitigation of negative environmental impact induced by their intensive use. As a consequence of exponential growth of researches in the polymers field, the macromolecular compounds' chemistry became a distinct science within organic chemistry, a well-defined domain, in connection with other natural sciences areas. The turning point in the development of macromolecular compounds chemistry as an exact science was the transition from semi-empiric researches to rigorous experiments marked in 1925–1930 period by Staudinger and

Carothers researches [1–4] that led to highlighting of polymer structures, defining the notions of macromolecule, macromolecular chain, homologous polymeric series, etc.

Polymers led to rapid development of plastic materials science due to their specific properties: chemical stability on acids, bases and solvents (higher than gold and platinum in some cases), high thermal stability, elasticity, plasticity, excellent mechanical strength, non-permeability to gases, low density, electrical insulation properties and also electroconductive properties. The latest characteristics referring to electroconductivity was thoroughly studied during the last 60 years, leading to the development of a new polymer chemistry subdomain, and dedicated to conductive polymers. The results, reported within research works focussed on conductive polymers, materialized in two Nobel prizes in chemistry [5]:

- Rudolph A. Marcus (1992) for his works on electron transfer theory in redox processes, with direct applications in biopolymers science and
- Alan J. Heeger, Alan G. Mac Diarmid and Hideki Shirakawa (2000) for development of intrinsic electroconductive polymers through researches related to polyacetylene.

At the same time, with macromolecular compound chemistry, a new domain related to the preparation and use of membrane-type advanced materials was developed. Membrane science developed continuously as an interdisciplinary science in which polymers have a central role. With the progress of research works dedicated to conductive polymers, researches on conductive polymers membranes preparation and their application were also initiated and developed.

## 1.2. Conductive polymers classes, specific characteristics, preparation methods and general applications

Classification of conductive polymers can be done based on various criteria. The most important criterion is related to the electric charge movement type which depends on the polymer chemical structure. Thus, two groups of electronic conductive polymers are formed:

- Redox polymers: polymers that possess redox potential within their structure groups (reduction/oxidation capacity);
- Intrinsic electroconductive polymers: polymers with conjugated π-π or p-π systems.

In the case of redox polymers, the movement of electrons is realized through reversible chemical reactions "donor-acceptor" type, in accordance with Eq. (1):

$$O_x + ne^- \rightleftarrows Red \tag{1}$$

where "*n*" is the number of transferred electrons.

The standard potential ($E^0$) is determined by Eq. (2):

http://dx.doi.org/10.5772/63560

$$E^0 = -\Delta G^0/nF \tag{2}$$

where $F$ is the Faraday constant and $\Delta G^0$ is the standard Gibbs free energy, calculated in accordance with Eq. (3):

$$\Delta G^0 = \Delta G^0_{Red} - \Delta G^0_{Ox} \tag{3}$$

In the case of redox polymers, the essential condition needed for continuous electron transport, which defines the "electronic conductive polymer" property, is that groups with redox potential have to be distributed one next to the other within the macromolecular structure (its spatial configuration) in such a way that electron jump between groups is possible (**Figure 1**).

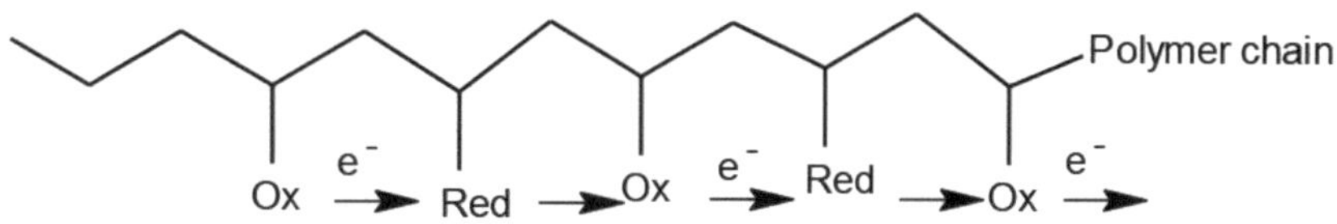


**Figure 1.** Electric charges movement within redox polymers.

Such polymers are those based on substituted nitro-styrene, quinones, dopamines and polymers which have within structure coordinative ligands based on Ir, Co, Re, Ru or Os [6].

The electrons transport principle of intrinsic electroconductive polymers consists of electrons transfer from π type bonds to nearby simple σ bonds, due to repulsion effect of same type charges.

The mandatory condition is that double bonds alternate with the simple ones (conjugated π-π systems), as shown in **Figure 2**.

**Figure 2.** Electric charges movement in "π-π" conjugated systems.

The same principle applies to the continuous transport of electrons in the case of polymers which contains heteroatoms N, S or O types within the macromolecular chain. Heteroatoms must be bonded to C atom that is involved in a double bond. Practically, non-participating p electrons of the heteroatom move to σ single bond, and through electrostatic repulsion effect, they further induce the movement of π electrons from the nearby double bond. That type of electrons transport is specific to conjugated p-π systems (**Figure 3**).

Intrinsic conductive polymers have the capacity to conduct electricity better than the majority of plastic materials which do not contain conjugated electron systems. For example, polyacetylene has a conductivity of $10^{-8}$–$10^{-7}$ S m$^{-1}$ in the form of "cis" isomer and $10^{-2}$–$10^{-3}$ S m$^{-1}$ in the form of "trans" isomer, compared with Teflon which has a conductivity of $10^{-16}$ S m$^{-1}$. However, compared with the conductivity of metals with best electric properties (Ag and Cu), which is of order $10^{8}$ S m$^{-1}$, the conductivity of these polymers is very low, having semiconductor properties. Through doping process, similar with that applied to classic inorganic semiconductors, the conductivity of intrinsic conductive polymers significantly grows, being closer to metals properties. Thus, polyacetylene doping with halogen vapour made (Cl, Br, I) the conductivity grow up to $10^{5}$ S m$^{-1}$ [5].

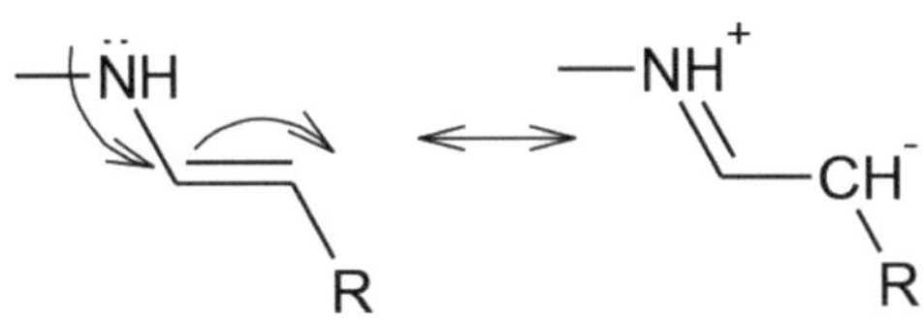


**Figure 3.** Electric charges movement in "p-π" conjugated systems.

Doping process involves the introduction of atoms capable of extracting or providing electrons to polymer's conjugated system, within its macromolecular chain. Introduction of such defects within the structure of macromolecule results in a more rapid jumping of electrons from created polar centres, assuring a better conductivity for the polymer. Usually, two distinct doping processes are applied [5]: oxidative doping (or p-doping) through which electrons are abstracted from the structure of polymer and reductive doping (or n-doping) through which electrons are introduced within the structure of polymer.

The two mentioned reactions are the processes for polyacetylene in Eqs (4) and (5):

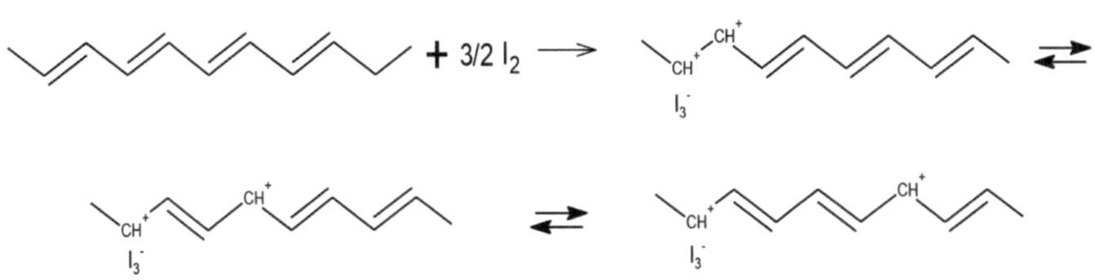


(4)

Oxidative doping

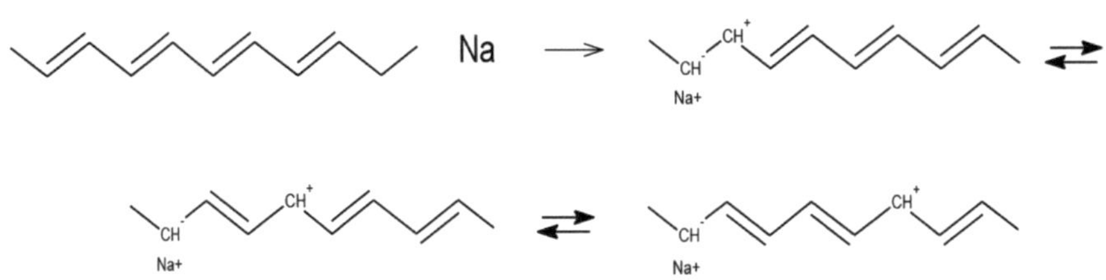


(5)

Reductive doping

In the first case, iodine is abstracting electrons from double bonds and in the second case, sodium atoms are releasing electrons to the double bonds bond resulting in their polarization. Both electric charges are migrating within the polymeric chain, adding the electric conductivity property, as shown in Eqs (4) and (5).

Doped intrinsic conductive polymers are obtained through one of the two main methods: chemical or electrochemical [7, 8]. Through chemical synthesis processes, monomers are polymerized/polycondensated by mixing with specific oxidative reagents such as ferric chloride or ammonium persulphate, at a specific pH. The method presents two main advantages compared with the electrochemical method: can be used to obtain polymeric powders or films that can be further processed and used at industrial level and can be used also in the synthesis of conductive polymers via electrochemical methods. Drawbacks of the methods are related to the fact that is highly sensitive, being dependent on reaction conditions (solvent nature, solvent and reagents purity, reagents' molar ratios, temperature, mixing mode and speed, reaction time, etc.). Electrochemical synthesis method is based on polymerization/polycondensation of monomers dissolved in a specific solvent through appliance of electricity between two electrodes. The monomer solution also contains the doping agent. Through this method, polymeric films (nanometers order thickness) with controlled/predefined structure are obtained. Electrochemical synthesis can be realized through three techniques: galvanostatic, potentiostatic and potentiodynamic [8]. The method presents the following disadvantages: limited polymer doping, conductive polymer quantity and polymeric film size are limited by electrode geometry and surface, difficulties in appliance for composite materials preparation (compared with the chemical synthesis).

| No. | Chemical name | Abbreviation |
|---|---|---|
| 1 | Polyacetylene | PAc |
| 2 | Polyaniline | PANI |
| 3 | Polypyrrole | PPy |
| 4 | Polythiophene | PTh |
| 5 | Poly(p-phenylene) | PPP |
| 6 | Polyazulene | PAZ |
| 7 | Polyfuran | PFu |
| 8 | Polyisopren | PIP |
| 9 | Polybutadiene | PBD |
| 10 | Poly(isothianaphtene) | PITN |
| 11 | Poly($\alpha$-naphthylamine) | PNA |

**Table 1.** A list of representative polymers and their abbreviations.

At the 2014 year level, more than 25 intrinsic conductive polymers [8] were known and used in various applications: electrochemical sensors [9–11]; gas sensors [12]; biosensors for medicine, food industry and environmental monitoring [13–15]; functionalized biomaterials with application in medicine [8, 16]; corrosion inhibitors [17]; fuel cells [18], etc. One of the main applications of conductive polymers is the development of optoelectronic devices based on electroluminescence phenomenon (field-effect transistors, FET; photodiodes; and light-emitting diodes, LEDs) [19]. The most representative polymers from those classes and their abbreviations are presented in **Table 1**.

For each basic polymer, a larger number of derivatives with electroconductive properties suitable for various applications were studied, and the research results were reported within the literature.

Many of the listed conductive polymers can be used for the preparation of polymeric membranes used in advanced separation processes, which is presented in the next section.

## 2. Membranes based on conductive polymers and their applications

### 2.1. General data on membranes and membrane processes

Membranes are advanced materials used for separation of compounds of sizes between 0.1 nm and 1 μm (from suspensions, dissolved macromolecules to simple or complex ions) from liquid and gaseous mixtures. It is difficult to find a membrane definition that covers simultaneously the issues related to its structure, separation mechanism and utilization domain. A generally accepted definition presented within the literature is as follows: *membrane is a selective barrier which actively or passively participates to the mass transfer between the phases separated by it* [20]. The membrane selectivity is determined by the material from which it is made, its structure (form, dimensions and pores distribution) and the force responsible for the separation process.

Membranes are classified based on material type and nature, structure and application domain. Based on these criteria, membranes are as follows:

- *Depending on material nature*: natural or synthetic;
- *Depending on material type*: polymeric or inorganic;
- *Depending on structure*: porous or dense (non-porous);
- *Depending on utilization domain*: microfiltration (MF), ultrafiltration (UF), nanofiltration (NF), reverse osmosis (RO), dialysis (D), electro-dialysis (ED), membrane distillation (MD), pervaporation (PV) and electro-osmosis (EO).

Considering the pores' form and distribution within the porous or non-porous membranes, they can be classified as

- *Symmetric structure:* straight or inclined cylindrical pores, uniformly distributed of mono-disperse microspherulites;

- *Asymmetric structure:* micropores with variable diameters forming a very thin layer 0.1–1 μm named active layer and non-regulated macropores forming the macroporous layer 100–200 μm;
- *Composite structure:* an active compact symmetric or asymmetric layer, an intermediary layer and a macroporous layer.

Depending on the separation surface, geometry membranes are plane, tubular hollow fibre type (inner diameter <0.5 mm), tubular capillary type (0.5 mm < inner diameter < 5 mm) and tubular type (inner diameter > 5 mm).

There are five main methods for membrane preparation: sintering, lamination, irradiation, phase inversion and deposition on thin layers [21]. Among them, phase inversion is the most commonly used method and is applied for membrane preparation at both laboratory and industrial level. Phase inversion concept was introduced in the literature by Kesting [22], and the concept implies transformation of one homogeneous polymeric solution in a two-phase system: one rich in polymer which forms the continuous part of porous membrane and other lacking polymer which fills the pores from membrane structure. The process has three main stages: polymer solubilization in a suitable solvent, skinning of polymer solution on a plane or tubular surface and polymer precipitation (phase inversion). The stage responsible for membrane structure is the precipitation one. Frequently, one of the following techniques for polymer transformation from liquid to solid phase is used: *vapour phase precipitation* [23], *controlled evaporation precipitation* [24], *thermal precipitation* [25] and *immersion precipitation* [26, 27]. Through phase inversion and immersion-precipitation techniques, membranes from various polymers are obtained: polysulphone, polyether sulphone, nylon 6,6, cellulose derivatives, polycarbonate, polyphenylenoxide, polyimides, polyamides, etc. These are used in both base form and modified (functionalized) form via various chemical reactions [28, 29].

Membrane processes are determined by the membrane structure and differentiated after the transport mechanism of chemical species through the membrane. Passing of one component from one side of the membrane to the other is defined as *permeation*, which is dependent on the driving force that generates it. After this criterion, membrane processes are classified as follows:

- Microfiltration, ultrafiltration and reverse osmosis (*driving force—pressure gradient*);
- Pervaporation, gas permeation, dialysis, separation through liquid membranes (*driving force—concentration gradient*);
- Thermo-osmosis and membrane distillation (*driving force—temperature gradient*);
- Electrodialysis and electro-osmosis (*driving force—electric potential gradient*).

Practical realization of membrane separation processes involves specific installations for each application field, in which the core element is represented by the equipment that contains the membrane, named *membrane module*. Depending on membrane shape there are two types of modules: *plane* and *tubular*. Plane membranes can be used in various geometries within modules similar to classic plate filters. Plane membranes can also be used within filter modules with both spiral and folded configurations. Tubular membranes can be used within tubular, capillary or hollow fibre modules.

Membranes and membrane process applications cover a vast area: treatment of water intended for human consumption from various sources (surface water including marine water, ground water), wastewater treatment, separation and concentration of proteins and enzymes from various natural and biosynthesis media, preparation of ultrapure water, development of medical devices (artificial kidneys, artificial lungs), development of sensors and biosensors with multiple applications, gas separations, fuel cells, ultrapure compounds via membrane distillation and pervaporation, etc.

Recent works in the field of membranes and membrane processes are focussing on obtaining cost reduction, improvement of separation characteristics (flow rate, selectivity, limitation of clogging) and extension of application domains at industrial scale [30]. Among these domains, the use of conductive polymer-based membranes is envisaged.

## 2.2. Membranes based on conductive polymers and their applications

Membranes based on conductive polymers represent a new class of advanced materials that can be used for separation or for interphase transfer processes of some chemical species based on their electrical properties. These membranes can be obtained through the previously described processes for preparation of classic membranes, their particularities being linked to polymers' "doping" methods in order to improve their conductive properties. Membranes of this type are obtained mainly from the polymers mentioned within **Table 1**, but considering the number of citations from the literature, the most used polymers are polyaniline (PANI) and polypyrrol (PPy). Apart from these polymers, during the last period, the use of functionalized polyetheretherketone (PEEK) as conductive polymer and of a considerable number of other polymers in fuel cells were thoroughly researched. Based on these facts, three main classes of conductive polymer-based membranes are presented. All the above-mentioned polymers have physical-chemical properties that do not allow the preparation of membranes using only the specific polymer but only in combination with other polymers. Such membranes usually lack conductive properties but have excellent mechanical strengths. Therefore, the majority of conductive polymer-based membranes are "composite membranes".

### *2.2.1. Membranes based on polyaniline (PANI) and their applications*

Polyaniline (PANI) is a macromolecular compound obtained through oxidative polymerization of aniline in accordance with Eq. (6):

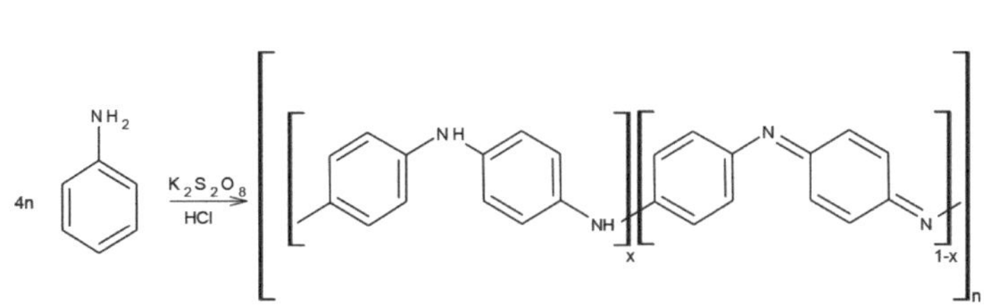


(6)

In its structure, there are structural units formed from benzene rings linked through aminic groups (-NH-) and structural units formed from a benzene ring linked with a quinone diimine. The ratio of the two structural units within polymeric chain is varying depending on its degree of oxidation. There are three particular structures of the polyaniline, which differentiate among them the function of the two structural units' ratio (**Figure 4**).

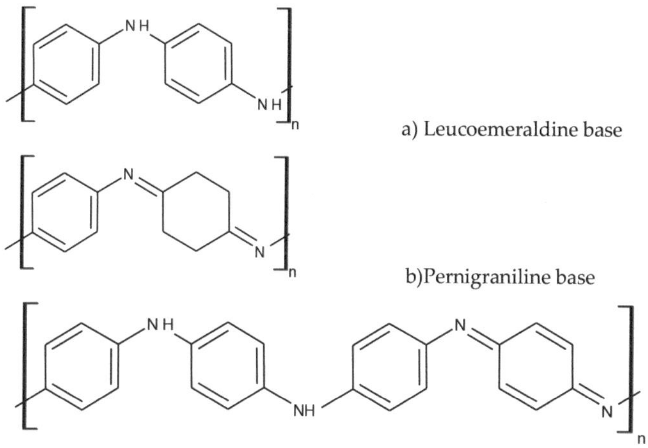


**Figure 4.** Structural PANI forms. (a) Leucoemeraldine base, (b) pernigraniline base and (c) emeraldine base.

Thus, in the limit case of PANI containing only structural units formed from benzene rings linked through aminic groups ($x = 1$), the compound is named *PANI-leucoemeraldine base* (**Figure 4a**), and in the limit case in which within the structure are only structural units formed from one benzene ring linked with a quinone diamine ($x = 0$), the compound is named *PANI-pernigraniline base* (**Figure 4b**). The third particular case is that in which within the macromolecular compound structure, the proportions of the two structural units are equal ($x = 0.5$), the compound being named *PANI-emeraldine base* (**Figure 4c**). Through treatment of these forms with acids or bases ("doping"), the polymer reversible passes from one form to other and gains electroconductive properties.

PANI in the base form has properties that do not allow obtaining simple membranes (low solubility in the majority of solvents commonly used for membranes preparation, low plasticity, thermal instability at temperatures above 160°C, etc.). For this reason, PANI-based membranes are obtained through blends with other polymer (usually chemically inert) suitable for membranes preparation. Composite membranes based on PANI with conductive properties are obtained. There are a large number of inert polymers used for PANI-based composite membranes preparation, the most used being cellulose and its derivatives, polysulphone, polystyrene and polypropylene.

PANI-based composite membranes are used in the majority of domains mentioned for conductive polymers, mainly for selective separation processes of some chemical species from complex liquid solutions, selective separation of gases, development of biosensors, electric and electronic devices (LED, photovoltaic cells), anticorrosive films and fabrication of antistatic textile materials.

The most recent scientific researches on preparation and specific applications of PANI-based composite membranes are presented in the following paragraphs.

Fibre-type cellulose was used in nanocomposites fabrication through oxidative in situ polymerization of aniline within fibres microstructure [31]. Polymerization was made in oxidative conditions using ammonium peroxydisulphate in hydrochloric acid aqueous solutions in which cellulose fibres impregnated with aniline are suspended. From the fibres separated at the end of the process, which contain PANI in their microporous structure, polymeric films were obtained via phase inversion process. Similarly, a composite material using nanofibrils bacterial cellulose as support material for PANI was obtained [32]. Research works showed the growth of PANI content within the composite material at the same time with the increase of its electric conductivity, with the prolonged reaction time from 30 to 90 min. Prolonging reaction time more than 90 min resulted in a decrease of electric conductivity due to aggregation of PANI particles and creation of discontinuities within nanocomposite structure. At the optimum time, a nanomaterial with the best conductivity of cca. 5.0 S $m^{-1}$ was obtained. Using this composite material, a flexible film (conductive membranes) that synergetically combines PANI conductive properties with mechanical strength (Young's modulus is 5.6 GPa and tensile strength is 95.7 MPa) provided by bacterial cellulose is obtained. The membranes obtained are applied in the field of electrochemical sensors, flexible electrodes and flexible displays. Compared with the processes in which first a nanocomposite material is obtained and then is used for conductive membrane preparation following a classic technique (phase inversion through immersion-precipitation technique), research works were conducted with the aim to obtain PANI-based composite conductive membranes following the next sequence: first, a semipermeable cellulose membrane is obtained and then on its surface, a thin layer of PANI is applied [33]. Deposition of the thin layer at membrane interface is realized via in situ polymerization of aniline with oxidative mixture containing ammonium peroxydisulphate in hydrochloric acid aqueous solutions. An aniline conversion of 80% was obtained after 24 h reaction time. At the end of the process, residual aniline was found on the active side of PANI membranes and secondary reaction products (ammonium hydrogen sulphate) obtained from ammonium peroxydisulphate were found on both sides of the membrane.

Cellulose esters are representing another class of polymeric materials used as support in preparation of PANI-based composite membranes. Investigations on PANI deposition on the surface of some microporous cellulose ester membranes were performed using two distinct techniques: deposition of PANI layer on membrane surface through in situ aniline polymerization in liquid phase or aniline polymerization in vapour phase [34, 35].

In situ polymerization of aniline in liquid phase was performed through dipping of one membrane from cellulose esters mixture in an aniline solution (PANI monomer) and $FeCl_3$ as oxidant [34]. In other experimental variant, aniline polymerization in liquid phase is made by immersion of pre-formed cellulose esters membrane in a solution that contains aniline and HCl followed by the addition (at a certain time) of oxidative agent such as ammonium

http://dx.doi.org/10.5772/63560

peroxydisulphate solution [35]. Aniline polymerization in vapour phase is done by soaking cellulose ester-based membrane in a solution of aniline and HCl and maintaining it in a closed tank with an oxidizing agent (ammonium peroxydisulphate and HCl solution)-saturated vapour atmosphere heated at 65–70°C [34, 35]. Cellulose acetate—PANI composite membranes with electric conductivities from $10^{-3}$ and 11 S $m^{-1}$ (using the liquid phase polymerization) [34] and respectively 98 $Sm^{-1}$ (using the vapour phase polymerization) [35] were obtained.

Polysulphone (Psf) is another frequently used polymer in the preparation of polymeric membranes used in membrane processes based on gradient pressure driving forces (microfiltration, ultrafiltration, reverse osmosis). This diversity of uses is due to polymers physical-chemical characteristics (chemical inertness, very good plasticity, excellent solubility in usual solvents used within phase inversion process, good mechanical strength, etc.). Psf is used as a substrate for the preparation of Psf–PANI composite membranes designed for advanced separation of compounds with polar groups from various mixtures, capitalizing the conductive properties of PANI from their structure.

Within the Psf – PANI composite membranes, PANI is present in the whole membrane's microporous structure, not only on surface. Preparation method is also a specific one and is differentiating from those already described. Thus, research works aimed to prepare Psf–PANI composite membranes through simultaneous formation of Psf-base membrane and aniline polymerization in oxidative conditions within membranes under formation pores [36]. Practically, the process consists of solubilization of Psf polymer within a specific solvent (N-methyl-pyrolidone or dimethylformamide) and aniline (PANI monomer), skinning of polymeric solution on a plane surface and immersion of polymeric film within an oxidative coagulation solution (ammonium peroxydisulphate and HCl). As the phase inversion process advances, Psf membrane is formed and within its pores PANI resulted from aniline polymerization in oxidative conditions.

Six types of composite membranes were obtained, using three polymeric solutions with 10, 12 and 14% Psf and two types of coagulants (distilled water and distilled water with 1.9% aniline). In all polymeric solutions, Psf was dissolved in a mixture of N-methylpyrrolidone and aniline. Obtained membranes were characterized from the point of view of flow and electroconductive properties through flows determination for solutions with variable pH (1, 3, 5, 7, 9 and 11), and selective separation properties were emphasized via determination of retention degree for standard proteins (albumin from bovine serum—BSA). BSA separation experiments proved that membranes obtained through coagulation from water and aniline solution present higher flows and retention degree compared to the membranes obtained by coagulation with only distilled water. For example, membranes obtained from 10% solution coagulated with water and aniline present a flow of 151.2 L/$m^2$h at pH = 4.9 and 196.3 L/$m^2$h at pH = 7.4 compared with membranes obtained from the same solution but coagulated in distilled water that showed a flow of 140.1 L/$m^2$h at pH = 4.9 and 189.4 L/$m^2$h at pH = 7.4. Retention degrees for membranes coagulated in water and aniline varied between 81.84 and 92.16% compared with 75.36–81.4% determined for membranes coagulated in distilled water.

Through this process composite membranes with conductive properties are obtained, having separation characteristics superior to those obtained using similar polymeric conditions that contains only Psf. Performed research emphasized the dependence of Psf/aniline ratio from polymeric solution and structural and hydrodynamic characteristics of Psf–PANI composite membranes.

In order to diminish the errors at laboratory level, mainly the manual skinning of polymeric solution and variation during the phase inversion process of the coagulation solution composition, Psf–PANI-based composite membrane preparation was studied in a steady-state installation [37]. That induces a modification of composite membrane preparation technology such that in one tank Psf membrane forming takes place. Psf membrane has in its pre-formed pores a certain quantity of aniline, and finalization of its structure is made in a reaction tank filled with oxidative mixture in which polymerization of aniline from pores occurs.

Using a 10% Psf solution (MW = 22,000 Da) dissolved in a mixture of N-methylpyrrolidone and aniline, membranes in a continuous system in the following working conditions were prepared: thickness of the polymeric film is equal to 0.2 mm, speed of the carrier in the tanks is equal to 1 m/min, temperature of the oxidative solution is equal to 25°C and reaction time is equal to 2 h. Characterization via distilled water flow for nine samples from the same membrane led to a maximum relative deviation of flow values of 2.45%, proving that through this approach, Psf–PANI composite membranes with reproducible hydrodynamic and conductive properties both in the entire surface and from one batch to other are obtained.

Other polymers studied as support materials in order to develop composite membranes based on PANI, with conductive polymers, are polystyrene [38, 39] and polypropylene [40]. Thus, from blends containing PANI and polystyrene in various ratios, dissolved in N-methyl-2-pyrrolidone, flexible polymeric films were obtained via phase inversion process, phase changing taking place through precipitation in vapour phase [38]. The particularity of the method consists in the fact that PANI was obtained in a separate oxidative polymerization process: a reaction media formed from aniline and alcoholic solution of 0.1 M $H_2SO_4$ in a volumetric ratio of 1/25 is cooled to −5°C; a solution containing ammonium peroxydisulphate as initiator is slowly added in a 2-h period within the reaction mixture; obtained PANI polymer is filtered and washed with acetone and a solution of 0.1 M $NH_4OH$ and mixed for 24 h; after that the polymer is filtered again, washed with distilled water and dried at 60°C for 24h. After drying, PANI and polystyrene are dissolved in N-methyl-2-pyrrolidone, the solution being coated on a support that is heated in an oven at 60°C for 24 h. Finally, the composite membrane obtained is removed from the support and subjected to doping process by immersing in a 5 M HCl solution for 9 min and dried afterwards [38].

Obtained polystyrene-PANI composite membranes present conductive properties depending on polystyrene/PANI ratio within the solution.

Polystyrene is used also in functionalized sulphonated form for the preparation of PANI-based composite materials [39]. Both polymers are obtained simultaneously in the same reaction environment consisting of aniline (PANI monomer), 4-styrene sulphonic acid sodium salt hydrate (sulphonated polystyrene monomer), ammonium peroxydisulphate and HCl (for oxidative polymerization).

The working procedure is the following: within a 3.47 mM aqueous solution of 4-styrene sulphonic acid sodium salt hydrate, heated at 80°C, the oxidant ammonium peroxydisulphate (aqueous solution 4.86 M) was added drop wise in a volumetric ration of 1/7 to styrene solution, under mixing for 1 h; after that an aqueous solution of aniline chlorhydrate 0.58 M (volumetric ratio of 1/8 to mixture) is added and after another 15 min a new quantity of 4-styrene sulphonic acid sodium salt hydrate (aqueous solution with a concentration of 0.72 M in a ratio of 1/9 to reaction media) was added drop wise; the reaction media is maintained under mixing at 80°C for 3 h, and afterwards the temperature drops to ambient temperature in a 24-h period. The composite polymer is precipitated in iso-propanol for 24 h without mixing and is washed with ethanol and dried at 70°C.

Obtained composite material contains in its structure PANI and sulphonated polystyrene macromolecular linked with chemical bonds (ionic bonds) through diimino protonated groups of PANI and sulphonic groups of sulphonated polystyrene. Due to these bonds, the conductive capacity of the composite material is lower compared with the above-presented composite materials, being practically a semiconductor. This composite material is used for polymeric membranes preparation through classic processes, dense films for antistatic packages, anticorrosive material or semiconductors.

Polypropylene is used as support material for preparation of PANI-based composite materials in the form of microporous membrane film through biaxial stretching technique [40]. PANI is formed within the polypropylene pores membrane through soaking of polymeric film in aniline followed by aniline oxidative polymerization using ammonium peroxydisulphate and HCl. Polypropylene—PANI composite membranes maintain the microporous structure of the support polymer, with various pores diameters and present conductive properties. These composite membranes are used in selective separation of chemical species from various liquid media through microfiltration, ultrafiltration, nanofiltration and reverse osmosis.

Besides polymeric materials, inorganics can be used as support for PANI-based composite materials. Thus, zeolites were used for new zeolite-PANI composite materials preparation due to their microporous structure and adsorption capacities [41]. Method consists of polymerization of aniline retained within zeolite matrix (Molecular Sieve 13X – SUPELCO Analytical) in oxidative conditions similar to those described previously (ammonium peroxydisulphate and HCl). The composite material is included within a quartz filter structure resulting in an inorganic-organic membrane with ultrafiltration flow rate properties. Conductive properties of these membranes (due to the PANI inclusion within their structure) allow separation of chemical species with high pollution potential from wastewater, such as heavy metals ions (Pb, Cu, Zn) and phenol derivatives (phenol, aminophenols or nitrophenols). Thus, retention degrees determined for heavy metal ions were 72.59% for $Pb^{2+}$, 87.48% for $Zn^{2+}$ and 99.55% for $Cu^{2+}$. Phenol derivatives from wastewater were removed using these composite materials with efficiencies above 98% (phenol, 98.15%; aminophenol, 99.78%; nitrophenol, 99.23%).

### *2.2.2. Polypyrrol-based membranes and their applications*

Polypyrrol (PPy) is a polymer obtained by pyrrol oxidation (Py) in the form of a black powder. This polymer has poor mechanical strength and thus low processibility. Its conductive properties in natural state are very low, and it is rapidly oxidized in contact with air, changing its properties. PPy conductivity is given by the existence within its structure of $\pi$ conjugated electron systems (from pyrrol ring) with p electrons available at N atom from pyrrol ring. This property is significantly improved by PPy "doping" with anions such as chloride, sulphate, perchlorate, dodecylsulphate and other organic compounds. Doped PPy is a polymer characterized by good chemical and thermal stability and better conductivity compared with other conductive polymers. Disadvantages related to PPy mechanical strength, plasticity and elasticity are improved both through doping process and by inclusion (as doped form, similar with PANI) within the polymeric and inorganic composite materials structure. PPy-based composite materials are frequently used as membranes within the processes in which driving force is no longer the pressure gradient but concentration gradient (gas separation from complex mixtures and pervaporation) and electric potential gradient (electro-dialysis).

During the last period, the research works related to PPy-based composite materials studies are focussed on their application in electro-analysis, medical field (systems for controlled release of drugs and use as biomaterial for artificial muscles) and antistatic and anticorrosive protection.

PPy polymerization in PPy-based composite materials is done through two methods: *polymerization through chemical oxidation* and *electrochemical polymerization* [42].

Ammonium peroxydisulphate, hydrogen peroxide and various compounds based on transitional metals salts ($Fe^{2+}$; $Cu^{2+}$; $Cr^{6+}$; $Mn^{2+}$; etc.) are frequently used as oxidative agents for the polymerization through chemical oxidation.

The process is presented in Eq. (7):

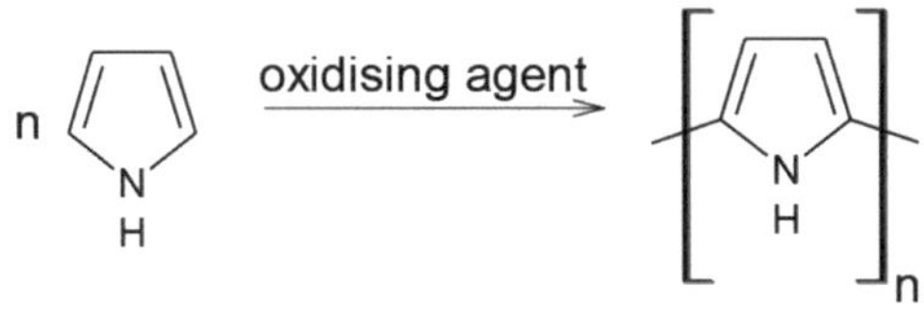


(7)

PPy polymeric chain can contain Py linked with three types of dimer sequences [8], presented in **Figure 5**.

Addition within the chemical oxidation reaction media of surfactants such as sodium dodecylbenzensulphonate and sodium alkylsulphonate and alkylnaphtalenesulphonate results in an increase of electric conductivity of the composite material and Py polymerization efficiency [42]. Electrochemical polymerization is performed in conditions mentioned within Section 1.2 based on application of an electric current between two electrodes immersed in a Py solution that contains also the dopant, in accordance with Eq. (8):

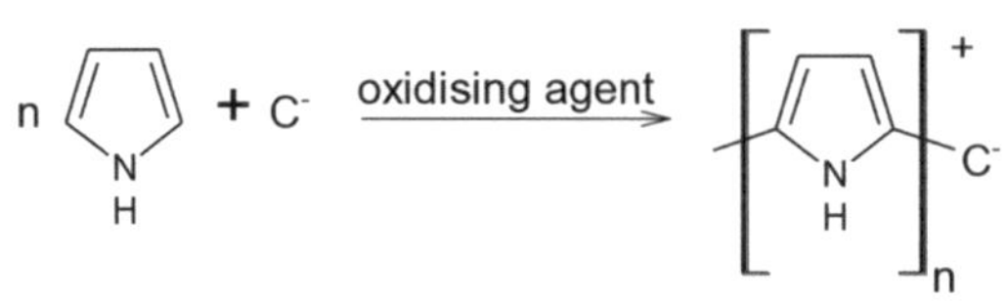


(8)

where $C^-$ is counterion.

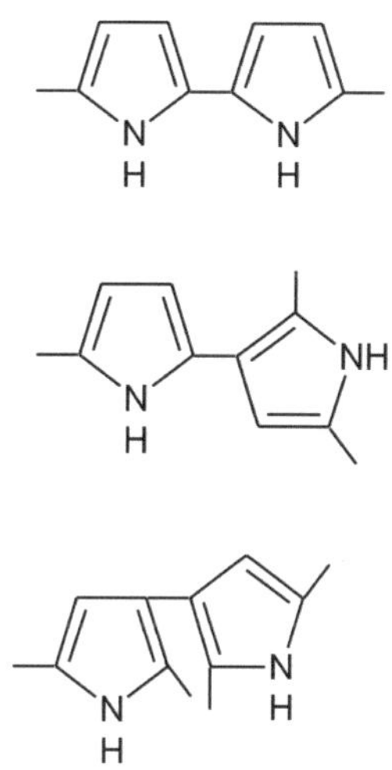

α,α′

α,β′

β,β′

**Figure 5.** Dimer sequences within PPy structure. (a) α,α′, (b) α,β′ and (c) β,β′

The Py chemical oxidative polymerization was applied to obtain membranes with high permeability for gas separation. Membranes were obtained through the technique of deposition in thin layers through interfacial polymerization [43, 44]. A freestanding polymeric film, 200–300 nm thick, was obtained by Py polymerization on an inert glass support through mixing of oxidative agent aqueous solutions containing ferrous chloride (0.4 M) and ferric chloride (0.5 M) with a solution of Py dissolved in an organic solvent (n-hexane) [43]. The membranes were prepared by pouring a solution of polydimethylsiloxane dissolved in n-hexane on PPy surface, after oxidant excess removal and polymeric film washing with methanol. The contact between components was maintained for 24 h and after that the composite membrane was annealed in air at 80°C for 15 min. Finally, the composite membrane is removed from the glass surface by simple washing with water. Obtained composite membranes present high selectivity for separation of oxygen and nitrogen from various mixtures, separation ratios $O_2/N_2$ of 17.2 being reported. The permeability for $O_2$ was 40.2 barrer. Using the same method of interfacial polymerization through chemical oxidation of Py or its derivatives (N-methylpyrrole), PPy-based conductive composite membranes with applications in the field of gas separation and pervaporation were obtained on a surface of microporous membrane supports [44]. Polymerization was performed at chamber temperature for 4 h using an aqueous 0.5 M Py solution and ferric chloride solutions (0.5, 1, 2 or 3M) as oxidative agent. The obtained membranes were washed with deionized water and stored in a 1 M HCl solution, and the operation was repeated daily for a week. In the final, membranes were stored in deionized water prior to characterization and use.

The Py chemical oxidative polymerization was used also for the preparation of PPy-based composite membranes having pre-formed membranes from sulphonated poly(styrene-co-divinylbenzene) as support, with its biotechnological applications and applications in wastewater treatment through electro-dialysis [45]. PPy polymer was formed through chemical oxidation within the structure of a commercial membrane with cationic exchange properties (Selemion CMT, manufactured by Asahi Glass Co.). The process consists of immersion of commercial membrane soaked in a ferric chloride oxidant solution in a Py aqueous solution at chamber temperature for about 4 h.

Similar properties related to ion exchange and possibility to apply in electro-dialysis also present PPy-based membranes obtained via Py chemical oxidative polymerization within the microporous structure of some inert polymeric supports [46] or inorganic supports [47]. In the first case [46], composite membrane is formed in one single stage through phase inversion techniques and chemical oxidation reactions. A solution of polysulphone and PPy dissolved in a N,N-dimethylformamide/methanol solvent system is coated on a plane surface and then immersed in a ferric chloride oxidant solution. At the same time, with formation of microporous polysulphone support, the chemical oxidative polymerization of Py within the preformed pores takes place. In the second case [47], the process consists of polymerization of adsorbed Py into silica through chemical oxidation, resulting in PPy inorganic-organic composite membranes.

PPy obtained through chemical oxidative polymerization is used as base material for medical devices [48, 49] due to its conductive properties. Py polymerization takes place in controlled conditions, a special attention being given to PPy doping and modification in order to be biocompatible. On PPy structure are engrafted biomolecules or cells (mammalian cells, endothelial cells, mesenchymal stem cells, etc.) through adsorption of covalent bonding [48]. Other studies emphasized the use of PPy incorporated in poly(ε-caprolactone) and gelatin nanofibres for cardiac tissues [49]. Performed experiments proved that increase of PPy concentration up to 30% within the composite material resulted in a reduction of average diameter of fibres from 239 ± 37 to 191 ± 45 nm, at the same time, with an increase of about six times of tensile modulus (7.9 ± 1.6 MPa initial and 50.3 ± 3.3 MPa after introduction of PPy).

Other applications of PPy-based composite membranes prepared through Py chemical oxidation is in the antistatic and anticorrosive materials field. Polyethylene used as natural [50] or modified [51] polymer constitutes an excellent polymeric material for the preparation of PPy composite membranes due to its properties (good mechanical strength, plasticity, elasticity) needed for membranes preparation. Thus, within the pores of polymeric films obtained through melt extrusion with subsequent annealing, uniaxial extension and thermal fixation, PPy was deposited through Py chemical oxidative polymerization [50]. Using another technique [51], polyethylene polymeric film is modified through engraftment within its structure of another polymer realized through acrylic acid irradiation with γ rays. Obtained material is highly hydrophilic which contributes to a better Py retention through surface adsorption that is further polymerized through chemical oxidation using iron chloride (III) or ammonium peroxydisulphate. Performed researches [51] were focussed on increase of electric conductivity of the composite material through increase of Py concentration within the reaction media from 0.3 to 0.9 M, but obtained results proved that this was insignificant (from 162 to 166 S $m^{-1}$). Introducing a new Py polymerization phase resulted in an increase of electric conductivity from 166 to 543 S $m^{-1}$.

http://dx.doi.org/10.5772/63560

PPy obtained via electrochemical method has similar application with those of PPy obtained through chemical oxidation.

Thus, PPy membrane formed on stainless steel net is used for separation of ethanol and cyclohexane mixtures through pervaporation [52]. Electrochemical polymerization takes place in a four-cell installation with porous glass walls and three electrodes. One of the electrodes is made from stainless steel net obtained through weaving of stainless steel fibre with 25 or 18 μm diameter. Within the cell that contains this electrode, a solution of 0.1M Py, acetonitrile as solvent and a doping agent ($Me_4NBF_4$ 0.05 M, $Bu_4NBF_4$ 0.1 M or $Bu_4NPF_6$ 0.1 M) are administered. Formed pervaporation membranes were tested in a specific installation and are proved to be selective for ethanol, permeation being dependent on Py polymerization degree and dopant type.

Using a similar method, electrochemical polymerized PPy deposited onto platinum sputter-coated polyvinylidene filters was obtained [53]. Process was realized through introduction within the cell containing filters of a Py aqueous solution containing as doping agents 8-hydroxyquinoline-5-sulphonic acid (HQS) or 2,9-dimethyl-4,7-diphenyl-1,10-phenanthroline disulphonic acid (BCS). PPy/BCS-type conductive membranes are permeable to a series of ions such as $Co^{2+}$, $Ni^{2+}$, $Zn^{2+}$, $K^+$, $Mg^{2+}$, $Ca^{2+}$, $Mn^{2+}$, $Fe^{3+}$ and $Cu^{2+}$. PPy/HQS conductive membranes are not permeable to all above-mentioned ions, significant results being obtained only for $K^+$, $Co^{2+}$ and $Cu^{2+}$. Moreover, ion fluxes are much higher (more than 10 times in case of $Cu^{2+}$) for PPy/BCS compared with PPy/HQS membranes. Conductive membranes with permeability for $Na^+$, $K^+$, $Ca^{2+}$ and $Mg^{2+}$ ions were obtained in similar conditions with those described above with the difference that PPy was deposited onto platinum sputter-coated polyvinylidene fluoride filters and polystyrenesulphonate/dodecylbenzenesulphonate (1%) or polyvinylphosphate/dodecylbenzenesulphonate systems were used for doping [54]. It was proved that ions transport fluxes are varying in the following order: $Na^+ > K^+ > Ca^{2+} > Mg^{2+}$.

Another domain that was thoroughly studied in the last period and uses PPy polymers obtained in the form of membrane films via electrochemical polymerization is that of electroanalysis. Through deposition of PPy membrane on the $Al_2O_3$ surface of one electrode, amperometric sensors with multiple uses in the field of analytical chemistry is obtained [55]. PPy membrane is prepared through electrochemical polymerization of Py on the electrode surface using an aqueous solution 0.2 M Py and 0.1 M KCl (dopant) or 0.1 M Py and 0.5 M $K_4[Fe(CN)_6]$ (dopant).

PPy based membranes applications within fuel cells domain should be mentioned also. The conductive polymers applications in the field of fuel cells are presented within the next section.

#### *2.2.3. Conductive polymers based membranes used for fuel cells*

Fuel cells are devices that generate electricity based on the free energy of a chemical reaction. A classical fuel cell consists of a porous anode fed with gas fuel that after oxidation led to electrons release; a porous cathode fed with oxidant, which generates protons and an electrolyte located between the two electrodes; and two bipolar plates and electric connectors that are linking electrodes through an exterior circuit. Chemical reactions for a classic combustion cell are presented in Eqs (9–11).

$$2H_2 + 4HO^- = 4H_2O + 4e^- \quad \text{Anode reaction} \tag{9}$$

$$O_2 + 2H_2O + 4e^- = 4HO^- \quad \text{Cathode reaction} \tag{10}$$

$$2H_2 + O_2 = 4H_2O + E + H \quad \text{Overall reaction} \tag{11}$$

where E = electrical energy and H = heat.

| Type of fuel cell | Operating temperature (°C) | Electrolyte | Reaction<br>At the anode (A)<br>At the cathode (C) |
|---|---|---|---|
| AFCs | 60–90 | KOH (liquid) | (A) $H_2 + 2HO^- = 2H_2O + 2e^-$<br>(C) $1/2O_2 + H_2O + 2e^- = 2HO^-$ |
| PEMFCs | 60–120 | Polymer–$SO_3H$ (solid) | (A) $H_2 = 2H^+ + 2e^-$<br>(C) $1/2O_2 + 2H^+ + 2e^- = H_2O$ |
| DMFCs | 60–120 | Polymer-$^+NR_3$ (liquid) | (A) $CH_3OH + 6HO^- = CO_2 + 5H_2O + 6e^-$<br>(C) $3/2O_2 + 3H_2O + 6e^- = 6HO^-$ |
| PAFCs | 160–220 | Phosphoric acid $H_3PO_4$ (liquid) | (A) $H_2 = 2H^+ + 2e^-$<br>(C) $1/2O_2 + 2H^+ + 2e^- = H_2O$ |
| MCFCs | 600–800 | Molten salt $Li_2CO_3/K_2CO_3$ (liquid) | (A) $H_2 + CO_3^{2-} = H_2O + CO_2 + 2e^-$<br>(C) $1/2O_2 + CO_2 + 2e^- = CO_3^{2-}$ |
| SOFCs | 800–1000 | Ceramic $ZrO_2/Y_2O_3$ (solid) | (A) $H_2 + O^{2-} = H_2O + 2e^-$<br>(C) $1/2O_2 + 2e^- = O^{2-}$ |

**Table 2.** Main combustion cell type's characteristics.

These devices present large spectra of applications due to the fact that global efficiencies obtained for electricity are higher than those of classical systems (thermic engines or hydro-electric turbines). At the same time, the effects induced upon the environment are less harmful than those produced by other electricity-producing systems such as fossil fuels burning.

Fuel cells can be classified based on two main criteria: electrolyte nature (charge carriers $HO^-$, $H^+$, $CO_3^{2-}$ or $O^{2-}$) and operation temperature. Based on the latter criterion, fuel cell types are: *low temperature*—alkaline fuel cells (AFCs—$T$ < 100°C), polymer electrolyte fuel cells (PEMFCs—*T* = 60–120°C), direct methanol fuel cells (DMFCs—T = 60–120°C), phosphonic acid fuel cells (PAFCs—*T* = 160–220°C) and *high temperature*—molten carbonate fuel cells (MCFCs—*T* = 600–800°C), solid oxide fuel cells (SOFCs—*T* = 800–1000°C).

**Table 2** is summarizing the main combustion cell types depending on operating temperature, electrolyte nature, anode (A) and cathode (C) reactions.

Conductive polymer membranes are thoroughly studied due to advantages offered by the fact that they function both as solid electrolytes and as selective separation barriers for the species implied in electricity generation within the fuel cells.

First applications were based on preparation and inclusion within the fuel cells structure of protons exchange membranes—Nafion, obtained from persulphonic acid and PTFE by Dupont Company 30 years ago. Nowadays researches are focussed on preparation of conductive polymers membranes with improved electric and mechanical properties. The most recent researches in preparation of both conductive membranes with protons exchange properties, applicable in PEMFCs, and conductive membranes applicable to AFCs are reviewed in the following paragraphs.

In a study dedicated to this domain [56] are emphasized the large number of composite membranes based on conductive polymers used for fabrication of high temperature proton exchange membrane fuel cells. Both organic composite membranes based on polymers with electric properties such as sulphonated poly (p-phenylene), sulphonated poly(ether ether ketone), sulfonated polysulfone, sulfonated poly (arylene ether sulfone), sulfonated poly(aryl ether ether nitrile), sulfonated poly(sulphide ketone), and organic-inorganic composite membranes such as fluorinated polymer/$SiO_2$, polyalkoxysilane/phosphotungstic acid, Nafion/PTFE/zirconium phosphate, Nafion/$TiO_2$ and Nafion/$SiO_2$ are reviewed.

One of the most studied polymers for organic composite membranes with applications in fuel cells is poly(ether ether ketone) (PEEK), which is used in base or modified form. Thus, from sulfonated poly(ether ether ketone) (SPEEK), asymmetric microporous membranes can be obtained via phase inversion method and immersion-precipitation technique. The SPEEK polymer was obtained by dissolving PEEK in concentrated $H_2SO_4$ added in a proportion of 5 wt%, at room temperature, reaction media being maintained by mixing for 24 h. From the resulted solution, a membrane was prepared by coating on a plane glass surface, followed by immersion in a coagulation bath containing distilled water (phase inversion method, immersion-precipitation technique). The obtained membrane is modified through in situ polymerization (within membrane pores) of Py doped with iron chloride and cerium sulphate [57]. Electroconductive properties of PPy from composite membrane structure (ionic conductivity of 0.34 S $m^{-1}$) made the membrane suitable for fuel cells.

Composite membranes are obtained, using SPEEK as base material, through inclusion within its structure of heteropolycompounds based on tungsten, molybdenum or wolfram [58]. Obtained composite membranes present a conductivity of 1 S $m^{-1}$ at chamber temperature and 10 S $m^{-1}$ at 100°C temperature, being used in PEMFC-type fuel cells.

Other studies on SPPEK applications for fuel cells showed that it can be functionalized with quaternary amine hydroxide and imidazolium hydroxide [59] (resulting membranes with $10^{-3}$ S $m^{-1}$ conductivity) or by PANI inclusion within membrane structure [60].

Polysulfone [61] and poly(1,4-phenylene ether ether sulfone) [62] are other polymeric materials that can be used to obtain composite membranes with conductive properties for fabrication of fuel cells. Using polysulfone, asymmetric membranes can be obtained via classical process of phase inversion and afterwards functionalized through incorporating acrylamide-based ionomers having proton-conducting sulphonics groups. Incorporation process of the new polymer is based on photopolymerization [61]. At the surface of membranes prepared from poly(1,4-phenylene ether ether sulfone) modified through addition of tungstophosphoric acid, a layer of PPy polymer is applied via chemical oxidation, resulting in composite membranes used for fabrication of DMFC [62]-type fuel cells.

Fuel cells protons exchange composite membranes, which can be used at temperature above 100°C, were obtained from poly(2,6-dimethyl-1,4-phenylene oxide), N-(3-aminopropyl)-imidazole and metal – organic frameworks [63].

Composite membranes with applications at high temperatures, for fuel cells, are obtained also from bi-functionalized copolymer prepared through radical copolymerization, having $SiO_2$ [64] within in its structure.

Another point of interest within the literature is represented by preparation of membranes for alkaline fuel cells. Studies performed in this domain [65, 66] classify conductive membranes for preparation of AFCs in heterogeneous and homogeneous membranes, each with their specific polymers and preparation methods. One of the most recent researches related to membranes for alkaline fuel cells is focussing on preparation of high ionic conductivity membrane from crosslinked poly(arylene ether sulphones) [67].

Besides traditional methods for preparation of polymeric membranes with conductive polymers (phase inversion, lamination, irradiation, etc.), a new technique was recently developed—plasma techniques both for plasma polymerization and for plasma modification of membrane surfaces [68]. Proton exchange membrane for PEMFCs and membranes for alkaline fuel cells (AFCs) can be obtained using this technique.

## 3. Conclusions

Conductive polymers themselves are not forming membranes that can be used in various processes due to low mechanical strength, lack of elasticity and plasticity. For this reason, conductive polymer-based membranes are mainly composite membranes.

Preparation methods of composite membranes based on conductive polymers are similar to those used for simple membranes (sintering, lamination, irradiation, phase inversion, deposition on thin layers) for the formation of support polymer that confers mechanical strength and elasticity. Chemical oxidation polymerization and electrochemical polymerization are used for the inclusion within the support structure of a polymer with conductive properties. There are three ways for preparation of composite membranes presented within the literature:

- Conductive polymer is formed at the same time with support membrane; in this case, the composite membrane contains the conductive polymer in all its structure;

- Conductive polymer is formed after support membrane preparation and its soaking in monomer solution followed by polymerization through chemical oxidation; in this case, composite membrane contains conductive polymer in all microporous structure;

- Conductive polymer is formed only after preparation of support membrane through deposition on its surface of conductive polymeric film; in this case, composite membrane contains in its structure two different layers—sandwich type.

Conductive properties of polymers that contain conjugated electrons systems (п-п or p-п) are low compared with metals, being at the level of semiconductors. In order to obtain polymers with better conductive properties, "doping" technique is used, through introduction within polymeric chain of atoms or groups of atoms that creates "defects" within macromolecule structure as a result having more rapid "jumping" of electrons between polar centres. Doping process takes place simultaneously with conductive polymer formation within the composite membrane structure, through dopants addition within monomer solution. Doping process takes place with conductive polymer preparation in some rare cases.

Conductive polymer-based composite membranes are used in membrane processes that generally use concentration gradient (pervaporation and gas separation) and electric potential gradient (electro-dialysis) as driving forces. There are also cases in which these membranes are used in processes that uses pressure gradient (MF, UF, RO). Many researches are focussed on conductive polymer-based composite membranes' use for fuel cells.

## Author details

Gheorghe Batrinescu*, Lucian Alexandru Constantin, Adriana Cuciureanu and Mirela Alina Constantin

*Address all correspondence to: batrinescugh@gmail.com

National Research and Development Institute for Industrial Ecology – ECOIND, Bucharest, Romania

## References

[1] H. Staudinger. Uber die makromolekulare Chemie. (English: On macromolecular chemistry). Angewandte Chemie.1936; 49: 801–813.

[2] W.H. Carothers. Studies on polymerization and ring formation. I. An introduction to the general theory of condensation polymer. Journal of American Chemical Society. 1929;51(8):2548-2559.

[3] W.H. Carothers. Polymerization. Chemical Reviews. 1931;8(3):353-426.

[4] W.H. Carothers. Collected Papers of Wallacce Hume Carothers on High Polymers Substances. New York: H. Mark & G.S. Whitby; 1940.

[5] The Royal Swedish Academy of Science, Retrieved on http://www.nobelprize.org/nobel_prizes/chemistry/ accessed: January 16, 2016

[6] R. Seeber, F. Terzi, C. Zanardi. Functional materials in amperometric sensing, In Redox Polymers and Metallopolymers, Springer Berlin Heidelberg. 2014. p. 59–97

[7] K. Kaneto. Chapter 7. In: K. Asaka & H. Okuzaki, editors. Conducting Polymers. Japan: Springer; 2014. DOI: 10.1007/978-4-54767-9_7

[8] R. Balint, N.H. Cassidy, S.H. Cartmell. Conductive polymers: towards a smart biomaterial for tissue engineering. Acta Biomaterialia. 2014;10:2341-2353.

[9] J. Janata, M. Josowicz. Conducting polymers in electronic chemical sensors. Nature Materials. 2003;2(1):19-24.

[10] C. Li, H. Bai, G. Shi. Conducting polymer nanomaterials: electrosynthesis and applications. Chemical Society Reviews. 2009;38(8):2397-3409.

[11] X. Ma, M. Gao, G. Li. Morphology tailoring of nano/micro-structured conductive polymers, composites and their applications in chemical sensors. Recent Patents on Nanotechnology. 2010;4(3):150-163.

[12] F. Selampinar, L. Toppare, U. Akbulut, T. Yalcin, S. Suzer. A conductive composite of polypyrrole II. As a gas sensor. Synthetic Metals. 1995;68(2):109-116.

[13] M. Gerard, A. Chaubey, B.D. Malhotra. Application of conducting polymers to biosensors. Biosensors and Bioelectronics. 2002;17(5):345-359.

[14] H. Yoon. Current trends in sensors based on conducting polymers nanomaterials. Nanomaterials. 2013;3:524-549.

[15] K. Arshak, V. Velusamy, O. Korostynska, K. Oliwa-Stasiak, C. Adley. Conducting polymers and their application to biosensors: emphasizing on foodborne pathogen detection. IEEE Sensor Journal. 2010;9(12):1942-1951.

[16] M. Muskovich, C.J. Bettinger. Biomaterials-based electronics: polymers and interfaces for biology and medicine. Advanced Healthcare Materials. 2012;1(3):248-266.

[17] N. Elhalawany, M.A. Mosad, M.K. Zahran. Novel water based coatings containing some conducting polymers nanoparticles (CPNs) as corrosion inhibitors. Progress in Organic Coatings. 2014;77(3):725-732.

[18] J. Peron, Z. Shi, S. Holdcroft. Hydrocarbon proton conducting polymers for fuel cell catalyst layers. Energy & Environmental Science. 2011;4:1575-1591.

[19] L. Persano, A. Camposeo, D. Pisignano. Active polymer nanofibers for photonics, electronics, energy generation and micromechanics. Progress in Polymer Science. 2015;43:48-95.

[20] M.C. Porter. Handbook of Industrial Membrane Technology. Westwood, New Jersey, USA: Noyies Publications; 1990.

[21] M.H.V. Mulder. Basic Principles of Membrane Technology. Dordrecht: Kluver Academic Publishers; 1991.

[22] R.E. Kesting. Synthetic Polymeric Membranes. New York: McGraw-Hill; 1971.

[23] R.J. Kay, W.B. Krantz, R.L. Sani. Linear stability theory model for finger formation in asymmetric membranes. Journal of Membrane Science. 1985;23(2):155-182.

[24] F.W. Altena, C.A. Smolders. Calculation of liquid-liquid phase separation in a ternary system of a polymer in a mixture of a solvent and a nonsolvent. Macromolecules. 1982;15(6):1491-1497.

[25] J.G. Wijmans, J.P.B. Baaij, C.A. Smolders. The mechanism of formation of microporous or skinned membranes produced by immersion precipitation. Journal of Membrane Science. 1983;14(3):263-274.

[26] G. Nechifor, N. Luca, G. Popescu, M. Nechifor. Preparation of membranes by immersion-precipitation method. Revue Roumaine de Chemie. 1989;34(11–12):2047-2056.

[27] G. Nechifor, G. Popescu. Asymmetric membranes prepared by immersion-precipitation technique. Revue Roumanine de Chemie. 1990;35(7–9):899-908.

[28] D. Garganciuc, G. Batrinescu, G. Nechifor, M. Olteanu. Functional polymers of polysulphone and polyphenylenoxide type used to obtain affinity membranes. Materiale Plastice. 2008;45(1):29-33.

[29] D. Garganciuc, G. Batrinescu, O. Popa, M. Olteanu. Preparation of some affinity membranes from functional polymers based on polysulphone and brominated polyphenylenoxide. Materiale Plastice. 2008;45(2):167-171.

[30] H.S. Muralidhara. Challenges of membrane technology in the XXI century. In: Z.F. Cui, H.S. Muralidhara, editors. Membrane Technology. Oxford: Elsevier; 2010. p. 19-32.

[31] L.H. Mattoso, E.S. Medeiros, D.A. Baker, J. Avloni, D.F. Wood, W.J. Orts. Ellectrically conductive nanocomposite made from cellulose nanofibrils and polyaniline. Journal of Nanoscience and Nanotechnology. 2009;9(5):2017-2022.

[32] W. Hu, S. Chen, Z. Yang, L. Liu, H. Wang. Flexible electrically conductive nanocomposite membrane based on bacterial cellulose and polyaniline. Journal of Physical Chemistry B. 2011;115(26):8453-8457.

[33] N.V. Blinova, J. Stejskal, M. Trchova, G. Cirici-Marjanovic, I. Sapurina. Polymerization of aniline on aniline membranes. Journal of Physical Chemistry B. 2007;111(10): 2440-2448.

[34] A.A. Qaiser, M.M. Hyland, D.A. Patterson. Effects of various polymerization techniques on PANI deposition at the surface of cellulose ester microporous membrane: XPS and electrical studies. Synthetic Metals. 2012;162(11–12):958-967.

[35] M.A. Shehzad, A.A. Qaiser, A. Javaid, F. Saeed. In situ solution-phase polymerization and chemical vapor deposition of polyaniline on microporous cellulose ester membranes: AFM and electrical conductivity studies. Synthetic Metals. 2015;200:164-171.

[36] A. Cuciureanu, G. Batrinescu, N.N. Badea, D.A. Radu. The influence of changing the polyaniline and polysulphone ratio on composite Psf–PANI membranes performances. Materiale Plastice. 2010;47(4):416-420.

[37] G. Batrinescu, M.A. Constantin, A. Cuciureanu, G. Nechifor. Polysulfone-polyaniline type membranes obtained in a steady state system: structural and hydrodynamic characteristics. Polymer Engineering and Science. 2014;54(7):1640-1647.

[38] I. S. Guimaraes, A.A. Hidalgo, H.N. Cunha, L.M. Santos, J.A.V. Santos, J.R. Santos Jr. Thermal and morphological characterization of conducting polyaniline/polystyrene blends. Synthetic Metals. 2012;162(7):705-709.

[39] M.A. Moussa, M.H. Abdel Rehim, Sh.A. Khalry, M.A. Soliman, A.M. Ghonaim, G.M. Turky. Electrical investigations of polyaniline/sulfonated polystyrene composite using broadband dielectric spectroscopy. Synthetic Metals. 2015;209:34-40.

[40] J. Yang, J. Hou, W. Zhu, M. Xu, M. Wan. Substituted polyaniline-polypropylene film composite: preparation and properties. Synthetic Metals. 1996;80(3):283-289.

[41] A. Ivan, D.L. Ghindeanu, V. Danciulescu, A. Raducu, A.C. Nechifor. Composite polyaniline-zeolite membrane material for wastewater ultrafiltration. Optoelectronics and Advanced Materials – Rapid Communication. 2012;6(11–12):1134-1138.

[42] L.X. Wang, X.G. Li, Y.L. Yang. Preparation, properties and applications of polypyrroles. Reactive and Functional Polymers. 2001;47:125-139.

[43] W.I. Son, J.M. Hong, B.S. Ki. Polypyrrole composite membrane with high permeability prepared by interfacial polymerization. Korean Journal of Chemical Engineering. 2005;22(2):285-290.

[44] C.R. Martin, W. Liang, V. Menon, R. Parthasarathy, A. Parthasarathy. Electronically conductive polymers as chemically-selective layers for membrane-based separations. Synthetic Metals. 1993;57(1):3766-3773.

[45] R. Scherer, A.M. Bernaders, M.M.C. Forte, J.Z. Ferreira, C.A. Ferreira. Preparation and physical characterization of a sulfonated poly(styrene-co-divinylbenzene)

and polypyrrole composite membrane. Materials Chemistry and Physics. 2001;71:131-136.

[46] C. Muscalu, R. David, S.A. Garea, A.C. Nechifor, D.I. Vaireanu, S.I. Voicu, G. Nechifor. Polysulfone-polypyrrole ionic conductive composite membranes synthesized by phase inversion with chemical reaction. Semiconductor Conference, CAS 2009 International. 2009;2:557-560.

[47] C. Baicea, A. Ivan, C. Trisca-Rusu, A.C. Nachifor, D.I. Vaireanu, S.I. Voicu, G. Nechifor. Ionic conductive silica-polypyrrole composite obtained by in-situ polymerization. Semiconductor Conference, CAS 2010 International, Sinaia, Romania. 2010;2:359-362.

[48] Z.B. Huang, G.F. Yin, X.M. Liao, J.W. Gu. Conducting polypyrrole in tissue engineering applications. Frontiers of Materials Science. 2014;8(1): 39-45.

[49] D. Kai, M.P. Prabhakaran, G. Jin, S. Ramakrishna. Polypyrrole-contained electrospun conductive nanofibrous membranes for cardiac tissue engineering. Journal of Biomedicals Materials Research A. 2011;99(3):376-385.

[50] G.K. Elyashevich, I.S. Kuryndin, E.Yu. Rosova. Composite membranes with conducting polymer microtubules as new electroactive and transport systems. Polymers for Advanced Technologies. 2002;13:725-736.

[51] N.S. Ilicheva, N.K. Kitaeva, V.R. Duflot, V.I. Kabanova. Synthesis and properties of electroconductive polymeric composite material based on polypyrrole. ISRN Polymer Science. 2012; Article ID 320316, 7 pages.

[52] M. Zhou, M. Persin, W. Kujawski, J. Sarrazin. Electrochemical preparation of polypyrrole membranes and their application in ethanol-ciclohexane separation by pervaporation. Journal of Membranes Science. 1995;108:89-96.

[53] V. Misoska, J. Ding, J.M. Davey, W.E. Price, S.F. Ralph. Polypyrrole membranes containing chelating ligands synthesis, characterization and transport studies. Polymer. 2001;42(21):8571-8579.

[54] J.M. Davey, S.F. Ralph, C.O. Too, G.G. Wallace. Electrochemically controlled transport of metal ions across polypyrrole membranes using a flow-through cell. Reactive and Functional Polymers. 2001;49(2):87-98.

[55] A. Michalska, S. Walkiewicz, K. Maksymiuk. Amperometric ion sensing using polypyrrole membranes. Electroanalysis. 2003;15(5–6): 509-517.

[56] S. Bose, T. Kuila, T.X.H. Nguyen, N.H. Kim, K. Lau, J.H. Lee. Polymer membranes for high temperature proton exchange membrane fuel cell: recent advances and challenges. Progress in Polymer Science. 2011;36:813-843.

[57] C. Baicea, A.C. Nechifor, D.I. Vaireanu, O. Gales, R. Trusca, S.I. Voicu. Sulfonated poly (ether ether ketone) – activated polypyrrole composite membranes for fuel cells.

Optoelectronics and Advanced Materials-Rapid Communications. 2011;5(11): 1181-1185.

[58] S.M.J. Zaidi, S.D. Mikhailenko, G.P. Robertson, M.D. Guiver, S. Kaliaguine. Proton conducting composite membranes from polyether ether ketone and heteropolyacids for fuel cell applications. Journal of Membranes Science. 2000;173(1):17-34.

[59] X. Yan, G. He, X. Wu, J. Benziger. Ion and water transport in functionalized PEEK membranes. Journal of Membrane Science. 2013;429:13-22.

[60] R.K. Nagarale, G.S. Gohil, V.K. Shahl. Sulfonated poly(ether ether ketone)/polyaniline composite proton-exchange membrane. Journal of Membrane Science. 2006;280(1–2): 389-396.

[61] P. Radonavic, M. Kellner, J. Matovic, R. Liska, T. Koch. Asymmetric membranes with interpenetrating proton-conducting morphology made by a combination of immersion precipitation and photopolymerization. Journal of Membrane Science. 2012;401–402: 254-261.

[62] S. Neelakandan, P. Kanagaraj, R.M. Sabarathinam, A. Nagendran. Polypyrrole layered SPEES/TPA proton exchange membrane for direct methanol fuel cells. Applied Surface Science. 2015;359:272-279.

[63] B. Wu, L. Ge, X. Lin, L. Wu, J. Luo, T. Xu. Immobilization of N-(3-aminopropyl)-imidazole through MOFs in proton conductive membrane for elevated temperature anhydrous applications. Journal of Membrane Science. 2014;458:86-95.

[64] A.K. Thakur, M. Manohar, V.K. Shahi. Bi-functionalized copolymer-sulphonated $SiO_2$ embedded with aprotic ionic liquid based anhydrous proton conducting membrane for high temperature application. Journal of Membrane Science. 2015;490:266-274.

[65] G. Merle, M. Wessling, K. Nijmeijer. Anion exchange membranes for alkaline fuel cells: a review. Journal of Membrane Science. 2011;377:1-35.

[66] G. Couture, A. Alaaeddine, F. Boschet, B. Ameduri. Polymeric materials as anion-exchange membranes for alkaline fuel cells. Progress in Polymer Science. 2011;36:1521-1557.

[67] Y.Z. Zhuo, A.N. Lai, Q. G. Zhang, A.M. Zhu, M.L. Ye, Q.L. Liu. Highly ionic-conductive crosslinked cardo poly(arylene ether sulfones) as anion exchange membranes for alkaline fuel cells. Journal of Membrane Science. 2015;491:138-148.

[68] Z. Jiang, Z.J. Jiang. Plasma techniques for the fabrication of polymer electrolyte membranes for fuel cells. Journal of Membrane Science. 2014;456:85-106.

# Chapter 2

# Conducting Polymers in Sensor Design

Jadwiga Sołoducho and Joanna Cabaj

Additional information is available at the end of the chapter

http://dx.doi.org/10.5772/63227

**Abstract**

Conducting polymers (CPs) as well as conducting polymer nanoparticles seem to be very applicable for the development of various analyte-recognizing elements of sensors and biosensors. This chapter reviews mainly fabrication methods as well as application of conducting polymers in sensors. Conducting polymers (CPs) have been applied in the design of catalytic and affinity biosensors as immobilization matrixes, signal transduction systems, and even analyte-recognizing components. Various types of conducting and electrochemically generated polymer-based electrochemical sensors were developed including amperometric catalytic and potentiodynamic affinity sensors. A very specific interaction of analyte with immobilized biological element results in the formation of reaction products.

**Keywords:** conducting polymers, conducting polymer nanoparticles, fabrication of conducting polymer nanomaterials, sensors, biosensors

## 1. Introduction

A high number of nanostructured materials including conducting polymers (CPs) become extremely essential in sensor and biosensor design. Conducting polymers as artificial convenient materials seem to be very applicable for the development of different analyte-recognizing parts of sensors and biosensors.

Conducting polymers (CPs) have been applied in the design of catalytic and affinity biosensors as immobilization matrixes, signal transduction systems, and even analyte recognizing components. Different types of conducting and electrochemically generated polymer-based electrochemical sensors were developed including amperometric catalytic and potentiodynamic affinity sensors. A very specific interaction of analyte with immobilized biological element results in the formation of reaction products.

Sensors based on conducting polymer materials present a series of effective features depending on the conducting materials as well as on the conducting polymer fabrication/modification methods. Differences of CPs were broadly utilized in the generation of proper polymers. There are clear signs that the capability of conducting polymers to transfer electrical carrier will be exploited in the fabrication of molecular tools.

Due to the sensor implementations, the most important features of CPs are presented as: (i) CPs are readily synthesized by electrochemical and chemical processes; (ii) CPs are strongly sensitive to a wide range of analytes at ambient conditions (i.e. temperature); (iii) a response is expected from their inherent transport properties (i.e. electrical conductivity, rate of energy transfer); (iv) the polymer structure can be modified to display selective responses toward specific analyte; (v) the sensitivity is readily tunable by adjusting the synthetic variables such as incorporated counter ions or polymerization temperature; and (vi) sensor array may be formed through electrochemical deposition, which permits for miniaturization and mass production of sensor tools [1].

Conducting polymer materials of nanosizes have emerged as significant elements of high-performance transducer applications. They possess unique properties associated with high surface area, small dimensions, and bulky counterparts [2]. Conducting polymer nanomaterials have been employed to detect diverse analytes such as toxic gases, volatile organic compounds, as well as biological species. Well known from literature are sensors based on conducting polymer nanomaterial, consisting of polypyrrole (PPy), polyaniline (PANI), poly-(3,4-ethylenodioxythiophene) (PEDOT) [1].

This contribution is reviewing principally fabrication methods of conducting polymer nanomaterials for biosensorics. Major types of biosensors based on conducting polymer nanoparticles including catalytic biosensors and molecularly imprinted polymer-based affinity sensors are presented.

## 2. Fabrication of conducting polymer nanomaterials

Conducting polymer nanomaterials are potentially useful for the fabrication of miniaturized sensors that enable small sample volumes, portability, and high-density arrays. Nanostructured CPs are favorable for electron transfer and biomolecule stabilization due to their increased surface area and high surface-free-energy. Nanostructured CPs, such as polypyrrole, polyaniline, and polythiophene (PTh), offer excellent prospects for both interfacing biological recognition species and transducing electronic signals to design novel bioelectronic devices [3].

Recently, many procedures for their fabrication with controlled shape and size have been developed. These ranged from lithographic techniques to chemical methods [4]. Moreover, the utilization of nanosize particles with narrow size distribution meets the demands for transparent and flexible sensor platforms.

There are a variety of interesting and useful characteristics associated with chemical approach, for example, template synthesis. Probably the most useful feature of template synthesis is that it is a general procedure with respect to the types of materials that can be prepared as well as templates that can be involved. Depending on the properties of monomers to be polymerized, the chemistry of the template, and their mutual interactions, that is, whether the monomer interacts electrostatically or whether it is chemically bound to the template, different micro- and nanostructures can be formed. The use of porous membranes with uniform pores yields monodisperse nanocylinders or nanorods, whose dimensions are carefully controlled by the pore size. These tubular or fibrillar structures can remain within the template or they can be extracted from the membrane. Porous materials in which polymer structures are grown can be considered as "encapsulating" materials aimed at the isolation of the functional, conducting micro- or nanostructure from the surrounding.

The synthetic strategies for conducting polymer nanomaterials can be classified into hard-template synthesis, soft-template synthesis, and template-free synthesis [5].

Hard-template synthesis is favorable in tailoring the dimensions of nanomaterials. It is often used in the production of nanosize materials such as nanotubes, nanorods, and nanocapsules. This method has been used to prepare conducting polymer nanomaterials as colloidal nanoparticles, nanofibers, and porous membranes [6]. An easy route to polypyrrole (PPy) nanotubes has been developed with an alumina membrane template. These nanotubes were produced by vapor deposition polymerization coupled with template synthesis. Significantly, the wall thickness of the structures was controlled within the range of a few nanometers [7]. This accurate control on the wall thickness of nanotube indicated that the template-mediated procedure was a precise and effective route for producing polymer nanotubes.

The soft-template approach has occurred as an alternative route to fabricate polymer nanomaterials with a good result. Soft template concerns surfactants, polyelectrolytes, liquid crystals, block copolymers, as well as biomolecules [8]. One of the examples of using this method for obtaining polymer nanomaterials was the formation of cylindrical micelles of bis(2-ethylhexyl)sulfosuccinate, through a co-interaction with iron cations in a solvent of low polarity [9].

Template-free synthesis is very accessible without using some specific templates. But this procedure is restricted to individual precursor material. For example, the template-free synthesis of polyaniline (PANI) nanofibers is systematically reported since the first report of Huang et al. [10]. A dispersion polymerization for the mass production of polyaniline nanorods was also presented by Jang's group [1].

Other templating method depends on the use of self-assembled monolayers (SAMs). These layers provide an opportunity to define their chemical functionality and applications with molecular precision and become increasingly valuable tools in the design of surface-confined synthetic routes to complex structures. The use of self-assembled monolayers with suitable functional groups protruding toward the monomer-containing solution can lead—via the suitable synthetic method—to covalent or electrostatic attachment of monomers to these functionalities. Consecutive polymerization results in one-dimensional (1D) or two-dimensional (2D) polymeric structures (brushes, "wires", surfaces), depending on the polymerization method [11, 12].

### 2.1. Electrochemical template method—polymeric native films

Here we present some applications of monomolecular setups for templated growth of polymeric materials. The thicker films on the electrode can also be used to govern the polymer growth. One of benefits of this procedure is that the thicker layers (i.e. insulating polymers) may block more effectively than monomolecular layers the access of the monomers to the electrode surface. In this situation, the polymer can be grown within defects. There are several works on deposition of conducting polymers onto thicker coating layers, but this results in the formation of composite structures where the coating layer acts as a support rather than a template [13, 14].

An interesting strategy of using polymeric coatings to prepare polypyrrole nanosize structures was reported by Jerome et al. [15–18]. This approach requires the use of an insulating polyethyl acrylate layer deposited on the glassy carbon (GC) electrode onto which pyrrole is electrochemically oxidized. When a suitable solvent for polyethyl acrylate is utilized, the insulating matrix is swollen and incorporates monomer molecules. The oxidized form of monomer creates within the coating some amount of nuclei that in the next step grow to create channels in the layer. These channels permit the PPy to occur from the polyethyl acrylate film in the form of nanowires. Despite the fact that different insulating layers might be used to template nanowire creation (ca. 350-4000 nm in diameter and hundreds of micrometers in length), films of polyethyl acrylate appear to give the best results [4].

The critical condition for nanowire preparation is the use of solvents with high donor numbers which are known to impart limited conductivity to PPy [19]. As a consequence, the monomer oxidation does not appear at the PPy nanowire, but simply between the electrode surface and foot of the growing polypyrrole wire, due to which the already created nanostructure is pushed away from the electrode.

In solvents with low donicity, PPy nanowires are not created due to the high conductivity of the resulting polymer that grows in cauliflower-like shape as an alternative [4]. Wang et al. [20] have proposed a little different route for fabrication of polypyrrole nanowires. The nanowires were fabricated by voltammetric or potentiostatic oxidation of pyrrole on GC electrode coated with a thin film of paraffin. It was investigated clearly that the nucleation process of PPy is of a 2D type, while the growth of the structure is a 1D process.

The 2D polymeric surfaces modified biosensor electrodes were also close to our earlier experiences [21, 22]. There were published sensing tools, where platinum electrode was modified with polymers obtained in the process of electrochemical polymerization. An example is thin film (550 nm) built of 3-methylthiophene/3-thiopheneacetic acid/2,7-bis[2-(3,4-ethylenedioxythiophene)]-*N*-nonylacridone [21]. The electropolymerized layer exists in the conducting oxidized state. The overall charge of the polymer was neutral due to the doping anions which are incorporated into the polymeric matrix during the electropolymerization. The optimal current density for copolymer deposition was observed as 12 $mA/cm^2$, and the polymerization time 1 min [21]. CPs porous film was used as suitable matrix for enzyme immobilization. The immobilization of laccase was performed in a one-step process. It was carried out for 10 min under galvanostatic conditions applying a current density of 3 $mA/cm^2$.

The optimal conditions presented in the experimental part were obtained considering a high signal-to-noise ratio, and stability of biosensor response [21].

Another CP (poly(*N*-hexyl-2,7-bis(thiophene) acridone)) was exploited by us in ceramic optical biosensor designed for the permanent monitoring of water solutions, **Figure 1** [22].

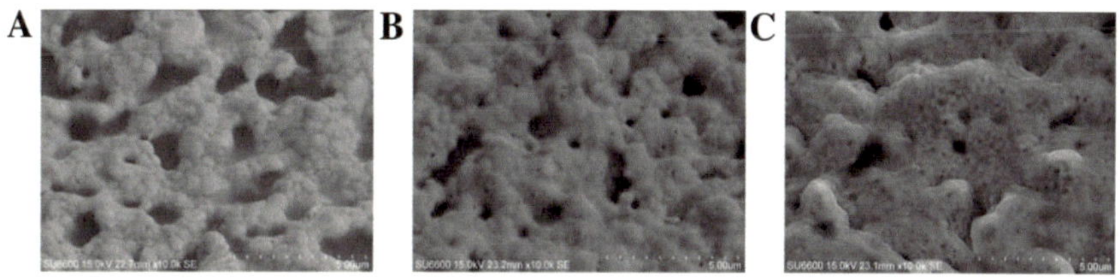


**Figure 1.** SEM images of: A—platinum bare electrode, B—electrode modified with polymer, C—electrode modified with polymer and laccase. Adapted with permission from Jędrychowska et al. [22].

### 2.2. Nanoparticle monolayers

A smart method of preparation of well-ordered honeycomb polymeric structures was reported by Han et al. [23]. They used monolayers of polystyrene nanoparticles supported on gold to template PANI growth into interstitial voids of the colloid. After dissolution of the template, the products form a negative pattern reflecting the ordered geometry of the particle monolayer (**Figure 2**). The colloidal assembly of polystyrene particles (diameter 600 nm) was prepared by sandwiching a gold substrate between two Teflon elements and adding dispersed polystyrene particles into the solution well of the top block.

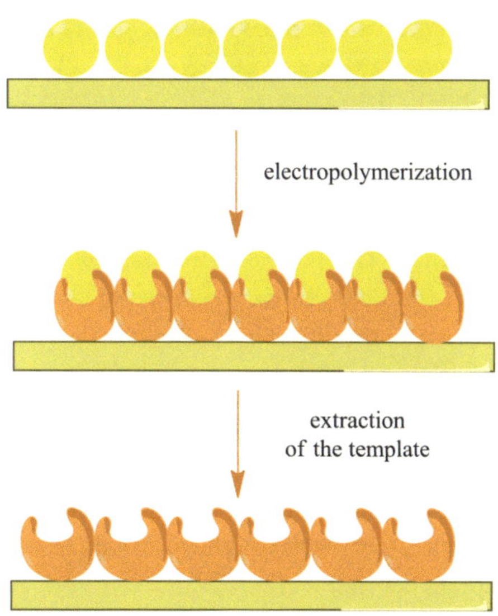


**Figure 2.** Procedure of formation honeycomb structures with nanoparticle monolayer as template.

The polymerization was performed by potential cycling of the electrode in the acidic solution of aniline. After the process of polymerization the deposits were placed in toluene to dissolve the polystyrene template. The above procedure yields a well-ordered 2D polyaniline honeycomb structure on the gold surface. It was found that the rigidity of polyaniline structure and long-distance order is greatly improved when polystyrene nanoparticles are pre-coated with polyelectrolyte multilayers (polydiallyldimethyl-ammonium chloride and polysodium-4-styrene-sulfonate).

Polyelectrolyte assemblies may also be used to create polymeric models. Briseno et al. also reported [24] truncated eggshell structures of the polyelectrolyte to template the PANI growth. The method involves assembling a film of polystyrene particles covered with a few polyelectrolyte layers onto a gold area. Extraction of the polystyrene templates results in the rupture of the apexes of the spherical polyelectrolyte shells and an expansion of the shells along the surface. The resulting structure consists of truncated eggshell features and retains the hexagonal arrangement of the original polystyrene particles. The polyelectrolyte shells are prepared by alternate assembling of cationic and anionic layers. The outermost layer is anionic and therefore it can assemble cationic anilinium cations from the solution. It was found that aniline is adsorbed onto eggshell structures and infiltrates into the void spaces within the polyelectrolyte multilayers. After immobilization the monomers can be polymerized by electrochemical oxidation in acidic water solution yielding composite polyaniline-polyelectrolyte truncated eggshell assembly.

### 2.3. Nanoparticle 3D orders

Another approach yielding 3D conducting polymer assemblies was reported by Sumida et al. [25]. This pathway is essentially the same as presented earlier for nanoparticle monolayers. The main dissimilarity is that the particles create 3D lattice rather than a monomolecular layer, consequently the resulting polymer is 3D. The synthetic pathway engages the use of colloidal particles generating "crystalline" lattice organized on the electrode surface. The required polymeric structure is settled onto the template which yields 3D network of the polymer. The main duty of the particles is to mechanically reduce the growth of the polymer to interstitial voids. The polymer creates interconnected macropores linked to each other via a symmetrical network of smaller pores. After the polymeric structure is generated the template may be removed by dissolution (**Figure 3**). The particle array template is usually fabricated of silica or polystyrene monodispersed colloids. The diameter of particles used is found as ca. 200 nm–1 μm. The colloid crystal phase was prepared by sedimentation of particles onto the electrode [25] or by vertical lifting from the colloidal suspension [26].

Due to an increase in mechanical stability of silica templates, the samples are sintered, which results in necking between neighboring particles [25]. The particle lattice is generated on the electrode, followed in the next step by electro-polymerization of the required monomer. The polymer is grown (electrochemical synthesis) by the template from conductive metal electrode surface onto which the template is organized.

Several polymers were electrochemically deposited onto colloid lattice templates including polypyrrole, polyaniline, polythiophene, polybithiophene, and so on and various electro-

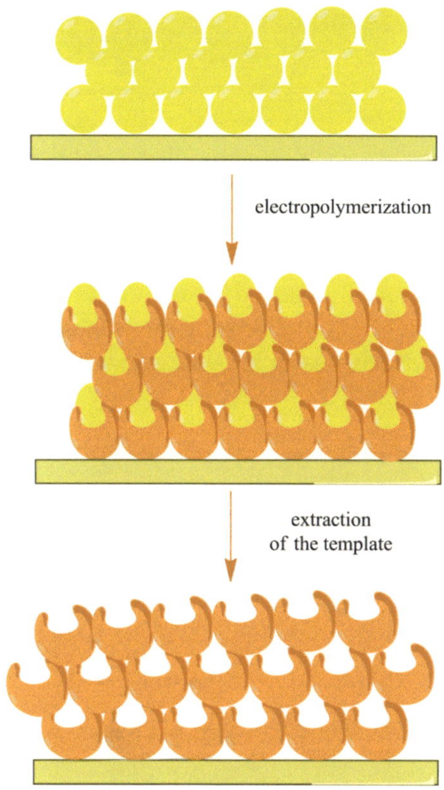


**Figure 3.** Procedure of formation polymeric inverted opals using 3D colloid crystals.

chemical techniques were employed to polymerize monomers on the template [11]. The most general is potentiostatic deposition, which permits credible monitoring over the structure of the growing polymer. The chronoamperometric transients expose at the start rather low current densities adequate to growth of polymer within the empty spaces of the template. In the next step, there is a rapid growth of current due to an increase in electrochemical process area when the new material reached the membrane/bulk solution interface. Polymerization above this transition point results in the subsequent growth of a cauliflower-like structure on the membrane surface; thus, effective preparation of replicas of colloidal particles requires avoiding deposition times that last too long [25].

Similar transient points are observed on potential-time dependence in galvanostatic deposition. After filling the template with polymer, a quick change in the potential is found when the growth front extends above the template/bulk solution interface [26]. The polymerization can also be processed by potentiodynamic methods, that is, cyclic voltammetry. Supposing the regulation over the thickness of the coating layer is more difficult, the inversed polymeric opals are characterized by improved quality originating from well-controlled polymerization process. This appears to fil efficiently the interstices in the colloid crystal and to generate a compact and defect-free structure [27].

## 3. Sensor based on transduction

### 3.1. Amperometric sensor

In case of amperometric sensors, signal is proportional to the concentration of an analyzed species. Specific target species are electroactive ones, that are able to be oxidized or reduced, when the oxidation or reduction potential is zero. The best known to this moment amperometric sensor is the Clark Oxygen Cell (developed in 1956). In this type of biosensors, concentration of a biocatalyst substrate is detected by the $O_2$ consumption by oxidase catalyzed process, or by the generation of $H_2O_2$. Electrochemical biosensors (**Figure 4**) are based on mediated or unmediated electrochemistry for electron transfer [27].

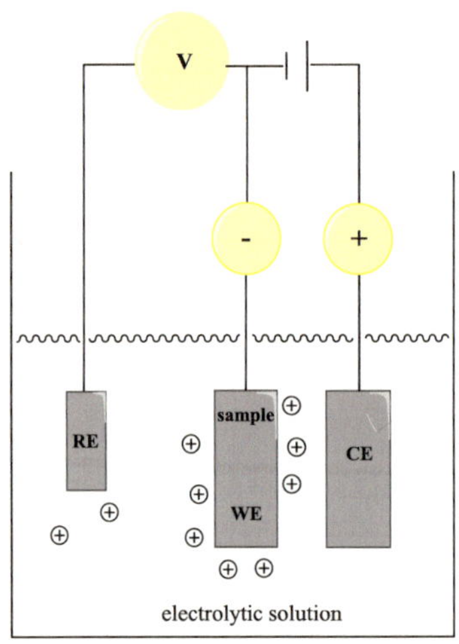


**Figure 4.** Scheme of amperometric sensor.

Direct electrical communication between redox enzyme and electrode was presented earlier by us [21]. However, usually this contact is prohibited because of a donor-acceptor separation distance - a major factor controlled the electron transfer rates. The most redox proteins (diameter 80–150 Å, laccase 50 Å) have the redox centers deeply embedded and therefore electrically insulated. That is, redox enzymes that form a donor-acceptor pair with an electrode support lack electrical contact with this support. In these systems, the biofunctions of the biocatalysts are electrochemically stimulated by electron transfer [21]. The electrical contact may be improved significantly by the application of CPs.

### 3.2. Piezoelectric sensor

An operating of this type of sensors is based on an acoustic wave expanded by an applied alternating current between two electrodes or interdigitated electrode fingers deposited on a piezoelectric solid that is, quartz. The types of piezoelectric transducers are based on the way of acoustic wave propagated between the electrodes. These tools are mainly used for gas phase monitoring, where $H_2S$, $CO_2$, $O_2$, $NO_2$, Hg, toluene as well as acetone sensors have been created [27, 28].

For the last few years, scientists have explored the possibility of using composite-material sensors by combining both silicon and polymers, examples of which include embedding of silicon sensing elements in polymer skins [29], packaging of silicon-based sensing devices in protective casing of polymer layer [29], and so on. Silicon-based tactile sensors have proven to provide high sensitivity, high spatial resolution, and ease of integration into electronic devices. There was also reported a piezoelectric polymer-based, skin-like tactile sensor which was selectively sensitive to stress and shear forces [29].

### 3.3. Optical sensor

Optical sensors are developed as tools based on the measurements of light absorbed or emitted as a consequence of a biochemical reaction. These sensors may be used for determination of pH, $O_2$, $CO_2$, and so on. [27]. Classically, these tools incorporate a material at the tip or on the side of the cable, which can generate an optical signal related to the concentration of analyte in the sample [28].

## 4. Sensor based on application mode

Different types of sensors have been fabricated using conducting polymers in various transduction modes. The transduction modes may be classified into five classes based on the operating rules, namely conductometric, potentiometric, amperometric, colorimetric, and gravimetric procedures [30, 31]. The conductometric mode utilizes changes in electrical conductivity as a result of a target species interaction. The conductivity of conducting polymers bridging the gap between two adjacent electrodes is mainly estimated as a function of analyte concentration, and it may also be monitored with a fixed potential in a sample.

### 4.1. Chemical sensors

There are many chemicals of concern that have to be determined, including toxic gases, volatile organic species, alcohol, and humidity. Different types of conducting polymer nanostructures have been used as elements of various detectors. Chemically synthesized polymer nanoparticles suspended in solvents can be simply deposited on a prefabricated electrode by drop casting to construct a sensor tool. Strongly sensitive chemiresistive sensors based on polypyrrole nanotubes were developed to determine toxic gases as well as volatile organic compounds [32].

PPy nanomaterial was transported onto a polydimethylsiloxane solid by a dry-transfer process and then micro-patterned Au electrodes were deposited onto the nanotubes by thermal evaporation. Different CPs such as PPy nanoparticles, PEDOT nanorods, and PEDOT nanotubes were introduced as well into electrode matrix to cause discriminative reactions toward individual target species in a sample. The lowest measured concentration of the sensors was ca. 0.01 ppm for ammonia, selectively. Electrochemical polymerization process permits, in a governable manner, deposition of the polymer on an electrode.

Recently, PPy nanowires (grown electrochemically on a microelectrode) were used as conductometric transducers to detect hydrogen gas at room temperature (**Figure 5a**) [33]. It was observed that the polypyrrole nanowires were assembled at high density (**Figure 5b**); moreover, a few of the nanowires bridged the gap between the electrodes (**Figure 5c**). The sensor was exhibited to various amounts of $H_2$ at room temperature in a limited space, during which the resistance of the polypyrrole nanowire electrode was verified. The resistance of the nanosized wires was supposed to reduce during exposition to $H_2$ (reducing gas). The sensors were characterized with a linear detection of ca. 600–2500 ppm for the gas. The sensitivities were observed depending on the quantity of the settled nanowires.

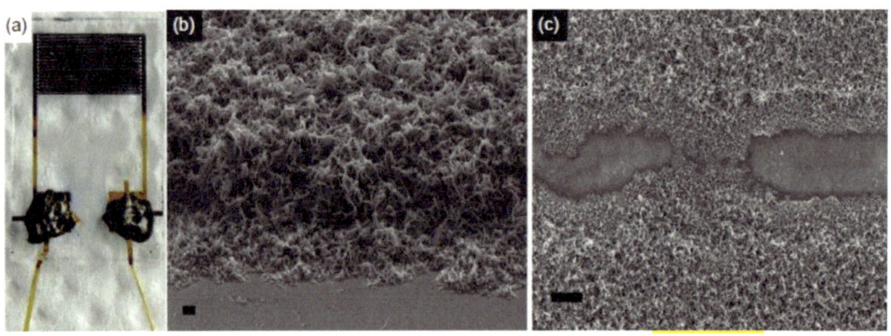


**Figure 5.** (**a**) Picture of a gas sensor electrode, (**b**) SEM images of polypyrrole nanowires settled on the electrode, and (**c**) top view of PPy nanowires bridging the insulating gap between the gold electrodes. Adapted with permission from Yoon [36].

Moreover, it was observed that PANI is able to interplay with $H_2$. There was reported immediate interaction of $H_2$ with doped polyaniline nanofibers to provoke a modification in the conductivity of the nanofibers [34]. The direct mass uptake of $H_2$ by polyaniline nanofibers was found as well by a quartz crystal microbalance (ca. 3% relative to the nanofiber mass). A reliable program of the $H_2$/PANI reaction provokes $H_2$ interacting with doped PANI at the charged amine nitrogen, proceeded after the dissociation of $H_2$ and the creation of N–H bonds at the amine nitrogen of the polyaniline [34, 35]. Afterward, charge transport between adherent amine nitrogen atoms regains the polyaniline back to its classical doped, emeraldine-salt pattern with a deliverance of $H_2$.

The nanoparticle-modified sensor for volatile organic compounds (VOCs) detection was reported by Vaddiraju et al. [37, 38]. They used, at first, oxidative chemical vapor deposition (oCVD) method as a comprehensive and substrate-independent technique for the formation of well-adhered conducting polymer layers which may be used as a platform to anchor the Au nanoparticles and ameliorate the robustness of the sensor. The resistive answer of the nanoparticle/polymer layers is used to determine the inherence of VOCs. The metal nanoparticles were grafted onto the surface of the functionalized conducting copolymer films using thiol as the linker molecule [39].

A system of electrochemical and electrical nanosensor for determination of nitroaromatic explosives vapors (TNT – 2,4,6-trinitrotoluene) was reported by Aguilar et al. [40]. The integrated sensor system consists of two elements with a thin layer coating of liquid crystal 1-butyl-3-methylimidazolium hexafluorophosphate (BMIM-PF6): (i) PEDOT nanojunctions for conductometric determination and (ii) electrodes for electrochemical detection (**Figure 3**). Conducting polymers can be integrated with many species to form composites that are sensitive to various chemicals. Polyaniline nanofibers treated with $CuCl_2$ exhibited the capacity to determine hydrogen sulfide with change in resistance by 4 orders of magnitude [41]. Moreover, water-soluble metal salts were to be readily included into polyaniline nanofibers, such as $CuCl_2$, $CuBr_2$, $CuF_2$, $Cu(O_2CCH_3)_2$, $Cu(NO_3)_2$, $EuCl_2$, $NiCl_2$, $FeCl_3$, and $CoCl_2$ [36]. The salts were sieved to find the best candidates for determination of arsine.

One of the habitual toxic gases is sarin (an organophosphorous derivative). Flexible nerve element sensors based on hydroxylated poly(ethylenedioxythiophene) nanotubes (HPNTs) equipped with unusual surface architectures were also reported [42]. Other gas detecting system was reported by Airoudj et al. as a *multilayer integrated optical sensor* (MIOS) using polyaniline as a sensitive material and demonstrated the detection of ammonia gas [43].

### 4.2. Biosensors

Biosensor is an analytical tool based on biological recognition element integrated with a signal transduction system. Undoubtedly, biosensors are one of the main issues of nanobiotechnology due to the fact, that commonly biosensing devices consist of various nanostructures meeting also nanosized biomolecules; as well as bio-bio and/or bio-non-bio reactions at nano-level which are consulted during the construction process. Sometimes miniaturization toward nano-dimension permits growth of selectivity as well as sensitivity of tools that are the main benefits of these analytical instruments. According to the recognition principals, biosensors can be classified into two clearly different groups as catalytic biosensors and affinity biosensors [36]. Conducting polymers have been applied in construction of both types of biosensors as immobilization matrixes [28, 44], signal transduction systems, and even analyte recognizing components [36].

In comparison to metals and ceramics, CPs are more compatible to biological structures. Inorganic nanosized materials have been included into biosensor setups by lithography and focused ion beam methods.

Nevertheless, the integration of CP nanomaterials into biosensors has been limited because of their incompatibility with the customary microfabrication procedures. However, the conducting polymers are subjected to chemical instability, weak mechanical properties, and poor biocompatibility, which may negatively affect the performance of the biosensors based on these polymers.

Aiming to solve the above-mentioned problems associated with conducting polymers, composite structures were fabricated by coating conducting polymers on rigid nanomaterials [45]. The composites are expected to maintain the merits and discard the shortcomings of the single component and act as excellent biosensor immobilization scaffolds. For example, graphene-PANI composite film was successfully synthesized in large scale by a facile electrodeposition method and horseradish peroxidase (HRP) was entrapped onto the film modified electrode to construct a $H_2O_2$ biosensor [46]. The biosensor exhibited a fast amperometric response after less than 5 s, a good linear range from 1 to 160 μM, and a low detection limit of 0.08 μM.

The incorporation of CNTs into conducting polymers can lead to new composite materials with improved properties due to synergic effects of the two. The π-conjugated structure of the nanotubes allows them to interact with organic aromatic compounds through π-π electronic and hydrophobic interactions [47]. Recently, Lete et al. presented a dopamine sensor based on poly(ethylenedioxythiophene) – PEDOT as well as CNTs. The tyrosinase-based biosensor prepared at low frequency range displayed the lowest limit of detection of 2.4 μM dopamine, the best repeatability of 4.9%, and a recovery of 100.9% for dopamine determination in the presence of catechol [48].

Xu et al. [49] fabricated glucose biosensor based on carbon nanotubes coated with PANI and dendrimer coated Pt nanoparticles. Xian et al. [50] chemically synthesized a composite of PANI nanofibers and Au nanoparticles. Chemical synthesis required several different chemicals, thereby increasing the complexity of the growth process.

Sun et al. have recently reported polypyrrole-Pluronic F127 nanoparticles (PPy-F127 NPs), a novel conducting polymer nanomaterial [51]. These novel nanomaterials were employed in the design of biosensors, which were further used in biochemical assays of L-lactic acid in pig muscle samples. Pluronic F127 was used because it is difficult to obtain PPy nanoparticles with the desired small size and good dispersion. F127 can be used as a template and dispersant for preparing nanoparticles with special shapes [52, 53] as well as according to its good biocompatibility. The reported biosensor had good electrocatalytic activity toward L-lactic acid with a linear range of 15μM–37.5 mM and a low detection limit of 8.8 μM. The L-lactic acid biosensor had also a good anti-interference property toward uric acid, ascorbic acid, glucose, and cysteine [51].

An alternative strategy of using PPy nanoparticles was presented by Yoon et al. [54]. Covalent links between polypyrrole nanotubes and a microelectrode were formed to obtain convenient electrical communication in solution. PPy nanotubes equipped with carboxyl groups were obtained and the carboxyl elements were bonded with the surface amino groups of the modified electrode. A liquid, ion-gated field-effect transistor (FET) sensor was successfully constructed by this procedure. The scheme of the FET sensor is almost the same as the normal metal-oxide-semiconductor FET, except for the gate, which includes the means of transduction from a chemical event to a voltage [55]. Several critical parameters determining the FET sensor response have also been investigated that is, Shirale et al. presented single PPy nanowire-based FET sensors for real-time pH determination. They also examined how the diameter of the nanowire affects the sensor activity [56]. The FET sensors had higher sensitivity with lower diameter and higher length [56].

Wang et al. reported an amperometric glucose biosensor based on a direct-electron-transfer mechanism [57, 58]. They electrochemically entrapped glucose oxidase (GOx) onto the inner wall of polyaniline nanotube orders fixed from an anodic aluminum oxide (AAO) pattern [59]. A pair of symmetric redox peaks was presented in the cyclic voltammogram, and a seeming constant electron transport rate of 5.8 ± 1.6 $s^{-1}$ was estimated. The biosensor exposed good sensing presentation with a violent response (3 s), a low detection limit (0.3 ± 0.1 μM), a linear range (0.01–5.5 mM), and high sensitivity (97.18 ± 4.62 μA $mM^{-1}$ $cm^{-2}$). The biosensor exhibited anti-interfering features versus common compounds such as ascorbic acid, uric acid, and 4-acetamidophenol.

Ramanathan et al. reported a single-step electrodeposition technique for the formation of PPy and PANI nanowires to bridge the gap between two electrodes on a silicon wafer [60, 61]. The deposition and growth of the nanowires are based on electrochemical oxidation polymerization.

To further improve the sensitivity and selectivity, a variety of biosensors based on CP composites have been developed, for example, the composites of CPs and carbon nanotubes, metal particles, metal oxide particles, metal salts, chitosan, and biocompatible materials [3].

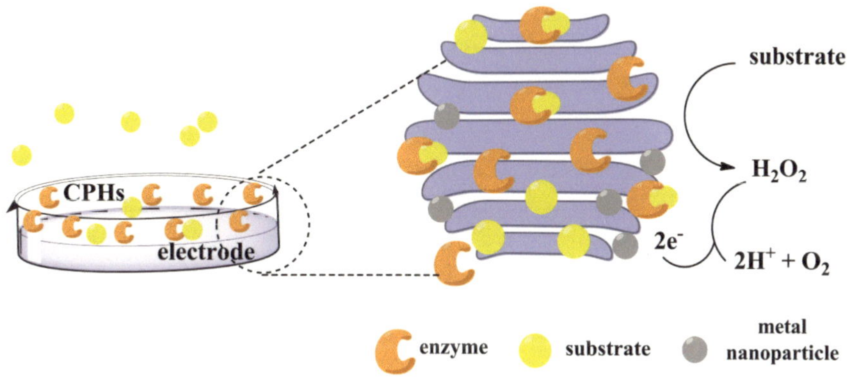


**Figure 6.** Classic example of electrochemical biosensor based on CPHs.

Recently, the growing attention is addressed to the synthesis of conducting polymer hydrogels (CPHs) and their introduction into functional biosensors. Different synthetic pathways, including templating against a hydrogel matrix [62], cross-linking by multivalence metal ions [63], and self-assembly, have been adapted to generate conducting polymer networks inside the water environment. Among them, there was reported a general CPH synthesis procedure by cross-linking CPs with multi-valence doping-acids [3]. Combining the features of both the CPs and hydrogels, these CPHs present great potential as advanced interface materials to be used in biosensors. Here is presented illustration of a reasonably constructed electrochemical biosensor based on CPHs (**Figure 6**). The CPHs are extremely important in enhancing the sensing properties by (i) spreading the planar electrode to the 3D organic matrix to enhance the efficient interface area, (ii) supplying links between soft and hard materials for sufficient enzyme immobilization, (iii) interfacing the ionic transferring phase and the electron transporting phase to lower the impedance, as well as (iv) ensuring high density loading with catalytic nanoparticles to provoke electron set [63]. Due to these, biosensors based on CPHs have displayed inviting features such as sensitivity, low detection limit, and quick response time. Until now, the development of CPH-based biosensors has been focused on the reasonable design of both the materials and instruments.

Heller as well as Asberg [64, 65] reported biosensors based on redox hydrogels and metal-ion-cross-linked CPHs, respectively, and suggested that the gels could be brilliant biosensor electrodes because of the benefits: (i) the specific charge transfer properties of CPs allow electron transport [66]; (ii) CPHs with three-dimensional porous nanostructures present a large surface area and short diffusion way; (iii) great biocompatibility [67]. Mano et al. also presented an electron-conducting cross-linked polyaniline-based redox hydrogel. The active film was created in a one-step process (by cross-linking a polymer acid-templated PANI with a water-soluble poly(ethylene glycol) diglycidyl ether (PEGDGE)) [68]. Glucose oxidase was immobilized in the hydrogel by simultaneously co-cross-linking, leading to the electrical wiring of the protein.

Asberg and Inganas investigated a general method to synthesize PEDOT hydrogels using multi-valence metal ions, such as magnesium, iron, and osmium to cross-link the poly(styrenesulfonate) (PSS) groups on PEDOT:PSS into a mesoscopic conducting network of hydrogels [65]. They also developed a biomolecule-enhanced electrode with the electroactive polymer hydrogel PEDOT:PSS in combination with poly(4-vinylpyridine) (P4VPy) cross-linked by osmium [65].

The integration of the benefits of organic conducting structures and hydrogels, CPHs present a hopeful interface for biosensor electrodes. The hierarchical structures of CPHs ensure an open porous system, a large surface area, and an organic matrix, which are profitable for enhancing the permeability of bio-species, increasing the interfacial area of electrodes, and immobilizing enzymes evenly and densely [3].

Another important issue in biosensor design concerns the relationship between receptors and transducers and the way they are coupled. Comprehensive receptors containing enzymes, antibodies, nucleic acids, and cells have been inserted into CPs through adsorption, entrapment, or covalent bonding. A well-known example concerns enzyme-based glucose determination for diagnostic of diabetes. Commonly, amperometric glucose determination has been reached with GOx electrodes. The GOx has been immobilized on electrodes by chemical cross-linking, electrodeposition, or electrostatic interactions [69]. The latest developments have suggested variants to the classic biocatalyst electrode [70, 71].

CPs nanomaterials have primarily been adapted as the conductive matrix with redox features. A FET-based sensor was reported by using GOx-attached PPy nanotubes as the conductive channel [72]. The polypyrrole nanotubes were fabricated by the chemical polymerization of pyrrole-3-carboxylic acid inside the cylindrical pores of an alumina membrane. A novel glucose sensor was developed by Wu and Yin, who utilize polyaniline-wrapped boron nitride nanotubes equipped with platinum nanoparticles as the electrode material [73]. Enzyme, according to the procedure, was mixed with the Pt/PANI/boron nitride nanotubes, which resulted in a cotton-like enzyme nano-hybrid. The nano-hybrid electrode has had good acid stability and heat resistance.

Molecularly imprinted polymers (MIPs) are tailor-made biomimetic materials that are capable of selective recognition toward target molecules. The selective recognition property, high stability, easy preparation, and low cost of these MIPs make them extremely attractive as sensing materials in biosensors and chemical sensors [74].

Molecularly imprinted sensors that demand no receptors have also been designed by CPs. For example, a simple procedure for the photoelectrochemical sensing of microcystin-LR (MC-LR) was obtained by utilization of PPy/titania nanotubes with MC-LR recognition places [75]. MC-LR is an agent that can induce structural as well as functional bothers of the liver, and is a potential cancer threat [76]. Due to its omnipresence and high toxicity, MC-LR is an agent of concern in water and environmental monitoring.

Zhou et al. reported an electrochemical sensor based on molecularly imprinted PPy for epinephrine recognition [77]. Pyrrole was electropolymerized with epinephrine on the surface of an electrode coated with silica nanoparticles and CNTs. After the removal of silica nanoparticles and epinephrine, a molecularly imprinted PPy/CNTs porous layer could be formed on the electrode. The amperometric answer of the PPy/CNTs electrode was recorded by adding of epinephrine in PBS. Also Li et al. [78] reported molecularly imprinted polymer arrays (3D MIP arrays) as the sensing materials. ZnO nanorods were deposited on indium tin oxide coated polyethylene terephthalate film and used as high-surface 3D scaffolds for 3D MIP arrays deposition. The obtained results indicated that this 3D MIP arrays-based electrochemical sensor exhibits high sensitivity and good selectivity for epinephrine.

Cai et al. prepared polyphenol nanocoating on the tips of CNTs in the presence of a template protein. Then, they relied on measurements of impedance to monitor the recognition. This nanosensor was sensitive and selective; moreover, it was also capable of detecting protein conformational changes [79]. Using $TiO_2$ NTs as the microreactor, Liu et al. fabricated MIPs onto the internal surface of $TiO_2$ nanotubes and used them as working electrode for photoelectrocatalytic and photocatalytic degradation of 9-AnCOOH [80]. Similarly, Tran et al. reported a novel photoelectrochemical sensor by surface modification of molecularly imprinted polymer on highly ordered and vertically aligned $TiO_2$ nanotube arrays which could be used for photoelectrochemical determination of perfluorooctane sulfonate [81].

## 5. Summary and outlook

Conducting polymers are favorable materials for both nanotechnology and biosensorics. The intensity of conducting polymer application in sensors and biosensors design increases permanently. For example, sensors based on conducting polymer-based nanocomposites show several inviting characteristics depend on the composite materials as well as on the CPs' formation procedure. Common utility of CPs is recently broadly utilized in design of suitable polymer-based nanocomposites. There were observed clear appearances that capability of CPs to transport electrical charge will be utilized in the design of different molecular tools. Furthermore, the conducting polymers will be adopted for integration nanosized elements and probably in the construction of single receptor molecule-based sensors within such molecular tools.

Frequent studies on conjugated polymer nanotubes and nanowires established that improvements in their molecular structure, crystallinity, and an increase in conjugation length are responsible for the increase in the room temperature conductivity when the diameter of such fibrillar structures decreases.

Sensors play a crucial role in environmental diagnostics, medical monitoring, industrial safety checking, as well as security. CP nanomaterials are supposed to possess much unexplored capability for sensor appliances. CP-based biosensors are likely to address the issues of biocompatibility for continuous monitoring of biological metabolites and drug dosages, and the possibility of *in vivo* sensing. Consequently, the forthcoming development into CP nanomaterials-based sensors may proffer large promise for the design of future-generation sensor instruments. Notably, it is envisaged that convenient and flexible high-performance sensors will soon be constructed by utilization of conducting polymer nanomaterials.

## Acknowledgements

The authors gratefully acknowledge the financial support from Wrocław University of Technology.

## Author details

Jadwiga Sołoducho* and Joanna Cabaj

*Address all correspondence to: jadwiga.soloducho@pwr.edu.pl

Wroclaw University of Science and Technology, Faculty of Chemistry, Wrocław, Poland

## References

[1] Yoon H., Jang J. Conducting-polymer nanomaterials for high-performance sensor applications: issues and challenges. Adv. Funct. Mater., 2009, 19, 1567–1576

[2] Hatchett D.W., and Josowicz M. Composites of intrinsically conducting polymers as sensing nanomaterials. Chem. Rev., 2008, 108, 746–769

[3] Li L., Shi Y., Pan L., Shi Y., and Yu G. Rational design and applications of conducting polymer hydrogels as electrochemical biosensors. J. Mater. Chem. B, 2015, 3, 2920–2930

[4] Mazur M., and Krysiński P. Electrochemical template synthesis of conjugated polymers. Curr. Top. Electrochem., 2007, 12, 15–31

[5] Jang J. Conducting polymer nanomaterials and their applications. Adv. Polym. Sci., 2006, 199, 189–260

[6] Jang J., and Bae J. Synthesis and characterization of polyaniline nanorods as curing agent and nanofiller for epoxy matrix composite. Adv. Funct. Mater., 2005, 15, 1877–1882

[7] Jang J., and Oh J.H. A facile synthesis of polypyrrole nanotubes using a template mediated vapor deposition polymerization and the conversion to carbon nanotubes. Chem. Commun., 2004, 882–883

[8] Jang J., and Yoon H. Multigram-scale fabrication of monodisperse conducting polymer and magnetic carbon nanoparticles. Small, 2005, 1, 1195–1199

[9] Yoon H., Chang M., and Jang J. Formation of 1D poly(3,4-ethylenedioxythiophene) nanomaterials in reverse microemulsions and their application to chemical sensors. Adv. Funct. Mater., 2007, 17, 431–436

[10] Huang J., Virji S., Weiller B.H., and Kaner R.B. Polyaniline nanofibers: facile synthesis and *chemical* sensors. J. Am. Chem. Soc., 2003, 125, 314–315

[11] Mazur M., and Krysinski P. Covalently immobilized 1,4-phenylenediamine on 11-mercaptoundecanoic acid-coated gold: effect of surface-confined monomers on the chemical in situ deposition of polyaniline and its derivatives. Langmuir, 2001, 17, 7093–7101

[12] Tagowska M., Mazur M., and Krysinski P. Covalently and ionically immobilised monomers on the gold surface. Synth. Met., 2004, 140, 29–35

[13] Neves S., Fonseca C.P., Zoppi, R.A., and de Torresi, S.I.C. Polyaniline composites: improving the electrochemical properties by template synthesis. J. Solid State Electrochem., 2001, 5, 412–418

[14] Roux S., Audebert P., Pagetti J., and Roche, M. Electrochemical growth of conducting polymer into zirconium oxopolymers sol-gel coatings. J. Sol-Gel Sci. Technol., 2003, 26, 435–439

[15] Jerome C., and Jerome R. Electrochemical synthesis of polypyrrole nanowires. Angew. Chem. Int. Ed., 1998, 37, 2488–2490

[16] Jerome C., Labaye D., Bodart I., and Jerome R. Electrosynthesis of polyacrylic polypyrrole composites: formation of polypyrrole wires. Synth. Met., 1999, 101, 3–4

[17] Jerome C., Demoustier-Champagne S., Legras R., and Jerome R. Electrochemical synthesis of conjugated polymer wires and nanotubules. Chem. Eur. J. 2000, 17, 3089 —3093

[18] Jerome C., Labaye D.E., and Jerome R. Electrochemical formation of polypyrrole nanowires. Synth. Met., 2004, 142, 207–216

[19] Ouyang J., and Li Y. Effect of electrolyte solvent on the conductivity and structure of as-prepared polypyrrole films. Polymer, 1997, 38, 1971–1976

[20] Ge D., Wang J., Wang Z., and Wang S. Electrochemical synthesis of polypyrrole nanowires on composite electrode. Synth. Met., 2002, 132, 93–95

[21] Jędrychowska A., Cabaj J., Świst A., and Sołoducho J. Electrochemical laccase sensor based on 3-methylthiophene/3-thiopheneacetic acid/bis(3,4-ethylenedioxythiophene)-Nnonylacridone as a new polymer support. J. Electroanal. Chem., 2014, 720–721, 64–70

[22] Jędrychowska A., Malecha K., Cabaj J., and Sołoducho J. Laccase biosensor based on low temperature co-fired ceramics for the permanent monitoring of water solutions. Electrochim. Acta, 2015, 165, 372–382

[23] Han S., Briseno A.L., Shi X., Mah D.A., and Zhou, F. Polyelectrolyte-coated nanosphere lithographic patterning of surfaces: fabrication and characterization of electropolymerized thin polyaniline honeycomb films. J. Phys. Chem. B, 2002, 106, 6465–6472

[24] Briseno A.L., Han S., Rauda I.E., Zhou F., Toh C.S., Nemanick E.J., and Lewis N.S. Electrochemical polymerization of aniline monomers infiltrated into well-ordered truncated eggshell structures of polyelectrolyte multilayers. Langmuir, 2004, 20, 219–226

[25] Sumida T., Wada Y., Kitamura T., and Yanagida S. Electrochemical preparation of macroporous polypyrrole films with regular arrays of interconnected spherical voids. Chem. Commun., 2000, 1613–1614

[26] Tian S., Wang J., Jonas U., and Knoll W. Inverse opals of polyaniline and its copolymers prepared by electrochemical techniques. Chem. Mater., 2005, 17, 5726–5730

[27] Gupta N., Sharma S., Mir I.A., and Kumar D. Advances in sensors based on conducting polymers. J. Sci. Ind. Res, 2006, 65, 549–557

[28] Janata J. (Ed.), Principles of chemical sensors. Plenum Press, New York, 1984

[29] Nambiar S., and Yeow J.T.W. Conductive polymer-based sensor for medical application. Biosens. Bioelectron., 2011, 26, 1825–1832

[30] Bartlett P.N., and Ling-Chung S.K. Conducting polymer gas sensors part II: response of polypyrrole to methanol vapors. Sens. Actuat. B, 1989, 19, 141–150

[31] Adhikari B., and Majumdar S. Polymers in sensor applications. Prog. Polym. Sci., 2004, 29, 699–766

[32] Turyan I., and Mandler D. Selective determination of Cr (VI) by self-assembled monolayer-based electrode. Anal. Chem., 1997, 69, 894–897

[33] Canh T.M. (Ed.), Biosensors, Chapman & Hall, London, 1993

[34] Belanger D., Nadreau J., and Froteir G. Electrochemistry of the polypyrrole glucose oxidase electrode. J. Electroanal. Chem. Interfacial Electrochem., 1989, 274, 143–155

[35] Yabuki S., Shinohara H., Ikariyama Y., and Aizawa M. Electrical activity controlling system for a mediator-coexisting alcohol dehydrogenase-NAD conductive membrane. Chem. Interfacial Electrochem., 1990, 277, 179–187

[36] Yoon H. Current trends in sensors based on conducting polymer nanomaterials. Nanomaterials, 2013, 3, 524–549

[37] Vaddiraju S., Seneca K., and Gleason K.K. Novel strategies for the deposition of -COOH functionalized conducting copolymer films and the assembly of inorganic nanoparticles on conducting polymer platforms. Adv. Funct. Mater., 2008, 18(13), 1929–1938

[38] Vaddiraju S., and Gleason K. Selective sensing of volatile organic compounds using novel conducting polymer-metal nanoparticle hybrids. Nanotech., 2010, 21(12), 125503–125511

[39] Laforgue A., and Robitaille L. Production of conductive PEDOT nanofibers by the combination of electrospinning and vapor-phase polymerization. Macromol., 2010, 43(9), 4194–4200

[40] Aguilar A.D., Forzani E.S., Leright M., Tsow F., Cagan A., Iglcsias R.A., Nagahara L.A., Amlani I., Tsui R., and Tao N.J. A hybrid nanosensor for TNT vapor detection. Nano Lett., 2010, 10, 380–384

[41] Virji S., Fowler J.D., Baker C.O., Huang J.X., Kaner R.B., and Weiller B.H. Polyaniline nanofiber composites with metal salts: chemical sensors for hydrogen sulfide. Small, 2005, 1, 624–627

[42] Kwon O.S., Park S.J., Lee J.S., Park E., Kim T., Park H.W., You S.A., Yoon H., and Jang J. Multidimensional conducting polymer nanotubes for ultrasensitive chemical nerve agent sensing. Nano Lett., 2012, 12, 2797–2802

[43] Airoudj A., Debarnot D., Beche B., and Poncin-Epaillard F. Design and sensing properties of an integrated optical gas sensor based on a multilayer structure. Anal. Chem., 2008, 80, 9188–9194

[44] Barik A., Solanki P.R, Kaushik A., Ali A., Pandey M.K., Kim C.G., and Malhotra B.D. Polyaniline-carboxymethyl cellulose nanocomposite for cholesterol detection. J. Nanosci. Nanotechnol., 2010, 10, 6479–6488

[45] Kesik M., Kanik F.E., Hizalan G., Kozanoglu D., Esenturk E.N., Timur S., and Toppare L. A functional immobilization matrix based on a conducting polymer and functionalized gold nanoparticles: synthesis and its application as an amperometric glucose biosensor. Polymer, 2013, 54, 4463–4471

[46] Feng X.-M., Li R.-M., Ma Y.-W., Chen R.-F., Shi N.-E., Fan Q.-L., and Huang W. One-step electrochemical synthesis of graphene/polyaniline composite film and its applications. Adv. Funct. Mater., 2011, 21, 2989–2996

[47] Carvalho R.C., Gouveia-Caridade C., and Brett C.M.A. Glassy carbon electrodes modified by multiwalled carbon nanotubes and poly(neutral red): a comparative study of different brands and application to electrocatalytic ascorbate determination. Anal. Bioanal. Chem., 2010, 398, 1675–168

[48] Lete C. Lupu S., Lakard B., Hihn J.-Y., del Campo F.J., Multi-analyte determination of dopamine and catechol at single-walled carbon nanotubes – conducting polymer – tyrosinase based electrochemical biosensors. J. Electroanal. Chem., 2015, 774, 53–61

[49] Xu L., Zhu Y., Yang X., and Li C. Amperometric biosensor based on carbon nanotubes coated with polyaniline/dendrimer-encapsulated Pt nanoparticles for glucose detection. Mater. Sci. Eng. C, 2009, 29, 1306–1310

[50] Xian Y., Hu Y., Liu F., Xian Y., Wang H., and Jin L. Glucose biosensor based on Au nanoparticles–conductive polyaniline nanocomposite. Biosens. Bioelectron., 2006, 21, 1996–2000

[51] Sun C., Wang D., Zhang M., Ni Y., Shen X., Song Y., Geng Z., Xu W., Liu F., and Mao C. Novel L-lactic acid biosensors based on conducting polypyrrole-block copolymer nanoparticles. Analyst, 2015, 140, 797–802.

[52] Bakshi S., Kaura A., Bhandari P., Kaur G., Torigoe K., and Esumi K. Synthesis of colloidal gold nanoparticles of different morphologies in the presence of triblock polymer micelles. J. Nanosci. Nanotechnol., 2006, 6, 1405–1410

[53] Mao C., Chen X.B., Hou X.M., Shen J., Zhu J.J., and Zhao W.B. Synthesis of rambutan-like hybrid nanospheres of Au-P123. Gold Bull., 2009, 42, 215–218

[54] Yoon H., Kim J.H., Lee N., Kim B.G., and Jang J. A novel sensor platform based on aptamer-conjugated polypyrrole nanotubes for label-free electrochemical protein detection. ChemBioChem, 2008, 9, 634–641

[55] Lee C.S., Kim S.K., and Kim M. Ion-sensitive field-effect transistor for biological sensing. Sensors, 2009, 9, 7111–7131

[56] Shirale D.J., Bangar M.A., Chen W., Myung N.V., and Mulchandani A. Effect of aspect ratio (length: diameter) on a single polypyrrole nanowire FET device. J. Phys. Chem. C, 2010, 114, 13375–13380

[57] Wang Z.Y., Liu S.N., Wu P., and Cai C.X. Detection of glucose based on direct electron transfer reaction of glucose oxidase immobilized on highly ordered polyaniline nanotubes. Anal. Chem., 2009, 81, 1638–1645

[58] Ruzgas T., Csoregi E., Emneus J., Gorton L., and Varga G.M., Peroxidase-modified electrodes: fundamentals and application. Anal. Chim. Acta, 1996, 330, 123–138.

[59] Ramanathan K., Bangar M.A., Yun M.H., Chen W.F., Mulchandani A., and Myung N.V. Individually addressable conducting polymer nanowires array. Nano Lett., 2004, 4, 1237–1239

[60] Yun M.H., Myung N.V., Vasquez R.P., Lee C.S., Menke E., and Penner R.M. Electrochemically grown wires for individually addressable sensor arrays. Nano Lett., 2004, 4, 419–422

[61] Green R.A., Baek S., Poole-Warren L.A., and Martens P.J. Conducting polymer-hydrogels for medical electrode applications. Sci. Technol. Adv. Mater., 2010, 11, 014107–014120

[62] Dai T.Y., Jiang X.J., Hua S.H., Wang X.S., and Lu Y. Facile fabrication of conducting polymer hydrogels via supramolecular self-assembly. Chem. Commun., 2008, 4279–4281

[63] Zhao Y., Liu B.R., Pan L.J., and Yu G.H. 3D nanostructured conductive polymer hydrogels for high-performance electrochemical devices. Energy Environ. Sci., 2013, 6, 2856–2870

[64] Heller A. Electron-conducting redox hydrogels: design, characteristics and synthesis. Curr. Opin. Chem. Biol., 2006, 10, 664–672

[65] Asberg P., and Inganas O. Hydrogels of a conducting conjugated polymer as 3-D enzyme electrode. Biosens. Bioelectron., 2003, 19, 199–207

[66] Guo B.L., Finne-Wistrand A., and Albertsson A.-C. Simple route to size-tunable degradable and electroactive nanoparticles from the self-assembly of conducting coil-rod-coil triblock copolymers. Chem. Mater., 2011, 23, 4045–4055

[67] Mao F., Mano N., and Heller A. Long tethers binding redox centres to polymer backbones enhance electron transport in enzyme "Wiring" hydrogels. J. Am. Chem. Soc., 2003, 125, 4951–4957

[68] Mano N., Yoo J.E., Tarver J., Loo Y.L., and Heller A. An electron-conducting cross-linked polyaniline-based redox hydrogel, formed in one step at pH 7.2, wires glucose oxidase. J. Am. Chem. Soc., 2007, 129, 7006–7007

[69] Wang J. Electrochemical glucose biosensors. Chem. Rev., 2008, 108, 814–825

[70] Yoon H., Ahn J.H., Barone P.W., Yum K., Sharma R., Boghossian A.A., Han J.H., and Strano M.S. Periplasmic binding proteins as optical modulators of single-walled carbon nanotube fluorescence: amplifying a nanoscale actuator. Angew. Chem. Int. Ed., 2011, 50, 1828–1831

[71] Barone P.W., Yoon H., Ortiz-Garcia R., Zhang J.Q., Ahn J.H., Kim J.H., and Strano M.S. Modulation of single-walled carbon nanotube photoluminescence by hydrogel swelling. ACS Nano, 2009, 3, 3869–3877

[72] Yoon H., Ko S., and Jang J. Field-effect-transistor sensor based on enzyme-functionalized polypyrrole nanotubes for glucose detection. J. Phys. Chem. B, 2008, 112, 9992–9997

[73] Wu J.M., and Yin L.W. Platinum nanoparticle modified polyaniline-functionalized Boron nitride nanotubes for amperometric glucose enzyme biosensor. ACS Appl. Mater. Interfaces, 2011, 3, 4354–4362

[74] Wackerlig J., and Lieberzeit P.A. Molecularly imprinted polymer nanoparticles in chemical sensing-synthesis, characterisation and application, Sens. Actuat. B, 2015, 207, 144–157

[75] Chen K., Liu M.C., Zhao G.H., Shi H.J., Fan L.F., and Zhao S.C. Fabrication of a novel and simple microcystin-LR photoelectrochemical sensor with high sensitivity and selectivity. Environ. Sci. Technol., 2012, 46, 11955–11961

[76] Xia Y.T., Deng J.L., and Jiang L. Simple and highly sensitive detection of hepatotoxin microcystin-LR via colorimetric variation based on polydiacetylene vesicles. Sensors Actuat. B, 2010, 145, 713–719

[77] Zhou H., Xu G.L., Zhu A.H., Zhao Z., Ren C.C., Nie L.L., and Kan X.W. A multiparous electrochemical sensor for epinephrine recognition and detection based on molecularly imprinted polypyrrole. RSC Adv., 2012, 2, 7803–7808

[78] Li H.-H., Wang H.-H., Li W.-T., Fang X.-X., Guo X.-C., Zhou W.-H., Cao X., Kou D.-X., Zhou Z.-J., and Wu S.-X. A novel electrochemical sensor for epinephrine based on three dimensional molecularly imprinted polymer arrays. Sens. Actuat. B, 2016, 222, 1127–1133

[79] Cai D., Ren L., Zhao H.Z., Xu C.J., Zhang L., Yu Y., Wang H.Z., Lan Y.C., Roberts M.F., Chuang J.H., Naughton M.J., Ren Z.F., and Chiles T.C. A molecular-imprint nanosensor for ultrasensitive detection of proteins. Nat. Nanotechnol., 2010, 5, 597–601

[80] Liu Y.T., Liu R.H., Liu C.B., Luo S.L., Yang L.X., Sui F., Teng R., Yang R.B., and Cai Q.Y. Enhanced photocatalysis on TiO2 nanotube arrays modified with molecularly imprinted TiO2 thin film. J. Hazard. Mater., 2010, 182, 912–918

[81] Tran T.T., Li J.Z., Feng H., Cai J., Yuan L.J., Wang N.Y., and Cai Q.Y., Molecularly imprinted polymer modified TiO2 nanotube arrays for photoelectrochemical determination of perfluorooctane sulfonate (PFOS). Sens. Actuat. B, 2014, 190, 745–751

Chapter 3

# Conducting Polymer Aerogels

Weina He and Xuetong Zhang

Additional information is available at the end of the chapter

http://dx.doi.org/10.5772/63397

**Abstract**

Conducting polymers are an important class of organic materials with electric conductivity and have experienced a rapid development. Meanwhile, as a novel class of porous nanomaterials, aerogels attract people's great interest for their ultra-low densities, large specific areas, rich open pores, etc. Thus, conducting polymer aerogels, combining the unique merits of aerogels with physicochemical properties relevant to conducting polymers, become a newly developed area. In this chapter, we give a brief introduction describing (1) synthesis strategies of conducting polymer (PEDOT, PPy, and PANi) aerogels through rational design for oxidant, cross-linker, soft template, sol-gel process, drying process; (2) advantages of these aerogels in physical and chemical performance, compared with the counterparts in bulk or membrane; and (3) their applications in energy storage, adsorption to metal-ions/dye-molecules, stress sensing, Joule heating. The chapter ends with a reflection on limitations of already proposed materials and a prospection of how conducting polymer aerogels developing in the future. As such, this chapter can act as a roadmap to guide researchers toward how conducting polymer aerogels produced and how these materials can be utilized, while also highlighting the current advancements in the field.

**Keywords:** conducting polymers, aerogel, sol-gel transition, energy storage, stress sensors

## 1. Introduction

### 1.1. Aerogels

"Aerogel" does not refer to a specific material with a set chemical formula but encompasses all materials with a specific geometrical structure, which is extremely porous with high connectivity between nanometers-sized branched structures. The aerogel contains little solid and up to 99.8% volume of air; hence, it is often referred to as "frozen smoke."

Aerogels, with merits of ultra-low densities, large specific areas, hierarchical open pores, elaborate 3D networks, etc., are a novel class of highly porous nanomaterials. These unique characteristics endow aerogels with interesting physical properties (low thermal conductivity, low sound velocity, etc.) and with potential applications (Cerenkov detectors, electronic devices, catalysis, etc.). The majority of aerogel is composed of silica, but carbon, metal, iron oxide, organic polymers, and semiconductor nanostructures can also form aerogels.

### 1.2. Conducting polymers

Intrinsically conducting polymers, also known as "synthetic metals," are polymers with a highly conjugated polymeric chain. They have been studied extensively due to their intriguing electronic and redox properties, along with the resulting numerous potential applications in many fields since their discovery in 1970s. Typical conducting polymers include polyacetylene (PA), polyaniline (PANi), polypyrrole (PPy), polythiophene (PTh), poly(para-phenylene) (PPP), poly(phenylenevinylene) (PPV), polyfuran (PF), etc. The chemical structures of these polymers are illustrated in **Figure 1**.

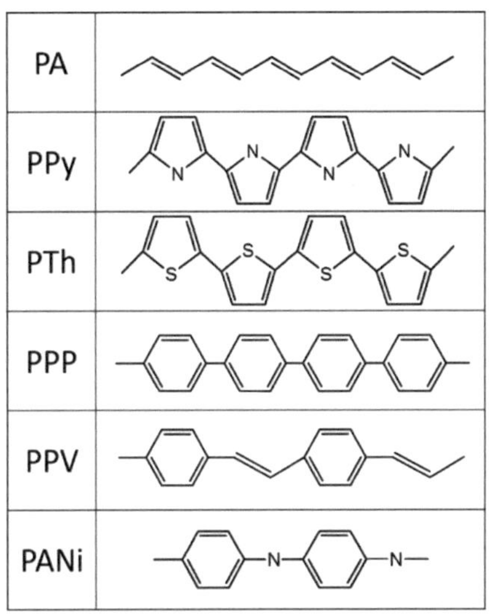


**Figure 1.** The chemical structures of typical conducting polymers.

These intrinsically conducting polymers exhibit adjustable electric conductivities, even across several orders of magnitude sometimes, depending on the level of doping or chemical/electrochemical treating. The reversibility of doping-dedoping endows these conducting polymer potential applications in actuators, sensors, etc., while the highly doped conducting polymers can be applied in nanoelectronic devices, corrosion protection coatings, microwave absorbing, etc., because of the high conductivity.

### 1.3. Conducting polymer aerogels

Aerogels can be divided into two categories according to their electrical properties: electric insulating aerogels and conducting ones. Conducting aerogels not only show ordinary properties as electric insulating aerogels but also exhibit excellent magnetic and electric-conducting performances. Therefore, conducting aerogels possess huge application potentials in energy storage, catalysis, sensing, electromagnetic shielding, and other fields. Heretofore, conductive aerogels are mainly obtained from (a) metal aerogel produced by sol-gel procedure, (b) carbonized product of organic aerogel, and (c) nanomaterial self-assembled by CNTs or G/GO sheets.

A new kind of conducting aerogels is produced by conducting polymers, named conducting polymer aerogels (CPAs). Theoretically, the inherent rigidity of most conducting polymers assist significantly in generating and maintaining permanent microporosity (pore diameter of <2 nm). Through the nanostructural design, CPAs combine the individual superiorities of conducting polymers and aerogels, providing high surface area for more efficient separation and energy storage, more acting sites for fast functionalization with various guest objects, etc.

However, there are still lots of challenges remaining. At first, as a result of the fast oxidation and the poor solvent solubility of intrinsically conducting polymers, there are great difficulties in preparing CPAs in large scale by far. Secondly, because of the inherent rigidity of the conjugated macromolecular chains resulted from the delocalized π-electron system along the conducting polymer backbone, it needs to design particular structures and explores new preparing methods to make CPAs either strong or elastic. Fortunately, improvements in the fragility and brittleness of CPAs announced recently by Zhang's research group means that a whole new world of applications may be opened up in the future, from solid 3D conducting network to flexible electronics. To date, conducting aerogels can be made from poly(3,4-ethylenedioxythiophene) (PEDOT), polyaniline (PANi), polypyrrole (PPy) conducting polymers. The progress in CPAs will be summarized in this chapter to inspirit many new explorations surrounding conducting aerogels in the further.

## 2. Synthesis of conducting polymer aerogels

In general terms, aerogel is produced by the synthesis of gel precursors and drying process. At first, the gel is created in a solution as a precursor via sol-gel transition, and then, the liquid component is removed through drying process, which removes liquid skillfully in order to maintain the nanostructures.

### 2.1. Synthesis of gel precursors

PEDOT, PPy, and PANi can be easily synthesized by chemical oxidation coupling with the presence of a series of oxidants, such as $(NH_4)_2S_2O_8$ (APS), $FeCl_3$, $Fe(NO_3)_3$, and so on. However, because of the fast reaction rate of oxidation and the poor insolubility of the outcomes, the products often precipitated as granules. So as to obtain the intact hydrogels as the precursors for aerogels, the oxidation rate should be controlled at first to let the reaction stop at the proper extent that the building blocks is neither too large to aggregated and separated as precipitates, nor too small to hardly support the framework or badly affect the electrical conductivity. The reaction rates are dependent on the concentration of monomers and oxidant or the reacting temperature. Thus, monomer suspension and oxidant solution with lower concentrations can be adopted, as well as lower reaction temperature provided by ice bath.

### *2.1.1. PEDOT gel*

As for nonporous conducting polymers, the PEDOT:PSS (PSS is the abbreviation for poly(styrenesulfonate)) complex possess the merits of high dispersibility in water, excellent film-forming performance through conventional solution processing, and unique physical properties for the resulting films (high transparency in the visible range, high mechanical flexibility, and excellent thermal stability). So it becomes the most commonly used conducting polymer in commercial applications. Since PEDOT:PSS complex is highly soluble, it is most likely to realize the preparation of conducting aerogels.

As expected, it was reported by Zhang's group [1] that PEDOT-PSS supermolecular hydrogel could be obtained for first time by polymerizing 3,4-ethylenedioxythiophene (EDOT) with excess ferric nitrate as oxidizing agent in the presence of PSS. The PSS plays two general roles: (1) acting as the source for the charge-balancing counter-ion, and (2) keeping the EDOT segments dispersed in the sol. Compared with the nonporous PEDOT-PSS synthesis, the oxidant amount and reaction temperature have been controlled effectively during PEDOT-PSS hydrogel preparation.

The supermolecular cross-linking mechanism of the hydrogel precursors has also been discussed. If APS was used as oxidizing agent, there was no hydrogel formed. Additionally, it was revealed by X-ray photoelectron spectroscopy (XPS) that excess iron existed in PEDOT-PSS aerogel samples. Taken the two aspects into account, it could be deduced that the small amount of iron ions cross-link PSS to form three-dimensional networks driven by the electrostatic interaction between the metal ions and the anions attached to PSS.

Although PEDOT:PSS hydrogel have been prepared, the presence of insulating PSS inevitably inhibits the electrical properties of hydrogel. The synthesis of conducting polymer hydrogels (CPHs) as precursors for CPAs is still a great challenge due to the poor solubility of conducting polymers in aqueous solution, originating from the lack of hydrophilic groups and stiff chains. Hence, Zhang's group synthesize sodium 4-(2,3-dihydrothieno[3,4-b][1,4]dioxin-2-yl)-methoxybutane-1-sulfonate (EDOT-S), which is one of the thiophene derivatives used as an amphiphilic monomer (**Figure 2a**) [2].

Thus, they prepared CPHs in one step through a combination of oxidative coupling polymerization and non-covalent cross linking of the amphiphilic thiophene derivative (**Figure 2b**) [2]. APS is a well-known non-crosslinking oxidant that can polymerize aniline, pyrrole, EDOT, etc., via oxidative coupling, but the corresponding conducting polymers are in the form of a precipitate due to the stiffness of their backbones and strong $\pi$-$\pi$ interactions among large conjugated units in their backbones. Unsurprisingly, PEDOT-PSS gels cannot be obtained

http://dx.doi.org/10.5772/63397

oxidized only by APS. However, the conducting polymer hydrogel was obtained for the first time starting from the amphiphilic EDOT-S using only APS as the oxidant. This indicates that the introduction of the hydrophilic ionic group to the hydrophobic monomer plays a key role in the synthesis of CPHs, which provides electrostatic interactions of the ionized polymer chains as extra cross-linking cites. By contrast, when $FeCl_3$ was used as the oxidant, the PEDOT-S hydrogels can be obtained under conditions of either a lower monomer concentration or a higher reaction temperature. This is probably due to the multivalent metal ions (excessive $Fe^{3+}$) offering an additional cross-linking force to the sulfonate groups attached to PEDOT-S backbones. Moreover, when the mixture of APS and $FeCl_3$ was used as the oxidant, a strikingly synergistic effect was achieved during synthesis.

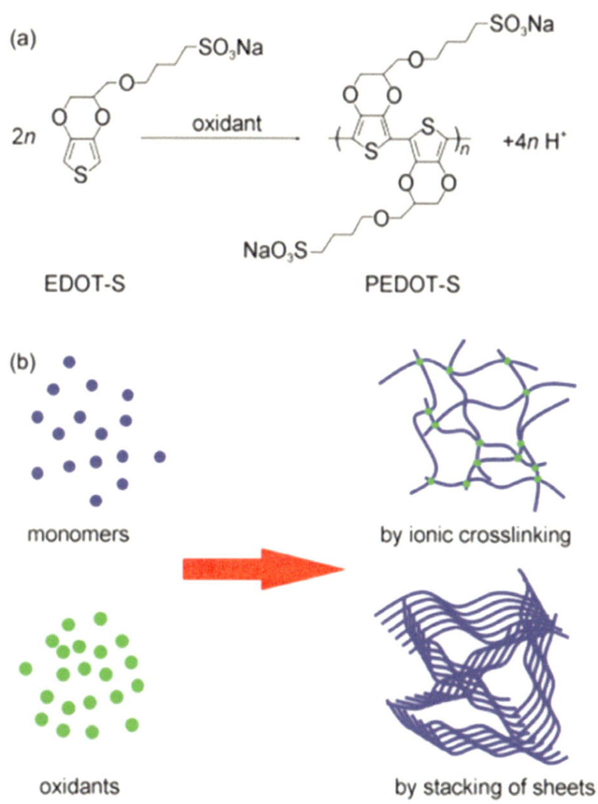


**Figure 2.** (a) Polymerization of EDOT-S into conducting polymer, and (b) schematic model for the formation of conducting polymer hydrogels through either ionic cross-linking or $\pi$-$\pi$ stacking. The oxidant used herein could be APS, $MCl_3$, or the mixture of them, where M represents Fe, La, Ce, Cr, or Sb [2].

Besides of the merits in gel forming, PEDOT-S can undergo significant 0D-2D transition during the polymerization and gelation process (**Figure 3**) [2]. The amphiphilic EDOT-S can self-assemble into sphere micelles in aqueous solution. However, when APS, $FeCl_3$, or APS-$FeCl_3$ was used as oxidant, PEDOT-S hydrogel was formed with the quasi-2D sheets, possessing an area-to-thickness ratio of much larger than 1000, serving as the building blocks.

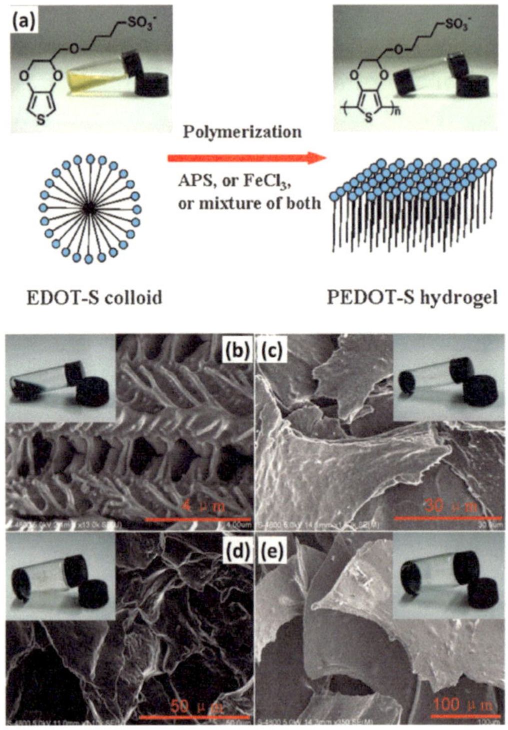


**Figure 3.** (a) A schematic diagram for synthesis of conducting polymer hydrogels. (b–e) SEM images of the resulting PEDOT-S products (see inset) synthesized with different oxidants: (b, c) using APS as the oxidant; (d) using $FeCl_3$ as the oxidant; (e) using a mixture of APS and $FeCl_3$ as the oxidant [2].

Considering the fact that the π-π interactions are negligible unless the number of aromatic rings is three or more, the authors attributed the 0D-2D morphological transition to the gradually increased π-π stacking effect. There is only one aromatic ring in each EDOT-S molecule, so the π-π stacking can be neglected. Combined with the difficult-happened overlapping of ring planes of the monomers due to the steric hindrance, EDOT-S easily forms spherical micelles. However, with the process of polymerization, overlapping of the ring planes of the macromolecules easily occurs because of the increasing π-π stacking among growing PEDOT-S macromolecules. Thus, initial EDOT-S spherical micelles gradually gather to form lamellar structure and ultimately convert into the PEDOT-S sheets.

Dimension evolution of the building blocks of the PEDOT-S hydrogel from 0D nanoparticles to 2D nanosheets did not only occur during polymerization and gelation process but was also observed by adjusting the reaction temperature or initial monomer concentration. It can be also explained by the enhanced π-π interaction of the conjugated length of polymer, which could surpass the hydrophobic interaction in original spherical micelles [3].

Except for the 0D-2D morphological transition of building blocks, the PEDOT-s hydrogel also exhibits chemical potential-dependent gel-sol transitions [2]. Chemicals with standard electrode potentials higher than 0.8 V [APS, $H_2O_2$, $K_2Cr_2O_7$, $HNO_3$, $Ce(SO_4)_2$, etc.] triggered

http://dx.doi.org/10.5772/63397

disbanding of the resulting conducting polymer hydrogels (called over-oxidation mechanism which has been confirmed via UV spectroscopy), promoting the occurrence of gel-sol transitions. It was evidenced that the gel-sol transition was driven by the decrease of conjunction length in the delocalized $\pi$ bond along the PEDOT-S chains caused by the strong oxidants.

Since EDOT-S can act as a reactive surfactant, it can be used to disperse EDOT to prepare PEDOT-S/PEDOT hydrogels through emulsion polymerization (**Figure 4**) [4]. The obtained hydrogels could also be an ideal precursor candidate for all-conducting polymer aerogels, which is the first all conducting polymer aerogel reported yet, with large BET surface areas, hierarchical pores, and abundant functional groups. It has been found that molar ratios of EDOT-S to EDOT have played a significant role in the stability of the EDOT-S stabilized EDOT emulsion—the higher molar ratio has led to the more stable colloid.

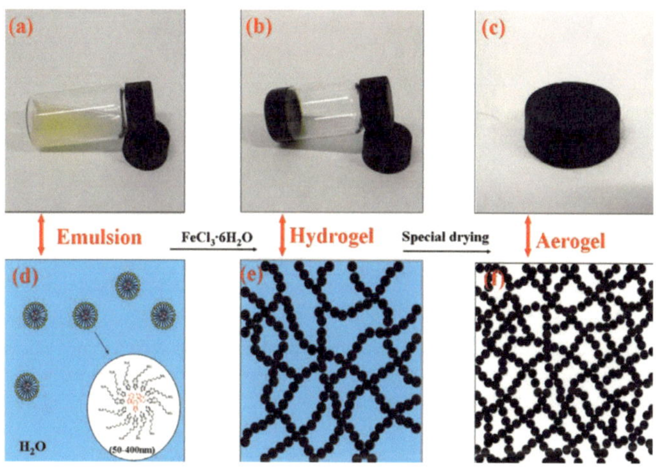


**Figure 4.** Emulsion-template synthesis strategy of all conducting polymer aerogels: digital photos of (a) EDOT emulsion stabilized by EDOT-S, (b) PEDOT-S/PEDOT hydrogel, (c) PEDOT-S/PEDOT aerogel and the corresponding schematic diagrams (d–f) [4].

### *2.1.2. PPy hydrogel*

Hydrogel could not only be fabricated by PEDOT and their thiophene derivative, could also be prepared by PPy, and, what is more important, PPy hydrogel can exhibit high elasticity through two-step synthesis strategy.

On account of inherent rigidity of the conjugated macromolecular chains originated from the delocalized $\pi$-electron system along the polymer backbone, it has been a huge challenge to make conducting polymer hydrogels elastic. Hence, Zhang's group turned to the two-step synthesis method, which successively contains a fast and a slow reaction procedure, to prepare elastic PPy hydrogels (**Figure 5**) [5]. The fast reaction procedure contributes to the formation of incipient network with low joint density to show certain flexibility. The next slow reaction process strengthens the incipient framework by forming conformal polymer coatings. The two-step synthesis method can be actualized by adopting deficient oxidant $Fe(NO_3)_3$ ($Fe^{3+}$ and $NO_3^-$ dominate the fast and slow oxidation procedures, respectively) as oxidant and aging for 30 days.

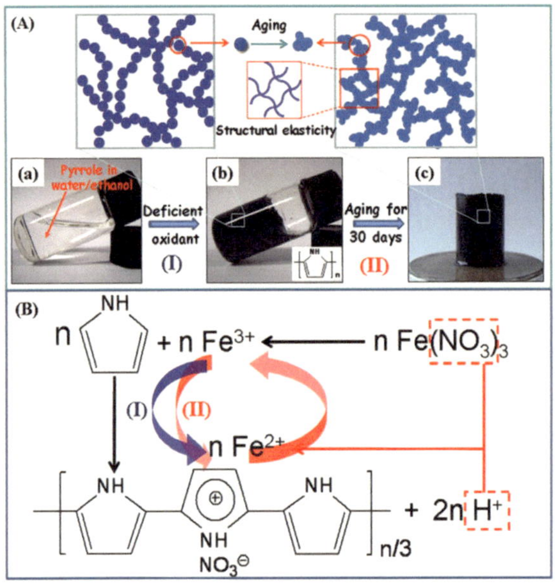


**Figure 5.** (A) The synthetic process of elastic polypyrrole hydrogel. Enlarged sketch maps indicated by squares in (A) have shown aggregated structural change of the polypyrrole hydrogel building blocks during polymerization. (B) The polymerization mechanism of the polypyrrole hydrogel [5].

To the best of our knowledge, it is the first report on the synthesis and properties of the elastic conducting polymer hydrogels. The work might also offer much inspiration to make more elastic conducting polymer hydrogels directly derived from PANi, PTh, etc., to prepare more elastic conducting aerogels.

### 2.1.3. PANi hydrogel

PANi hydrogels have been synthesized by embedding PANi into continuous matrix or using various non-conducting cross-linkers to cross-link PANi to form gels. However, both of the two strategies inevitably introduce non-conducting components, which impairs the performances of the resulting PANi hydrogels. Is it possible to make PANi hydrogels with continuous conjugated framework but without using any additional cross-linkers?

Herein self-cross-linked PANi hydrogels was synthesized for the first time via oxidative coupling reaction with APS as the oxidant and aniline hydrochloric salt as the precursor without any additional crosslinkers [6]. Aniline hydrochloric salt was chosen as the raw material on account of its water solubility and that the dissociated HCl could also dope PAni to ensure high conductivity. Aniline hydrochloric salt was oxidized by equimolar amount of

APS. With the polymerization proceeded, the PAni hydrogel formed. Subsequently, the hydrogel aged at ambient temperature to obtain an ideal precursor for PANi aerogels.

### *2.1.4. Composite conducting hydrogels*

Conducting hydrogels could be fabricated by typical conducting polymers (PEDOT, PPy, PANi) via well-designed synthesis strategies as maintained above. On this basis, conducting hydrogels can also composite with other functional materials to obtain more integrate performances for final aerogel products.

At first, by embedding carbon nanotubes into PEDOT-PSS supermolecular hydrogels in the presence of polyvinyl alcohol (PVA) as stabilizer for carbon nanotubes, Zhang's group [7] have prepared PEDOT-PSS/MWCNTs and PEDOT-PSS/c-MWCNTs composite gels with low densities of 0.04–0.07 g $cm^{-3}$.

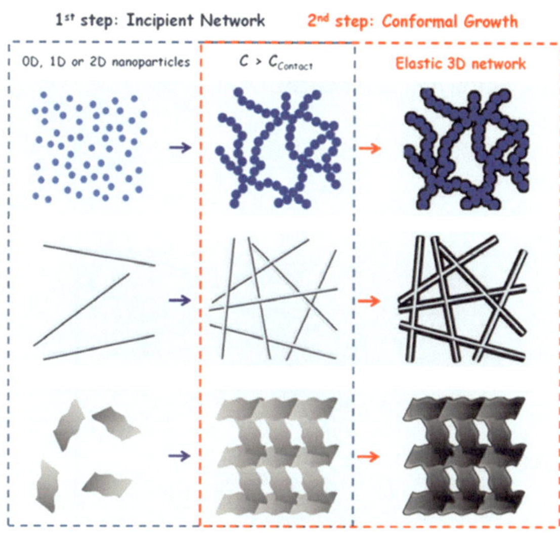


**Figure 6.** Schematic representation of incipient network conformal growth (INCG) technology [8].

On the basis of elastic PPy hydrogels, an incipient network conformal growth (INCG) technology was proposed to prepare hybrid and elastic porous materials (**Figure 6**) [8]. 0D, 1D, or 2D nanoparticles (NPs) are dispersed in solvent to form a uniform suspension at first. Once the concentration of suspension is within a proper range, the incipient network will form through NPs contacting with each other. The monomer of conducting polymer is then introduced and supposed to coating the incipient network conformally through polymerization. The conformal coatings not only endow NPs suspension with high dispersion but also offer composite conducting polymer with additional structural elasticity, meeting requirements for future generations of portable, compressive, and flexible devices.

To put INCG technology into practice, the fabrication process for PPy-Ag NW hybrid gels is demonstrated in **Figure 6** [8]. Ag nanowires (Ag NWs) were selected to contribute the incipient

3D networks considering easy synthesis and uniform diameters. Evaluated by Doi and Edwards theory, Ag NWs suspension with the volume friction between $\phi_1 = 8 \times 10^{-7}$ and $\phi_2 = 10^{-3}$ was appropriate to form an incipient 3D networks. Above $\phi_1 = 8 \times 10^{-7}$, the dispersed Ag NWs in suspension began to contact with each other. Pyrrole (Py) was then added and preferentially adsorbed to the surface of Ag NWs resulting from metal-$\pi$ strong interactions, to form core-shell structures. Once the oxidant was introduced in the next, Py was oxidized and polymerized in situ. Thus, the 3D network and the core-shell morphology were fixed, obtaining PPy-Ag NW coaxial nanowire aerogels.

Besides, a "wet" process to incorporate chiral polyaniline nanowires into the agarose was put forward, and hence, the chiroptical properties could be attached to PANi hydrogels [9]. The chiral polyaniline nanofibers were obtained by successively using potassium tetrachloroaurate (PTC) and APS to oxidize aniline in the presence of (1S)-(+)-10-camphorsulfonic acid [(S)-(+)-CSA] or (R)-(−)-CSA as a chiral-inducing agent. The optical activity of the chiral polyaniline is dependent on the mass ratio of PTC to APS, reaching peak values in the ratio range of 1.0–2.5. The chiroptical properties of the polyaniline nanofibers are retained within the composite hydrogel, although the degree of chiral organization of the polyaniline appears to be somewhat modified by the presence of the agarose matrix.

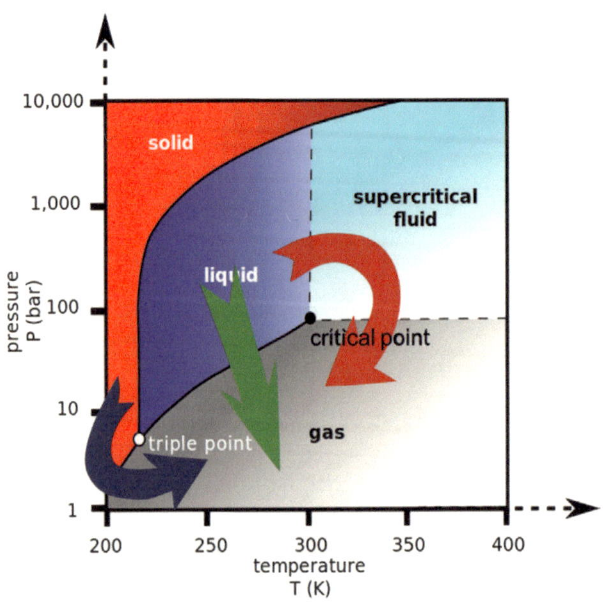


**Figure 7.** Phase diagram of pure materials and arrow illustrations of different drying process (supercritical drying—red arrow, ordinary drying—green arrow, and freeze-drying—blue arrow).

## 2.2. Drying of gel precursors

After the conducting polymer hydrogels were produced, the hydrogels should be processed into aerogels by replacing the liquid solvent with air. When a substance is dried via normal methods of applying heat and pressure at a finite rate, the substance passes through the liquid-gas barrier (green arrow in **Figure 7**), where the amount of capillary stress changes, causing the substance to deflate. To avoid this problem, there are supercritical drying (SCD), which dries a substance via high heat and pressure and goes beyond the critical point (red arrow in **Figure 7**), and freeze-drying, which goes through solid-gas barrier (blue arrow in **Figure 7**) to avoid the direct liquid-gas transition.

### *2.2.1. Supercritical drying*

Supercritical drying (SCD) is performed to replace the liquid in a material with a gas to isolate the solid component without destroying the material's delicate nanostructured porous network. Carbon dioxide ($CO_2$) is the most used as supercritical fluid, because it is supercritically extracted at a lower temperature (31.1°C) than an organic solvent without the risk of combustion. Through low-temperature SCD, most of the obtained conducting polymer hydrogels mentioned above, such as PEDOT-PSS, PEDOT-s/PEDOT, PEDOT-PSS/MWCNTs, PPy, PPy/Ag NW, PANi hydrogels, can be converted into corresponding aerogels successfully. Low-temperature supercritical drying for aerogel production conducts as follows:

A gel is prepared using sol-gel chemistry. The gel contains a mixture of organic solvent and water in its pores.

1. The gel is soaked in a pure organic solvent (ethanol, methanol, acetone, amyl acetate, etc.) and changes the solvent for several times over several days to remove the original solvent from its pores.

2. Finally, the gel is supercritically dried in a pressure vessel. The gel is placed in a pressure vessel, and the pressure vessel is then filled with liquid $CO_2$. Adjusting temperature and pressure to make the gel soaked in supercritical $CO_2$ and flushing through new $CO_2$ every 1–2 h. After the solvent in the pores of the gel has been completely replaced by supercritical $CO_2$, the vessel is then isothermally depressurized to give aerogels.

It is notable that when the solvent of the gel is exchanged for supercritical $CO_2$, the gel may also shrink slightly. This stems from favorable interactions between the two liquids, to put it simply, molecules take up smaller space in a mixture than molecules in either of the separate liquids. Solvents with low volume changes of mixing with supercritical $CO_2$ include acetone, amyl acetate, ethanol, and methanol. In general, aerogels made using supercritical drying from $CO_2$ may shrink up to 5%. Even so, low temperature SCD is still the best drying method to maintain deliberate nanostructures of aerogels so far.

### *2.2.2. Freeze-drying*

Although SCD is a mostly used drying method to maintain deliberate nanostructures of aerogels, it is not always successful for gel precursors, because (1) some conducting polymer gels dissolve in alcohol or acetone, like aforementioned PEDOT-s hydrogels [2, 3], (2) gels are prepared in solvent which cannot exchange with alcohol or acetone or supercritical $CO_2$ and (3) some metal-containing gels will react with the carbon dioxide to create metal carbonates.

In these circumstances, the freeze-drying is a common alternative for gel precursors to proceed into aerogels (also called cryogels sometimes).

In freeze-drying, the gel precursors were usually quickly frozen in liquid $N_2$ at first. Then, reduce the surrounding pressure to allow the frozen solvent in the gel pores to sublimate directly from the solid phase to the gas phase, leaving aerogels.

To avoid the structural damages caused by ice crystals of solvent produced and grown during the freezing process, the frozen process should finish within the shortest time, usually quenching the gel precursors in liquid $N_2$ with a very quick speed. Well, it is indeed a crazy idea to carry out in a diametrically opposite way, called ice-segregation-induced self-assembly (ISISA) [10]. The pore morphology of resulting cryogels can be well controlled via ISISA process by unidirectional freezing at a definite immersion rate (**Figure 8**). During unidirectional freezing, crystalline ices are produced, causing the original dispersed solute to be expelled to the boundaries between adjacent ice crystals gradually. After experiencing freeze-drying, the ordered-arranged solutes become "walls," enclosing the empty spaces that formerly occupied by ice crystals. Aligned or unaligned conducting polymer cryogels with 3D macroporous architectures have been prepared using the ice-segregation-induced self-assembly (ISISA) of different PEDOT-PSS freezing precursors as a dispersion or a formed hydrogel.

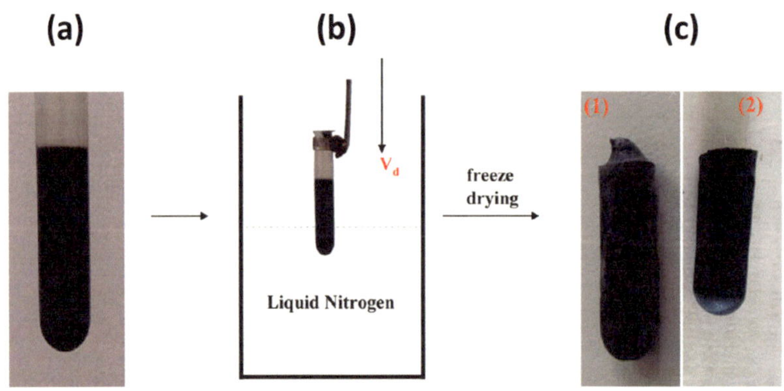


**Figure 8.** (a) An example of the PEDOT-PSS freezing precursor held in a plastic centrifugal tube. (b) A schematic diagram for unidirectional freezing of the PEDOT-PSS precursor by dipping vertically into liquid nitrogen at a rate of 3–50 mm $min^{-1}$. (c) Digital photos of the PEDOT-PSS cryogels starting from dispersion (left) and hydrogel (right) [10].

It is believed that ISISA is a facile method for producing hierarchically macroporous 3D monoliths of conducting polymers with many advantages as follows. (1) It is a green and inexpensive method for ice template was merely needed without any other organic solvents. (2) It is facile, reproducible, and readily controllable by adjusting immersion rate, solute concentration, etc. (3) It endows conducting polymers with unusual long-range ordered macropores, which may introduce many unexpected properties and thus may exploit some new applications in the fields of electronic components (including batteries and transistors, solar cells), tissue engineering, and next-generation catalytic and separation supports.

# 3. Properties of conducting polymer aerogels

## 3.1. PEDOT aerogels

The PEDOT-PSS aerogels dried through SCD show low density (0.138–0.232 g $cm^{-3}$), large surface area (170–370 $m^2$ $g^{-1}$), and hierarchical porous structures [1]. The resulting PEDOT-PSS aerogel (with molar ratio of 1:1 for PEDOT: PSS) has a conductivity of $10^{-1}$ S $cm^{-1}$, being comparable with the value of PEDOT-PSS thin film.

Processing PEDOT-PSS via ISISA endows the conducting polymers with novel properties. Well-ordered alignment of the macroporous structure could be observed along the longitudinal direction (i.e., the ice growth direction) for the resulting aerogels from SEM, which resulted from phase separation that occurs during the directional freezing process (**Figure 9**) [10]. Closer observations in different areas showed that there were several domains with random orientations over the whole macroporous monolith and that the boundaries of these domains could be easily recognized. More interestingly, unexpected fingerprint-like morphology can be observed from the cross section of the cone-shaped bottom of the cryogel (**Figure 9d**). This may be attributed to the super-cooling of the dispersion in the immersed portion, thus instead of crystalline ice, the amorphous ice grow and act as a template. These observations may indicate an efficient way to produce the man-made fingerprint for identifications and markings.

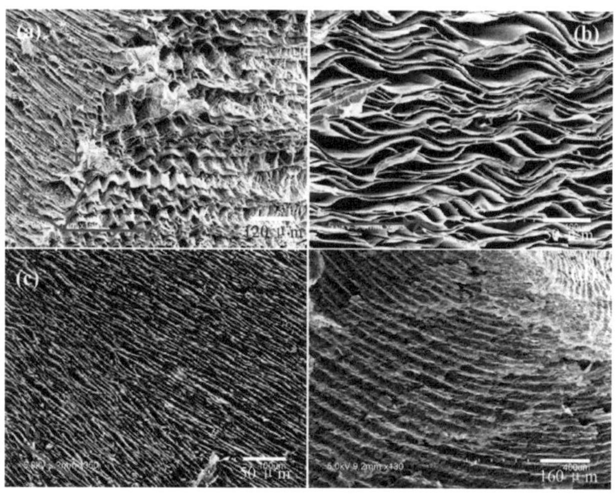


**Figure 9.** SEM images of PEDOT-PSS cryogel prepared via ISISA showing a well-ordered macropore structures from (a) the whole domain, (b) face $Zr_1$, (c) face $r_1r_2$, and (d) cross section of the cone-shaped bottom of the cryogel. (Arrow Z and r denotes the longitudinal and radial directions, $Z \perp r_1 \perp r_2$.) [10].

Besides, all conducting polymer PEDOT-S/PEDOT aerogels obtained by emulsion polymerization and dried through SCD have high conductivity with the level of 101 S $m^{-1}$ [4]. In comparison with any other conventional aerogels, the resulting PEDOT-S/PEDOT aerogels, without the necessity of further functionalization, show superb adsorption ability to various guest substances, such as dyestuffs and heavy metal ions, and enhanced electrochemical.

Furthermore, the specific surface areas, thermal stability, and electrical conductivities can be significantly enhanced by embedding MWCNTs into PEDOT-PSS aerogel matrix [7]. The resulting composite aerogels show low density (0.044–0.062 g $cm^{-3}$), large surface area (280–400 $m^2$ $g^{-1}$), high electrical conductivity (1.2–6.9 × $10^{-2}$ S $cm^{-1}$), and hierarchical porous structures.

### 3.2. PPy aerogels

The PPy hydrogel could be also readily converted into the lightweight, elastic, conductive, and aerogel through SCD. Thus, pure organic, electrically conductive (ca. 0.5 S $m^{-1}$), lightweight (0.07 g $cm^{-3}$) PPy aerogel was first obtained by Zhang's group (**Figure 10**) [5]. The obtained aerogels have still kept the excellent elasticity after drying of the PPy hydrogels, which could be compressed by ≥70% and recovered to its original shape in 30 s.

**Figure 10.** Digital photos of the lightweight, elastic, conductive, and organic PPy aerogels made from the resulting PPy hydrogels through supercritical fluid drying process located on a dandelion and connected to the LED bulb [5].

The authors attributed the elasticity of the PPy aerogels to the elaborate synthesis design and the resulting micro-structural changes. (1) At first, the reaction rate is slowed down and the reaction extent is limited by adopting the deficient oxidant, bringing an incipient network with the reduced joint density. Thus, the obtained PPy aerogels exhibit less stiffness. (2) Then, the initial network can be reinforced effectively by coarsening the framework conjunctions, benefited from the secondary growth during the slow reaction process. Thus, the obtained PPy aerogels can avoid the structural fracture when suffering compressions. (3) At last, the asymmetrically epitaxial growth of the original polymer particles during the secondary growth offers many new weak contacting joints. Single-point contacting for the building blocks of traditional PPy aerogels suffering compression is replaced by multi-point contacting or even face contacting. Thus, inside stress of the compressed PPy aerogels can be easily dissipated, preventing damage caused by stress concentration. In summary, the PPy aerogels can restrict the irreversible fracture of the hydrogel network when bearing compression.

http://dx.doi.org/10.5772/63397

Besides PPy aerogels, PPy-Ag NW aerogels prepared through INCG [8] also exhibited superb compressive elasticity, which could be compressed by large deformations (>90%) and return to its original shape in seconds once withdrawing the compression (**Figure 11a**, **b**). From the comparison stress-stain (σ-ε) curves for 50 compress-release circles along the loading direction with under a fixed max strain of 20%, PPy-Ag NW aerogel recovered their deformations with little mechanical failures (**Figure 11c**).

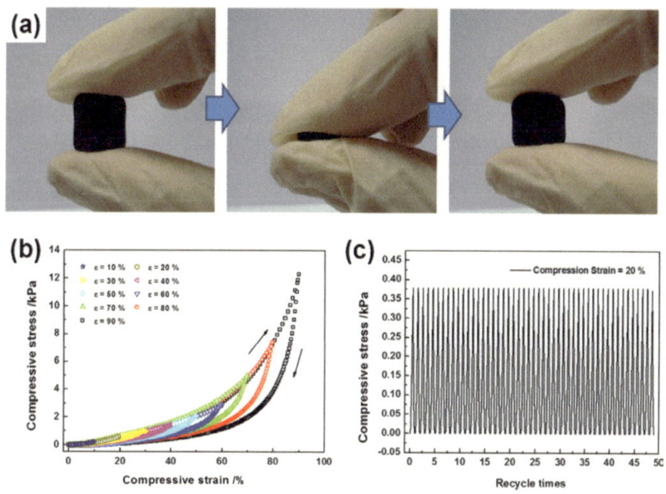


**Figure 11.** (a) Digital pictures showing a superb elasticity of PPy-Ag NW aerogel. (b) σ–ε curves for PPy-Ag NW aerogel along the loading direction during loading–unloading cycles (ε = 10–90%). (c) 50 consecutive compression tests for PPy-Ag NW aerogel at ε = 20% [8].

Similarly to PPy aerogels, the superb elasticity of PPy-Ag NW aerogels also comes from the rational-designed nanostructures. The resultant PPy-Ag NW aerogels possess: (1) strong and flexible 3D network contributed by ultra-long coaxial nanowires, giving aerogels resistance and elasticity to external compressions by bending their framework skeletons; (2) rich pores, making aerogels possible to dissipate the external compressive energy by shutting off the pores; (3) in situ welded junctions and strong metal-π interactions, avoiding the interfacial slippage resulting from the poor load transfer between adjacent coaxial nanowires or Ag NW and PPy coatings.

### 3.3. PANi aerogels

Self-cross-linked PANi hydrogels can be also converted to corresponding aerogels by SCD [6]. The obtained aerogels exhibit adjustable densities from 0.03–0.15 g $cm^{-3}$, coral-like micro-morphology, high BET surface area (39.54 $m^2$ $g^{-1}$), and comparative mechanical strength. The resulting PANi aerogels showed a coral-like micro-morphology, formed by branched nano-fibers with a diameter of 50–150 nm and a length up to tens of micrometers (**Figure 12**). This unique morphology insures higher electrochemical performance, because the charges can transport effectively in 3D porous network than the corresponding bulk materials.

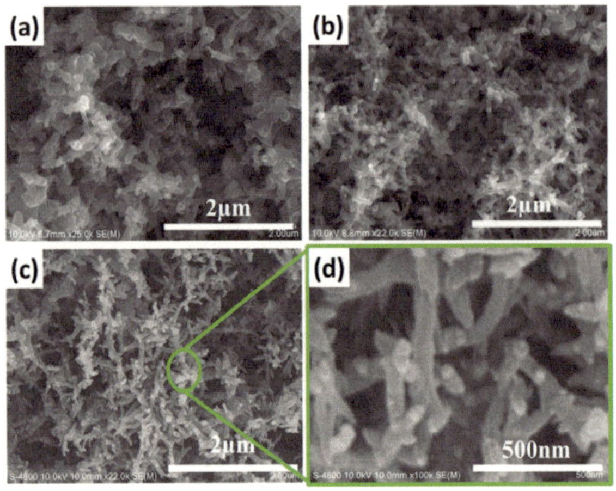


**Figure 12.** SEM images of PANi hydrogels with a density of (a) 0.05 g $cm^{-3}$, (b) 0.08 g $cm^{-3}$, (c) 0.11 g $cm^{-3}$, and (D) the magnified image of PANi hydrogel with 0.11 g $cm^{-3}$ [6].

As the obtained PANi aerogels exhibit high porosity from SEM detections, it is necessary to analysis their specific surface area and pore size distribution (PSD) by Brunauer-Emmett-Teller (BET) tests. The adsorption-desorption isotherm curve (**Figure 13**) of PANi aerogels presents the characteristic feature of the type-IV isotherm with a H3 hysteresis loop, indicating the existence of a large number of mesoporous (with a diameter of 2–50 nm). A further analysis shows that PANi aerogel possesses a BET specific surface area of 39.54 $m^2$ $g^{-1}$ and a unimodal pore size distribution peak around 28 nm. The high surface area and mesoporous structure are supposed to enlarge electrode-electrolyte contact surface and boost the electrolyte transfer obviously.

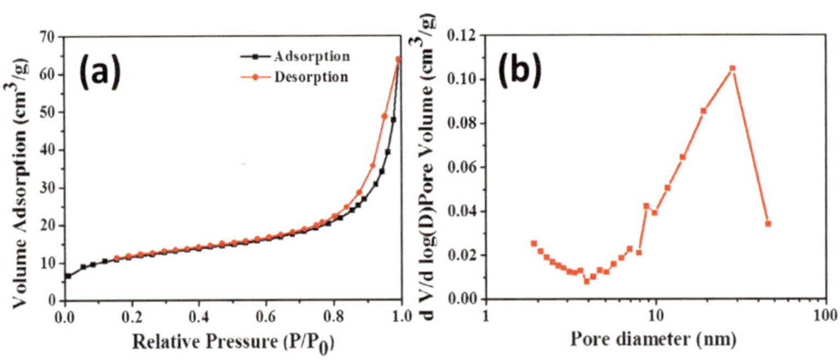


**Figure 13.** (a) Nitrogen adsorption/desorption isotherms of PANi aerogels analyzed by BET detections and (b) the corresponding pore size distribution [6].

Besides, PANi aerogels exhibit increasing mechanical strength with the increase of the PANi density. The comparative mechanical strength insures better integrity of the hydrogel network, which also contributes to better transfer of ions, charges, and electrons in redox reaction interface during electrochemical tests.

# 4. Application of conducting polymer aerogels

The large surface area and wide pore size distribution, together with their electro properties, would allow these conducting polymer aerogels to be applied in many fields with unexpected performance.

## 4.1. Electrochemical energy storage

The conducting polymer hydrogels and aerogels show superior electro-chemical properties including a fast electrochemical response, a high specific capacitance, and a low electric resistance, resulting from the combined merits of organic semiconductors and three-dimensional porous gels.

At first, the electrochemical capacitive behavior of the supercritically dried PEDOT-S/PEDOT aerogel has been investigated by cyclic voltammetry (CV) between 0.2 and 0.6 V in 1.0 mol $L^{-1}$ KCl solution [4]. Fixing the scan rate at 50 mV $s^{-1}$, the CV curves for both PEDOT-S/PEDOT and PSS/PEDOT aerogel electrodes show a rectangular-like shape with an obvious wide peak at the potential interval of −0.2 to 0.0 V, indicating the presence of a pseudo-capacitive effect caused by the redox of the conducting polymers. The electrochemical capacitance of the electrode materials can be calculated further from the integral CV area. For the PEDOTS/PEDOT aerogel, the electrochemical capacitance is 68.5 F $g^{-1}$, much more than the value of the PSS/PEDOT aerogel electrode (19.2 F $g^{-1}$). Furthermore, the good rate capability (slight decay of the electrochemical capacitance along with the increase of the scan rate) and high cycling stability (negligible changes of the electrochemical capacitance after 100th charge-discharge cycle) of the resulting PEDOT-S/PEDOT aerogel electrode have been further confirmed from CV detections. The excellent performance in electrochemical energy storage of the resulting PEDOT-S/PEDOT aerogel electrode might be ascribed to low ionic resistance and fast kinetics of the electrochemical process in an aqueous medium.

In the meanwhile, CV curves of the PEDOT-S gels were recorded in 1.0 mol $L^{-1}$ $Na_2SO_4$ aqueous solution [3]. The specific capacitance of the PEDOT-S hydrogels is estimated to be ca. 30–60 F $g^{-1}$, which is much higher than PEDOT:PSS gels. This can be ascribed to the higher proportion of active ingredients in PEDOT-S than PEDOT:PSS gels for PSS does not contribute electrochemical capacitance directly.

Besides PEDOT aerogels, PANi gels are proved to be a promising material for electrochemical energy storage [6]. For the self-cross-linking PANi hydrogel, functionalized electrodes have exhibited a new record of specific capacitance (750 F $g^{-1}$). At first, CV curves of the PANi hydrogel electrode were collected in 1.0 M $H_2SO_4$ electrolyte solution (**Figure 14**). The representative redox peaks, which belong to the transformation between leucoemeraldine base (LB), emeraldine salt (ES), and pernigraniline base (PB), persisted from 5 to 200 mV $s^{-1}$, indicating a high stability in redox electrochemical activity. Then, EIS analysis was performed to evaluate the charge transfer property of PANi hydrogel. The solution resistance of the PANi gel electrode is as low as 1.7 X. The charge-transfer resistance of electrode of the PANi hydrogel electrode is <0.1 X, which is remarkably low due to well-matched resistance between the PANi hydrogel (electronic conductivity) and liquid electrolyte phase (ionic conductivity), insuring good ion transport within 3D gel framework. At last, chronopotentiometry (CP) tests were carried out to further evaluate the electrochemical energy storage performance of PANi hydrogel electrode (**Figure 14**). 75.8% capacitance remained with the increasing current density from 2 A $g^{-1}$ (588 F $g^{-1}$) to 20 A $g^{-1}$ (446 F $g^{-1}$), indicating an excellent rate capability for high power performance. The specific capacitance of the PANi hydrogel electrodes is dependent on hydrogel density, reaching the peak level (750 F $g^{-1}$) at the density of 0.11 g $cm^{-3}$.

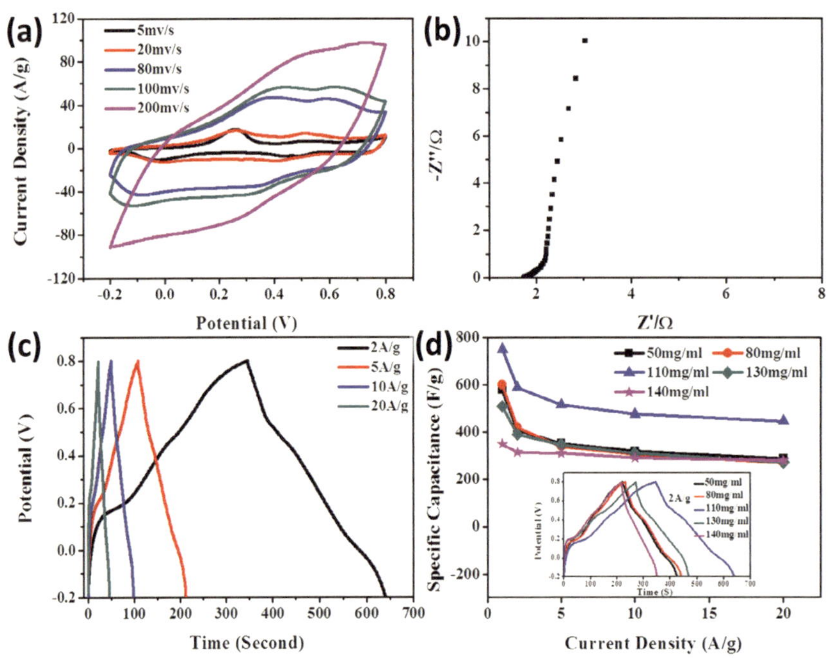


**Figure 14.** Electrochemical performance of PANi hydrogel electrodes: (a) cyclic voltammogram curves, (b) Nyquist plot, (c) galvanostatic charge/discharge curves, (d) specific capacitances of PANi hydrogel electrodes with various PANi densities, and (d-inset) chronopotentiogram curves of PANi hydrogel electrodes at current density of 2 A $g^{-1}$ [6].

### 4.2. Stress sensing

Piezoresistive sensors, transducing pressure message to resistance signal, have been rapidly developed and widely used for their advantages in high sensitivity, fast response, feasible fabrication, low cost, and easy signal collection. Conductive and elastic aerogels are an emerging kind of candidates for fabrication of piezoresistive sensors because of their combinational electric conductivities and mechanical flexibilities.

As mentioned above, the PPy hydrogel can be readily converted into the lightweight, elastic, conductive, and organic aerogels via SCD [5]. The obtained aerogels have still kept the excellent elasticity after drying of the PPy hydrogels and been made into stress sensors successfully. The decreases of the electrical resistance are directly proportional to the compression strain, reaching at 3% when the aerogel was compressed by 80%. It is not difficult to understand that the external responses in electrical properties are brought by the different contacting situations of the gel network upon compressive stresses. The resistance decreases of the compressed PPy aerogels could complete within 1.5 s and so was the electrical resistance returns to the original value once being released. It is because of the elasticity of PPy aerogels that their gel skeletons can bend and recover in seconds without structural fractures. Besides of the short response time, the electrical responses of PPy aerogels also exhibit high stability under compress and release circles.

Although pure conducting polymer aerogels show great potentials in stress sensing, they have not solved the "temperature drift," a common problem that the piezoresistive materials always meet caused by the temperature-dependent electrical resistances. Hence, self-temperature compensated sensors are highly desirable in direct and precise measurement. The most feasible solution is developing hybrid sensing materials with a low/nearly zero temperature coefficient resistivity (TCR) by offsetting the positive and negative temperature dependence of electric resistance. In this respect, the polypyrrole/silver (PPy/Ag) coaxial nanowire hybrid aerogel prepared via INCG [8] is a wise choice for Ag NWs show positive temperature-coefficient, while PPy show negative temperature-coefficient. Through the reasonable selection of the ratio of Ag NWs at 83.3%, the electric resistance of the hybrid aerogels can be immune from temperature influence, showing a nearly zero temperature coefficient (**Figure 15a**).

To process into a stress sensor, PPy-Ag NW aerogel was imbedded into two parallel copper electrodes, as shown in **Figure 15b**. The electric resistance decreased directly with the compression strain, changing by 83% at the strain of 80% (**Figure 15d**). The electric resistance response of PPy-Ag NW aerogel caused by compression could be completed within 1 ms and so was the electric resistance recovery after compression released. The detection minimum was as low as 4.93 Pa when the aerogel was compressed by 0.5%. Furthermore, the electrical response of PPy aerogel upon stress exhibited high stability, indicated by suffering hundreds of compression-release circles for each compressive strain (**Figure 15e**). The sensitivity S (= $\delta(\Delta R/R_0)/\delta\sigma$) was as high as 0.33 $kPa^{-1}$ for the nearly zero temperature coefficient sample, while S = 0.06 $kPa^{-1}$ for another sample with a Ag NW content of 2.7 wt% (**Figure 15f**). It implies that a series of sensing samples with different sensitivities can be prepared by programmed synthesis to meet the needs of applications. There was no deviation in sensing performance at different temperatures (15, 25°C) (**Figure 15f**) or after 3 V electric heating for 160 s, affirming the high sensing stability of the PPy-Ag NW aerogel sensor with nearly zero temperature coefficient.

### 4.3. Joule heating

A stress-triggered Joule heater can be fabricated by aerogels used for stress sensing in principle. Under compression, the electric resistance can be reduced and the Joule heating generated by electric current running through aerogels can be raised. It is possible that the thermal generation may be inconspicuous for aerogels in original state upon inputting a proper voltage, while for the compressed aerogels, the Joule heating can raise its temperature considerably, named stress-triggered Joule heating.

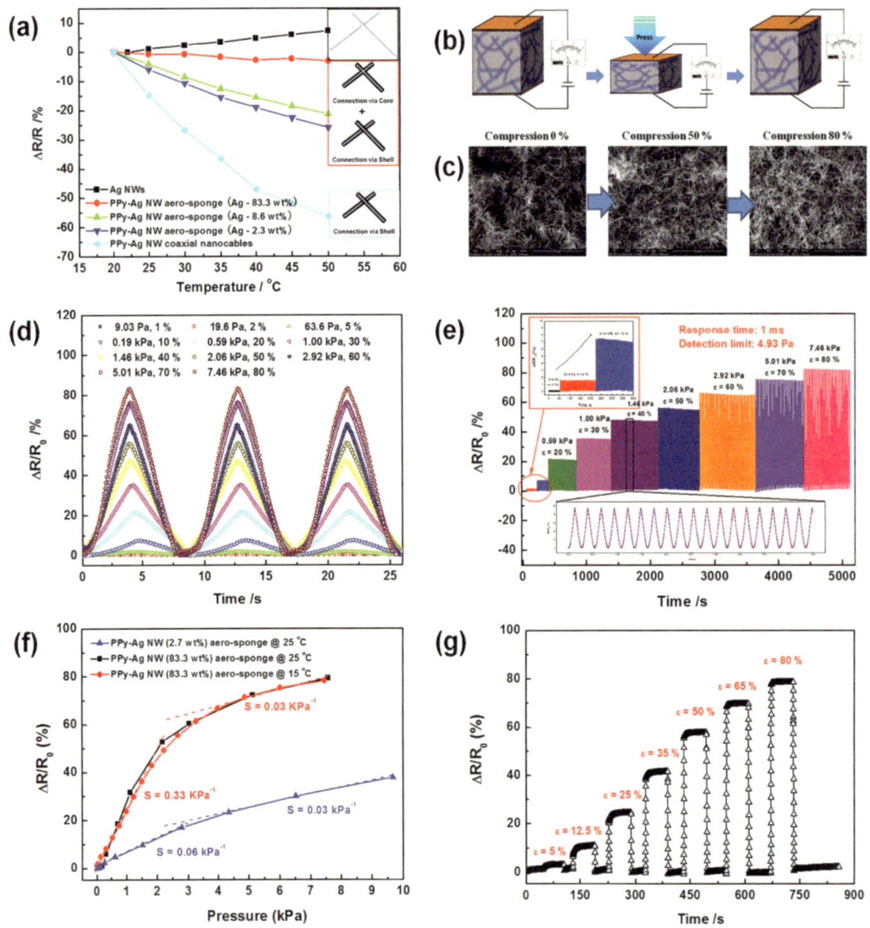


**Figure 15.** Performance of PPy-Ag NW hybrid aerogels in stress sensing. (a) Curves of how the electric resistance changes with temperature relatively (ΔR/R ~ T) for Ag NWs, PPy-Ag NW membrane deposited by nanocable suspension, and PPy-Ag NW hybrid aerogel, respectively. (b) Schematic illustration of how to prepare a stress sensor with PPy-Ag NW aerogel. (c) SEM images of PPy-Ag NW aerogel with compression. (d) Multiple-cycles electric resistance tests under a series of repeated loading-unloading pressures with 1 V input. (e) Reliability test of PPy-Ag NW hybrid aerogel under a series of repeated loading-unloading pressures with partial enlarged details in insets. (f) Comparison of how electric resistance changes with compressive pressure relatively (ΔR/R ~ p) for the nearly zero temperature coefficient PPy-Ag NW aerogel performed at 25 and 15°C. (g) Electric resistance response of the nearly zero temperature coefficient PPy-Ag NW aerogel at different strains by introducing a series of persistent compressive pressure [8].

The stress-triggered Joule heater can be fabricated by PPy-Ag NW hybrid aerogel, which is placed on thermal insulating blanket and covered by a reversible thermo-allochroic membrane (**Figure 16a**) [8]. Inputting a persistent voltage of 3 V, the Joule heating generated by electric current running through the original aerogel was mostly dissipated and will not causing any visible differences for thermo-allochroic membrane (**Figure 16b**). However, if imposing a persistent compressive pressure, the electric current was boosted by nearly four times. As a consequence, Joule heating increased, warmed the aerogel, and induced the color

http://dx.doi.org/10.5772/63397

changes of the thermo-allochroic membrane (the temperature point of color transition for thermo-allochroic membrane used in this work is about 43°C) (**Figure 16c**). Once the compressive pressure was released, the membrane returned its original color gradually, indicating the temperature recovery of the aerogel.

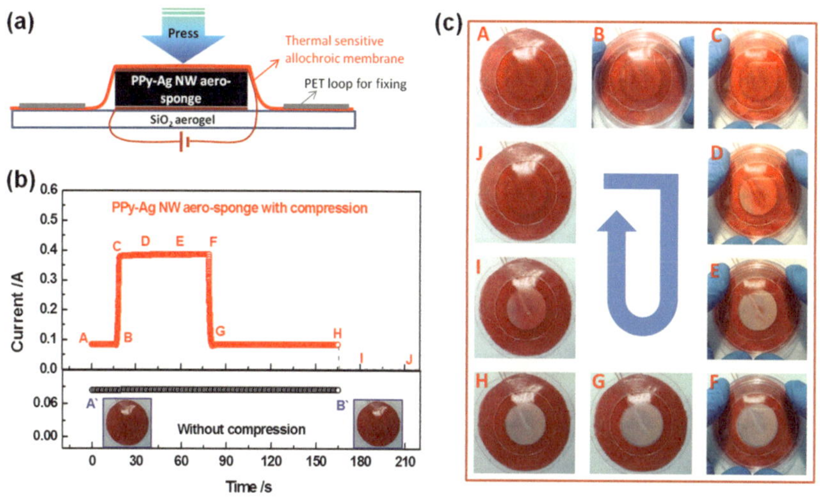


**Figure 16.** The performance of stress-triggered Joule heater. (a) Schematic illustration of how to construct a stress-triggered Joule heater with PPy-Ag NW aerogel. (b) Real-time I–t curves of the stress-triggered Joule heater under both free and compressive states with 3 V voltage input, and inset is the digital pictures of the Joule heater in free state. (c) Digital pictures of the intelligent Joule heater at the corresponding time point marked in (b) [8].

### 4.4. Adsorbing and separation

Hierarchical pores, a large surface area, and abundant conjugated chain or functional groups have endowed the conducting polymer aerogels with adsorption ability of some guest substances, such as heavy metals, dye molecules, etc.

For PEDOT aerogels, let us have a look at molecular structure of PEDOT-S at first. The conducting polymers possess linear π-conjugated system and lots of attached sulfonic acid groups. The π-conjugated system makes both of PEDOT-S and PEDOT-S/PEDOT aerogels show good affinity to the most of π-conjugated based dyestuffs via π-π stacking. But the adsorptive capacities of these aerogels to cationic and anionic dyestuffs exhibit a significant difference, contributed by the electrostatic interaction of the adsorbed dyestuffs and the sulfonic acid groups attached on the PEDOT-S polymer chains. The adsorptive capacity of PEDOT-s and PEDOT-S/PEDOT aerogels can reach around 164 and 170 mg $g^{-1}$ for basic fuchsine, while gels can only adsorb 39 and 160 mg $g^{-1}$ acid fuchsine, respectively [3, 4]. Moreover, for both of the two kinds of aerogels, the adsorbed basic fuchsin can be released by addition of a cationic surfactant cetyltrimethyl ammonium bromide (CTAB) and the more the CTAB added, the more the basic fuchsin released, which provides an efficient way to control the release of the adsorbed dye molecules.

In addition, the PEDOT-S/PEDOT aerogel also show certain adsorptive capacity to heavy metals ions caused by electrostatic interactions [4]. But the adsorptive capacity is not uniform towards different kind of metals, 184.1 mg $g^{-1}$ for $Hg^{2+}$, 111.1 mg $g^{-1}$ for $Ag^{+}$, 79.6 mg $g^{-1}$ for $Pb^{2+}$, and 28.3 mg $g^{-1}$ for $Cu^{2+}$. Author ascribed it to the different affinity of metal ions to alkoxy sulfonate groups attached on the main chains of the conducting polymer. On all accounts, these adsorption investigations have shown a very promising application of the PEDOT aerogels in the removal of color and heavy metal ions from wastewater, etc.

Although there is no ionized functional group attached on PPy chains, the π-conjugated structures can provide enough active cites for dyestuff adsorptions. The high specific surface area (ca. 12 $m^2$ $g^{-1}$), excellent mechanical strength, and strong π-π interaction between dye molecules and sorbent matrix allow PPy aerogels to extract dye molecules from the dye solutions in a very fast and effective way (**Figure 17**) [5].

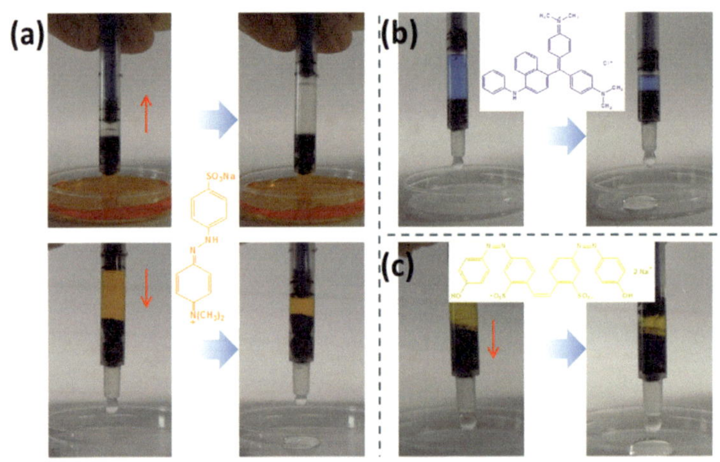


**Figure 17.** Digital photos showing fast removal of (a) methyl orange, (b) victoria blue, and (c) brilliant yellow from waste water with the elastic PPy hydrogels in syringes [5].

The PPy gels were placed in the tube of a syringe, and the plunger was pulled or pushed allowing take in or expel the dye solutions after permeating through the PPy hydrogels. For all three kinds of dye molecules, the colorful dye solutions changed into colorless in several seconds. PPy gel could absorb methyl orange (MO) with a capacity of 389.5 mg $g^{-1}$ and an efficiency of 99.99%. Besides, the used hydrogel can be refreshed by treating with NaOH aqueous solution with a concentration of 2 mol $L^{-1}$ at 80°C, and the refreshed gel could still adsorb MO with a capacity of 196.4 mg $g^{-1}$ and an efficiency of 99.89%.

## 5. Outlook: problems, prospects, and challenges

This chapter has focused on recent research progress in the development of conducting polymer aerogels, with specific focus on PEDOT-containing, PPy, and PANi aerogels discus-

http://dx.doi.org/10.5772/63397

sing from (1) synthesis strategy of conducting polymer (PEDOT, PPy, and PANi) aerogels through rational design for oxidant, cross-linker, soft template, sol-gel process, drying process; (2) advantages of these aerogels in physical and chemical performance, compared with the counterparts in bulk or membrane; and (3) applications in energy storage, adsorption to metal-ions/dye-molecules, stress sensing, Joule heating.

However, there are still lots of challenges remaining. At first, as a result of the fast oxidation and the poor solvent solubility of intrinsically conducting polymers, there are great difficulties in preparing conducting polymer aerogels in large-scale by far. Secondly, because of the common existing differences in physicochemical properties, it lacks universal methods to make widely used conducting polymer aerogels. Besides, lots of the exploited applications of conducting polymer aerogels in the present go back to the liquid surroundings, such as electrochemical energy storage, dyestuff absorption, liquid-phase catalysis, etc., making drying process futile to some extent.

As such, the future development of conducting polymer aerogels will focus on (1) exploring new universally applicable synthesis strategies on the base of INCG, emulsion polymerization, two-step synthesis, ISISA described in this chapter; (2) expanding new application areas of conducting polymer aerogels, especially applications in gas environment, such as gas phase catalysis, air purification, solar cells, electrostatic shielding, heat insulation combining with thermoelectric conversion functions, etc. Enlightened by the progress of typical PEDOT, PPy, and PANi aerogels presented in this chapter, we expect to witness the widespread use of many new conducting aerogels in the near future.

## Acknowledgements

The authors would like to thank the financial supports from National Natural Science Foundation of China (21504104) and the Natural Science Foundation of Jiangsu Province (BK20140391).

## Author details

Weina He[1] and Xuetong Zhang[2,3*]

*Address all correspondence to: xtzhang2013@sinano.ac.cn

1 Shanghai Jiao Tong University School of Medicine, Shanghai, People's Republic of China

2 Suzhou Institute of Nano-tech and Nano-bionics, Suzhou, People's Republic of China

3 Beijing Institute of Technology, Chinese Academy, Beijing, People's Republic of China

## References

[1] Zhang X.T., Chang D.W., Liu J.R., Yunjun Luo Y.J. Conducting polymer aerogels from supercritical $CO_2$ drying PEDOT-PSS hydrogels. Journal of Materials Chemistry. 2010;20(24):5080–5085. doi:10.1039/c0jm00050g

[2] Du R., Xu Y.Z., Luo Y.J., Zhang X.T., Zhang J. Synthesis of conducting polymer hydrogels with 2D building blocks and their potential-dependent gel–sol transitions. Chemical Communications. 2011;47(22):6287–6289. doi:10.1039/c1cc10915d

[3] Du R., Zhang X.T. Alkoxysulfonate-functionalized Poly(3,4-ethylenedioxythiophene) hydrogels. Acta Physico-Chimica Sinica. 2012;28(10):2305–2314. doi:10.3866/PKU.WHXB201209142

[4] Xu Y.Z., Sui Z.Y., Xu B., Duan H., Zhang X.T. Emulsion template synthesis of all conducting polymer aerogels with superb adsorption capacity and enhanced electrochemical capacitance. Journal of Materials Chemistry. 2012;22(17):8579–8584. doi: 10.1039/c2jm30565h

[5] Lu Y., He W.N., Cao T., Guo H.T., Zhang Y.Y., Li Q.W., et al. Elastic, conductive, polymeric hydrogels and s. Scientific Reports. 2014;4:5792. doi:10.1038/srep05792

[6] Guo H.T., He W.N., Lu Y., Zhang X.T. Self-crosslinked polyaniline hydrogel electrodes for electrochemical energy storage. Carbon. 2015;92:133–141. doi:10.1016/j.carbon.2015.03.062

[7] Zhang X.T., Liu J.R., Xu B., Su Y.F., Luo Y.J. Ultralight conducting polymer/carbon nanotube composite aerogels. Carbon. 2011;49(6):1884–1893. doi:10.1016/j.carbon.2011.01.011

[8] He W.N., Li G.Y., Zhang S.Q., Wei Y., Wang J., Li Q.W., et al. Polypyrrole/silver coaxial nanowire aero-sponges for temperature-independent stress sensing and stress-triggered Joule heating. ACS Nano. 2015;9(4):4244–4251. doi:10.1021/acsnano.5b00626

[9] Zhang X.T., Chechik V., Smith D.K., Walton P.H., Duhme-Klair A.K., Luo Y.J. Nanocomposite hydrogels-controlled synthesis of chiral polyaniline nanofibers and their inclusion in agarose. Synthetic Metals. 2009;159(19–20):2135–2140. doi:10.1016/j.synthmet.2009.08.002

[10] Zhang X.T., Li C.Y., Luo Y.J. Aligned/unaligned conducting polymer cryogels with three-dimensional macroporous architectures from Ice-Segregation-Induced Self-Assembly of PEDOT-PSS. Langmuir. 2011;27(5):1915–1923. doi:10.1021/la1044333

Chapter 4

# Coating of Conducting Polymers on Natural Cellulosic Fibers

Saiful Izwan Abd Razak, Nor Syuhada Azmi,
Khalida Fakhruddin, Farah Nuruljannah Dahli,
Izzati Fatimah Wahab,
Noor Fadzliana Ahmad Sharif,
Abdul Halim Mohd Yusof and
Nadirul Hasraf Mat Nayan

Additional information is available at the end of the chapter

http://dx.doi.org/10.5772/63304

**Abstract**

The process of combining natural cellulosic fibers with conducting polymers (CPs) is being pursued by scientist and researchers for their achievable synergistic electrical and biofriendly properties. CPs can be deposited on to a wide variety of cellulosic substrate and fibers, thus achieving good interactions between them. Various methods of deposition include in situ polymerization, physical coating, multilayering, and printing. Such materials are used for achieving more sustainable and low-cost CP-based applications.

**Keywords:** conducting polymers, cellulosic fibers, coating, paper, electrical conductivity, polyaniline, polypyrrole

## 1. Introduction

Since the discovery of conducting polymers (CPs), extensive researches have been carried out to explore the interesting properties of this relatively new and exciting material. Their ease of preparation, low cost, and good electronic properties have lead to the discovery of many CP-related applications. Due to their uniqueness and versatility, CPs are being deposited onto many type surfaces to induce their properties to the template or substrate material. Among those substrates, plant-based cellulosic fibers have gained wide scientific interest. In comparison with synthetic fibers, cellulosic fibers are eco-friendly and more abundant. Conductive cellulosic fibers produced by coating with CPs are being explored for various applications including supercapacitors and batteries, conductive paper and packaging, fillers in polymer composites, transistors and conductive wires, adsorbents, and actuators. Here, we discuss recent and past reports on the research activities regarding coating of CPs on cellulosic fibers and their directed applications.

## 2. Cellulosic fibers

Cellulosic fibers derived from plant have drawn much attention due to their sustainability and renewability. Plant fibers find applications in many fields. Conventional synthetic fibers like glass, carbon, and aramid can be produced with a definite range of properties, whereas the characteristic properties of natural fibers vary considerably. In general, cellulosic fibers may be bast, leaf, or seed including wood or nonwood types. The major constituents of lignocellulosic fibers are cellulose, hemicelluloses, and lignin. Cellulose is the most abundant natural polymer which contain D-anhydroglucose units (cellobiose) and are joined by β-1,4-glycosidic linkages at $C_1$ and $C_4$ positions. The chemical structure is shown in **Figure 1**. The main functional groups of cellulose are its hydroxyl groups (primary and secondary). These hydroxyl groups are involved in a number of intra- and intermolecular hydrogen bonds which result in various ordered crystalline arrangements. In fibers, cellulose chains are aligned along the fiber length, which make them high in flexural and tensile strength [1]. Hemicelluloses can consist of various monomeric units such as glucose, mannose, galactose, xylose, and arabinose. Depending on the monomeric unit, they can vary in structure and are highly branched with much lower degree of polymerization than that of cellulose giving them a non-crystalline structure. Hemicelluloses act as supportive matrix for cellulose fibers. Lignin is an amorphous, high molecular weight phenolic compound, which function as a structural supporting material in plant. Their structure varies and do not have predictable, continuous, and uniform properties. Though the exact chemical structure of lignin still remains unestablished, the high carbon and low hydrogen content suggested that the compound is highly unsaturated or aromatic [2].

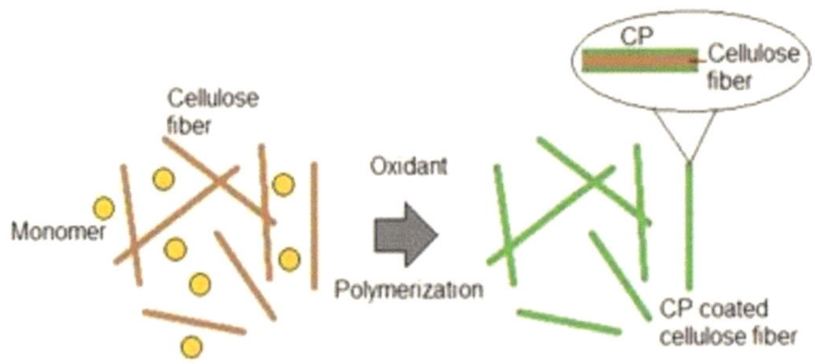


**Figure 1.** In situ polymerization of CP on cellulosic fibers.

http://dx.doi.org/10.5772/63304

## 3. Polyaniline, polyprrole, and PEDOT, PEDOT:PSS

Polyaniline (PANI), polypyrrole (PPY), poly (3,4-ethylenedioxythiophene) (PEDOT), and poly (3,4-ethylenedioxythiophene):poly (styrenesulfonate) (PEDOT:PSS) can be regarded as the most explored CPs up to date. PANI has the longest history among the CPs. It is one of the oldest artificial CPs and its high electrical conductivity among organic compounds has attracted continuing attention. The green protonated emeraldine salt has conductivity many orders of magnitude higher than that of common polymers, which ranges from $10^{-10}$ to $10^{1}$ S/cm but lower than that of typical metals [3]. PPY is another interesting CP with conductivity ranging from $10^{-4}$ to $10^{-2}$ S/cm. It has been studied in many electrochemical and sensor applications [4]. Another extensively explored family of CPs is PEDOT. It has high conductivity reaching 300 S/cm. Despite being described as more environmentally friendly compared to other CPs, they have the disadvantage being insoluble water and infusible [5]. PEDOT can be imbedded in poly(4-styrenesulfonic acid) (PSS) to form a more processable water-based PEDOT:PSS complex. This CP forms a stable dispersion that allow for coating process, which include spin coating. The conductivity can be from $10^{-5}$ to 1 S/cm depending on the PEDOT:PSS ratio [6].

## 4. CPs on cellulosic fibers

In situ polymerization is regarded the most popular method of depositing CPs on cellulosic fibers. The method varies widely depending on the types of oxidants, medium, dopants, monomers, the ratio between them, and the processing steps and parameters. In most cases, the in situ polymerization can be simplified as having the monomers of CPs being polymerized in a reactor or medium in the presence of the cellulose fibers (**Figure 1**). CPs will be formed on the bulk fibers or surface depending on the arrangement of the fibers.

Wood-based celluloses have been explored widely with the modification of CPs. PPYs deposited on crosslink cellulose have conductivity reaching 1.1 S/cm with good mechanical properties. The crosslink structure might keep the rigid PPY in a good conductive state due to the weaker interaction between the crosslinked cellulose and the PPY [7]. Aniline monomer has been described to penetrate all areas of wood veneer structures during in situ polymerization. In situ PANI formation was more prominent in cell walls and middle lamella. PANI modification resulted in the reduction of mechanical properties due to the prohibition of secondary interactions within the cell wall. The conductivity was anisotropic, mostly along the direction of wood fibers. Furthermore, the lignin component of the wood and PANI formed strong interaction as described by increase in glass transition temperature of lignin [8]. Lignin is a major component of wood; it contains the structures of phenol, carboxyl, and aldehyde groups, among others. Its hydrophobic character and presence of aromacity facilitates the interaction with the hydrophobic PANI. Blends of emeraldine base of PANI (36%) and lignin film have shown good homogeneous blend and good oxidation/reduction potential characteristics. The interaction occurs mainly between amine of PANI and carbonyl of lignin; imine of PANI and hydroxyl of lignin [9, 10].

Due to the vast growing effort of reducing wood-based resource, bamboo has been explored as the substitute to replace wood. Peeled bamboo veneers have shown increase in conductivity from $10^{-4}$ to $10^{-3}$ S/cm when coated with PANI, tuned by the phosphate acid doping concentration. Cellulose structure of bamboo was intact but with a slight reduction in the degree of crystallinity [11]. Curauá fiber, natural vegetal fibers of amazonic origin have shown increase in conductivity from $10^{-8}$ to $10^{-4}$ S/cm when in situ coated with PANI doped with sulfuric acid. PANI-modified fibers achieved conductivity changes of approximately 20,000% when tested for its compression sensitivities due to the high contacts. The modified fiber has potential applications as a pressure-sensing material by the function of its conductivity to applied pressure [12]. Similarly, mango fiber was reported to achieve conductivity by modification with PANI. The coated fiber was reported to achieve magnetic properties, being attracted by magnetic field [13]. It was shown that man-made cellulose-based fibers and textile, such as cotton, viscose, cupro, and lyocell, were able to be modified with CPs. This type modification is sought after in the applications of technical apparel, such as antistatic fabrics for work clothing, sportswear, heating, and cooling equipment [14]. Another method of coating the fabric with CPs besides the bulk in situ technique is the vapor phase polymerization. The cellulose fabric is impregnated with aqueous solution of oxidant and dopant, and followed by exposure to monomer in order to start the polymerization. The method revealed a partial penetration of PPY into the amorphous zones of fiber bulk [15]. Coating of CPs on protein fibers such as wool and silk has also been reported [16, 17]. Dip coating of PEDOT:PSS on wood-based microcrystalline celluloses showed that PEDOT was preferentially adsorbed rather than PSS. A strong interaction between the PEDOT:PSS and the cellulose was observed, implying a broad molecular distribution of the CP. As the pH of the solution increased, the amount of adsorbed PEDOT:PSS decreased due to the higher repulsive forces existing at higher pH levels [18].

## 5. CPs on bacterial cellulose

Bacterial cellulose (BC) is a straight chain polysaccharide with an ultrafine nanosized 3D fibrous network structure and is produced by certain microorganism (bacteria). The fibril diameter ranges between 10 and 100 nm with crystallinity up to 90%. It is highly hydrophilic due to neat hydroxyl groups of cellulose on the surface. These fibers are stronger than those of conventional natural fibers [19].

In situ coating of PANI on BC template showed high conductivity ($10^{-2}$ S/cm) when doped with organic acids such as HCl. The PANI forms a continuous nanosheath surrounding the BC due to the attraction force between the amine of aniline and abundant available hydroxyl of BC during the polymerization. The twisting-induced conductivity of the BC/PANI was reported to be activated at twisting angle of 200° [19]. Coating of BC with PANI doped with dodecylbenzene sulfonic acid (DBSA) depends on the amount of DBSA used. The highest conductivity ($10^{-5}$ S/cm) was at DBSA to aniline molar ratio of 1.5. Excess of DBSA dopant for the BC/PANI might avoid the penetration of PANI into the bulk BC with more preferable PANI formation at the surface of the BC that lead to the lowering of the conductivity [20]. It was later on reported that $FeCl_3$ oxidant can yield a better conductivity ($10^{-1}$ S/cm) to the BC/PANI [21]. A higher doping level was achieved by using p-toluene sulfonic acid dopant with good thermodynamic stability [22]. A further optimized reaction protocol was able to produce BC/PANI with conductivity of 5.1 S/cm and specific capacitance of 273 F/g. This was achieved due to the flake-shaped morphology of the BC/PANI facilitated by DMF (dimethylformamide) during the in situ polymerization [23]. Good redox properties were also observed for BC coated PPY [24].

## 6. CPs/cellulosic fiber as filler in polymer composites

These CP-coated cellulosic fibers have been explored as fillers in polymer composites. Incorporating fibers modified with CPs into insulating polymer matrices could induce some degree of electrical conductivity to the load-bearing host mainly for antistatic, electromagnetic, and pressure-sensing applications.

Curauá fiber modified with PANI was reported to be blended with polyamide-6 thermoplastic in a twin-screw extruder to form composites. The conductivity of composites was at $10^{-7}$ S/cm with a good polymer-fiber interface. The fibers provide easy conductive path (avoiding recombination of charge transport) for the PANI compared to PANI alone in the polymer matrix [25]. It was reported that epoxy resin could be cured with a PANI-coated kenaf fiber to form composites. The composite achieved conductivity in the range of antistatic applications. Mechanical properties of the composite could be improved by using suitable doping agent [26]. Similar approach was demonstrated using PANI-coated coconut fiber blended in polyurethane derived from castor oil. An increase in conductivity was observed when stress was applied to the composite [27].

## 7. CPs on paper and pulp fibers

Paper is a porous sheet made up of cellulosic fibers. The sheet usually has many pores generated by the fiber texture, depending on paper grades and manufacturing processes. Cellulose-based paper is intrinsically hydrophilic and contains strong hydrogen bonds after being dried. The main advantages of paper substrate are their lightweight, low cost, and environmental friendly characteristics which have potential to be modified with CPs.

Most of the earlier reports focus on the in situ polymerization of the CPs during the pulping stage or at the interface of the paper. The interest in conductive paper was initiated by a report by [28]. It was shown that the individual cellulose fiber of a filter paper was coated with 50–150 nm of PPY spheres. The spheres were fused together forming an integral layer of CPs with approximate thickness of one sphere. The pores within the paper matrix were not filled with PPY, indicating there is a significant bonding between the cellulose surface and the PPY. Furthermore, these pores open up for further modifications for new grades of paper. The sonication process for removal of impurities after the in situ polymerization did not remove any bounded PPY. The strong interaction was suggested to be an H bond between the free available OH groups of the cellulose and the amide groups of the PPY. Similar structure was observed by using PANI as the CP but with a slight reduction in the available pores. Low temperatures (0–25°C) during the in situ polymerization of PANI coating on filter paper yielded better electrical conductivity ($10^{-5}$ S/cm) compared to that of high temperature condition (50°C) ($10^{-8}$ S/cm) [29].

In situ coating of PPY on bleached kraft soft wood pulp indicated there is a concentration threshold of pyrrole monomer for the composite paper to achieve a certain degree of surface conductivity. It was reported that the monomer concentration should be more than 1.8 $g{\cdot}L^{-1}$ in order to achieve a stable conductive paper [30]. PPYs have low oxidation potential; the redox reaction of PPY is more sensitive to oxygen, thus making the resulting conductive paper composites less stable in air or oxygen-rich environment. Conductive PPY paper composite is electrically unstable when stored in air at an ambient temperature. The electrical conductivity decays with an increase in storage temperature. On the contrary, the decay was inhibited when stored in a nitrogen environment. This signified that the conductivity of the paper composite was related to the oxidation of PPY which in turn relates to the temperature. PPY particles were observed to become larger when aged at 100°C in air due to the thermal oxidation accompanied by a huge increase in resistivity [31]. Doping of PPY with amphiphilic cationic polyacrylates resulted in a more stable coating on the cellulose substrate. High-charge dopant induces easier PPY entrance into the fiber lumen and easier formation of bipolaron charge carriers. The mechanical strength of the paper was enhanced due to the good interfiber bonding induced by the polymeric dopant that was able to build connections between the fibers. The bonding between PPY-coated fibers (doping other than polyacrylate) is much weaker than H bond between cellulose fibers [32].

PEDOT coating on bleached softwood kraft pulp showed that the polymerization temperature and time is optimum at 60°C and 4 h, respectively, with good environmental stability due to the higher oxidation potential of PEDOT compared to other CPs. The volume resistivity was achieved at 5.9 Ω/cm with monomer concentration of 3 $g{\cdot}L^{-1}$ [33]. PEDOT in situ coating on a filter paper revealed a significant increase in tensile strength due to the interlocking between PEDOT and cellulose fibers. Furthermore, the PEDOT layer fixes and covers the fibrils tightly [34].

It was describe by X-ray photoelectron spectroscopy that the bond between PANI and cellulose in pulp paper existed in the form of H bonding and the lowering of mechanical properties is due to the coverage of hydroxyl groups of the fibers by PANI. p-toluene sulfonic acid doping of PANI achieved a conducting network and lowest resistivity at 30% of coated PANI [35]. PANI-coated commercial paper pulp has been described to be rigid and flexible enough to be used as paper folding artwork or origami. The electrical conductivity increased with a bend angle of 250°. The twisting process induced high contact between fibers in the sheet, resulting in the increase of electrical conductivity [36]. Pulping conditions of the paper played a significant role in PANI deposition and yield. It was shown that chemical pulping had higher amount of coated PANI and higher conductivity than those prepared from high-yield pulp. Furthermore, bleached chemical pulp had better paper properties. It was proposed that the increasing content of sulfonic group of the chemical pulp had a positive linear effect to the amount of coated PANI. Sulfonic groups in the pulp fibers could be dissociate and promote the adsorption of the positively charged PANI onto the fibers [37]. Interfacial polymerization was suggested to be the better method in protecting the fiber strength to some degree. The aniline polymerization reactions take place at the paper interface and not across the whole sample. Thus, multiple steps of polymerization are required. Polyamideamine-epichlorohydrin (PAE), a commercial wet strength agent in papermaking showed to be beneficial in improving the mechanical properties of multiple interfacials of cellulose paper in situ coated with PANI [38].

Other types of cellulosic paper sources have shown potential to be deposited with CPs. Coating of PANI on rice straw and bagasse pulp paper has shown potential for packaging applications. The main drawback of such paper substrate is that their mechanical properties such as breaking length, burst factor, and tear factor decreased with increasing PANI amount. A total of 10 wt % of PANI is needed to impart electrical conductivity to the paper composite [39]. Pineapple leaf fiber paper pulp was coated with PANI using the in situ polymerization method. The addition of PANI showed a marginal drop in mechanical properties (burst strength and tear index) and a reduction in moisture absorption with increasing PANI amount. The DC conductivity was achieved around $10^{-5}$ S/cm at the concentration threshold of PANI [40]. It was shown that the mechanical and electrical conductivity of PANI-coated kenaf paper could be optimized by controlling the PANI amount and dopant concentration [41]. Further studies have shown that PANI-coated kenaf fibers were able to blend with kenaf pulp during the pulping stage. The composite paper revealed enhancement in the electrical conductivity with improved mechanical properties [42]. Another effort is by coating pulp of wasted egg holder tray by PPY. The raw pulp was bleached and in situ coated with PPY, giving conductivity value of $10^{-4}$ S/cm in the frequency-independent region [43].

Paper is usually extremely rough, with peak-to-valley roughness values of up to hundreds of micrometers. The packing density of fibers also limits its overall mechanical properties. Nanosized fiber papers or nanopapers can provide excellent mechanical properties, provide smooth surface, and be optically transparent. Modification of nanopapers with CPs will open up many opportunities in electronic/optoelectronic devices [44]. It was displayed that at the percolation threshold, nanofibrillated cellulose/PANI composite paper had mechanical properties similar to its unmodified counterpart. The percolation was achieved at 4.57 vol% of PANI with conductivity of $10^{-5}$ S/cm. The nanosize of the cellulose opened more available OH for the in situ polymerization, thus increasing the PANI coating [45]. PEDOT:PSS can also be spin coated on nanofibrillar cellulose film [46].

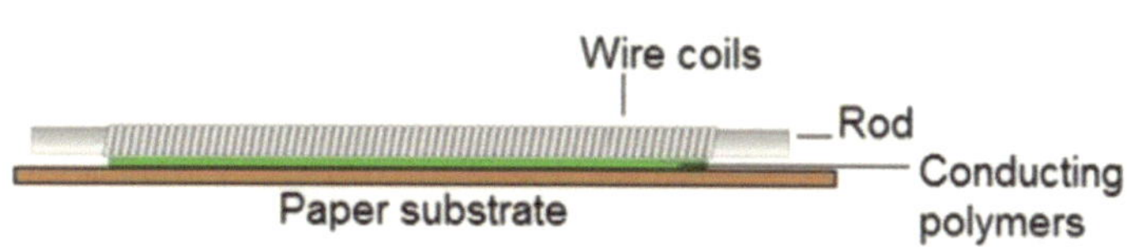


**Figure 2.** Mayer rod coating of CP on paper.

Mayer rod coating (**Figure 2**) is one of the most popular coating methods on a paper substrate. A Mayer rod is a stainless steel rod that is wound tightly with stainless wire of varying diameter. The grooves between the wire coils determine the precise amount of coating materials that will pass through as it moved along the web. Coating technology can avoid the complication faced with in situ polymerization technique and is a very feasible process for coating a layer of any material over paper substrate. Nevertheless, the viscosity of the CP solution should be low enough to allow easy flow between the wire windings.

Physical coating of PEDOT:PSS using a Mayer rod was achieved by using organic solvents such as dimethyl sulfoxide and N-methylpyrrolidone. The solvents induced plasticizing effect and conformation changes to the PEDOT molecules that lead to higher conductivity compared to PEDOT:PSS coating without any solvents. The paper conductivity level was at $10^{-3}$ S/cm with a slight increase in tensile strength [47]. It was reported that multiwalled carbon nanotubes could be dispersed in PEDOT:PSS solution without the use of any added surfactant. This mixtures were coated onto the paper substrate using a Mayer rod coater and heated using a specially design unidirectional ceramic heating board. The heating method revealed better conductivity compared to those dried using conventional drying. This was due to the good bonding between the solution mixture and the paper fiber during the drying stage [48]. Clay has been suggested to act as carrier for the CP for conductive paper. Though the reasons are unclear, the nanosize of clay minerals can provide good dispersion of the CP for the physical rod coating process [49].

## 8. Electroactive papers, supercapacitors, and adsorbents

Due to the interesting redox properties of CPs, they are being explored as electroactive polymers and actuators. They can create a mechanical response from an electrical stimulation. The response can be either a transverse bending or an axial expansion/contraction. Papers that are modified with CPs for their actuation properties can be classified as an electroactive paper. Electroactive paper has the advantages of low cost, lightweight, biodegradability, large displacement output, and low actuation voltages. Drawbacks of this material are that it suffers performance degradation over time and produces a low displacement output at a low humidity.

It was shown that PANI-cellophane actuator had better performance than PPY-cellophane in terms of bending displacement and ruggedness to humidity. This was described due to the relatively high conductivity of PANI and the crystalline nature of PANI which changes to amorphous structure after the actuation [50]. Electroactive paper of regenerated cellulose and PPY with dispersed hydrophobic ionic liquid resulted in reduced resistance and better durability under ambient humidity conditions. The movement of anions in the conducting system resulted in bending displacement [51]. In situ coating of PANI on cellulose solution using p-toluenesulfonic acid showed that the resulting paper had low Young's modulus and reduced thermal stability. The cellulose-PANI electroactive paper was fabricated by depositing very thin gold electrodes on both surfaces of the paper. A large bending displacement and long actuation stability was reported at ambient humidity and temperature [52].

The key technology that underlies the performance of supercapacitor materials has turned it into a research hotspot is the electrode material. Electrode materials mainly include carbon material, metal oxide, and CPs. Flexible electrode-based paper is very promising for low cost and lightweight energy storage applications. The rough and porous paper surface is advantageous when electrochemical active materials were coated onto cellulose fibers.

Cellulose extracted from green polluting *Cladophora* algae was in situ coated with PPY and revealed excellent ion-exchange capacity and cycling stability when used as a working electrode in a chloride-containing solution [53]. The group later on developed the *Cladophora*/PPY as aqueous batteries that can exhibit charge capacities 25 and 33 mAh $g^{-1}$. This was suggested to be due to the thin 50 nm coating layer of PPY on the individual cellulose fiber [54]. A sandwich of PPY/A4 paper and PVA/$H_3PO_4$ showed electrical conductivity of 15 S/cm and an energy density of 1.0 mWh $cm^{-3}$ [55]. Similar energy storage potential was shown by PEDOT:PSS dispersed with multiwalled carbon nanotubes coated on paper substrate [56]. It was shown that a regenerated cellulose film coated with PPY achieved conductivity of 0.59 S/cm at nearly 30 wt% of PPY. The specific capacitance was reported at 392–308 F/g [57].

Anion-exchange properties of PPY on wood saw dust have shown potential use as adsorbent for the uptake of Cr(VI) in wastewater solution [58]. Another study later on improved the uptake findings using industrially favorable hydrogen peroxide as the oxidant to minimize the cost of using ammonium peroxydisulfate oxidant [59]. Layers of copper metals were reported to be deposited electrochemically onto the outer surface of lightweight balsa wood coated with PPY and reduction of silver by suspending the coated samples in hydrazine hydrate solution. The conductivity was comparable to that of the metal alone [60]. Similarly, redox-active surface of CPs on kraft paper pulp was utilized for the reduction of silver metals [61].

## 9. Inkjet printing of CPs on cellulose paper

Recent research interest has been directed toward inkjet printing of CPs. The precision and flexibility of inkjet printer is ideal for printed electronics compared to the high cost and labor-intensive method of photolithography and vapor deposition. The main challenges of inkjet printing of CPs on paper substrate are their film homogeneity, wetting behavior, and nozzle clogging during printing.

Suspension of DBSA-doped PANI has been successfully printed using a commercial piezoelectric desktop inkjet printer. The surface tension was reported to be in a range suitable for inkjet printing with bulk conductivity as same magnitude as those of drop-coated film [62]. PANI has also been identified for printing conductive wires on paper substrates for the fabrication of transistors with a low operating voltage. It was shown that the resistance of the printed paper has linear correlation with the short length scale roughness of the print substrate [63]. PEDOT:PSS have been explored as inks for printing on paper substrates. Roughness of the paper substrate, additives, and the ink formulation itself play a role in the conductivity.

A more stable conductivity was reported to be achieved by adding single-walled carbon nanotubes into the PEDOT:PSS ink. Retention aids and fixation agents of the paper seem to interact with the PSS ions thus lowering the conductivity of the printed films [64]. Another interesting method of patterning CPs on to paper is the "pen-writing" method. Patterns or drawing of $FeCl_3$ oxidant were drawn on the paper substrate using an iridium nib pen and followed by exposure to pyrrole vapor that leads to quick interfacial polymerization. The paper can be served as sensors for ammonia gas, thermal heat, and NIR (near-infrared) irradiation [65].

## 10. Layer-by-layer assembly of CPs on paper

Cellulosic substrate or paper can be deposited by CPs using a layer-by-layer (LbL) multilayering technique. Surface properties of the paper substrate can be controlled according to electrostatic or hydrophobic interaction. It was described earlier that LbL polyelectrolyte coating can modify the surface charge to make the resulting paper stronger [66]. An LbL of wood microfibers, PEDOT:PSS, and small amount of multiwalled carbon nanotubes (0.2%) have shown conductivity up to 20 $S/cm^{-1}$ and electrical capacitance of $10^{-11}$ $F/in^2$ [67]. It was reported that unbeaten eucalyptus-bleached kraft pulp were coated with cationic polyelectrolyte polyethyleneimine and PEDOT:PSS. The conductivity of the resultant paper was in the range of $10^{-5}$–$10^{-4}$ S/cm without loss of paper strength. Increased contact between PEDOT:PSS, low salt concentration, and the calendering process contributed to the improved conductivity [68]. Multilayering of conductive paper sheets made of carboxymethylated fibers displayed a much higher conductivity than those made of non-carboxymethylated fibers when both pulps were coated with PEDOT:PSS and poly(allyl amine) polyelectrolyte [69].

## 11. Flame retardancy

CPs are good flame retardant agents. Flame retardant agents are added in textile or plastics to inhibit or delay the flaming process. Coating of CPs on to papers would open up many potential applications in functional papers by enhancing the thermal stability. It was reported that both forms of PANI, base and protonated (hydrochloride), give the flame-retardant performance to cellulose filter paper. The PANI-coated cellulose retains the original fibrillar morphology after burning. This was suggested due to the protection of the solid carbonaceous PANI that restricts the access of oxygen to cellulose [70]. Similar heat protection was reported by coating of PANI onto wood scantlings [71]. It has been investigated that the flame retardancy increased with the amount of PANI deposited in the paper pulp and the doping agent played a major role in the flame retardancy. It was suggested that organic sulfonic acids (p-toluenesulfonic acid and sulfosalicylic acid) were more suitable than inorganic acids (sulfuric acid and hydrochloric acid). This was due to the easy dedoping and low doping level of the latter dopants [72, 73]. A further study reported that phytic acid greatly enhanced the flame retardancy of the composite paper but the conductivity level was lowered [74].

http://dx.doi.org/10.5772/63304

## 12. Conclusion

Coating of CPs on cellulosic fibers show promise in wide variety of applications from conductive papers to smart actuating materials. The utilization of natural-based and low-cost resource for high-end applications is feasible by modifying the cellulosic fibers with CPs. CPs have been shown to impart their unique features to many types of cellulosic fibers such as pure cellulose, commercial grade papers, kenaf, bamboo, vegetal fibers, rice straw, and BC. In general, further investigation should be focused on producing more mechanically reliable and electrically stable CP-coated cellulosic fibers. Polyaniline, polypyrrole, poly (3,4-ethylenedioxythiophene):poly (styrenesulfonate), and their derivatives are currently being explored extensively for their applications with cellulosic fibers. The coating method of CPs such as in situ during pulping, physical deposition using a Mayer rod, spin coating, layer-by-layer assembly, or inkjet printing plays a major role in determining the targeted properties. It envisioned that the combination of CPs and cellulosic fibers could promote more sustainable electrically functional materials for the future.

## Acknowledgements

The authors would like to express their gratitude to Universiti Teknologi Malaysia for the financial assistance through research grants.

## Author details

Saiful Izwan Abd Razak[1,2*], Nor Syuhada Azmi[2], Khalida Fakhruddin[2], Farah Nuruljannah Dahli[3], Izzati Fatimah Wahab[2], Noor Fadzliana Ahmad Sharif[3], Abdul Halim Mohd Yusof[3] and Nadirul Hasraf Mat Nayan[4]

*Address all correspondence to: saifulizwan@utm.my

1 IJN-UTM Cardiovascular Engineering Centre, Universiti Teknologi Malaysia, Skudai, Johor, Malaysia

2 Faculty of Biosciences and Medical Engineering, Universiti Teknologi Malaysia, Skudai, Johor, Malaysia

3 Faculty of Chemical and Energy Engineering, Universiti Teknologi Malaysia, Skudai, Johor, Malaysia

4 Faculty of Engineering Technology, Universiti Tun Hussein Onn Malaysia, Batu Pahat, Johor, Malaysia

## References

[1] Park, S., Baker, J. O., Himmel, M. E, Parilla, P. A., Johnson, D. K. (2010). Cellulose crystallinity index: measurement techniques and their impact on interpreting cellulase performance. *Biotechnology for Biofuels, 3,* 10–20.

[2] Mohanty, A. K., Misra, M., Hinrichsen, G. (2000). Biofibres, biodegradable polymers and biocomposites: an overview. *Macromolecular Materials and Engineering, 276*(277), 1–24.

[3] Bhadra, S., Khastgir, D., Singha, N. K., & Lee, J. H. (2009). Progress in preparation, processing and applications of polyaniline. *Progress in Polymer Science, 34*(8), 783–810.

[4] Khomenko, V., Frackowiak, E., & Beguin, F. (2005). Determination of the specific capacitance of conducting polymer/nanotubes composite electrodes using different cell configurations. *Electrochimica Acta, 50*(12), 2499–2506.

[5] Groenendaal, L., Jonas, F., Freitag, D., Pielartzik, H., & Reynolds, J. R. (2000). Poly (3, 4-ethylenedioxythiophene) and its derivatives: past, present, and future. *Advanced Materials, 12*(7), 481–494.

[6] Stöcker, T., Köhler, A., & Moos, R. (2012). Why does the electrical conductivity in PEDOT: PSS decrease with PSS content? A study combining thermoelectric measurements with impedance spectroscopy. *Journal of Polymer Science Part B: Polymer Physics, 50*(14), 976–983.

[7] Yin, W., Li, J., Li, Y., Wu, J., & Gu, T. (2001). Conducting composite film based on polypyrrole and crosslinked cellulose. *Journal of Applied Polymer Science, 80*(9), 1368–1373.

[8] Hassel, B. I., Trey, S. M., Leijonmarck, S., & Johansson, M. (2014). A study on the morphology, mechanical, and electrical performance of polyaniline-modified Wood-A semiconducting composite material. *BioResources, 9*(3), 5007–5023.

[9] Rodrigues, P. C., Muraro, M., Garcia, C. M., Souza, G. P., Abbate, M., Schreiner, W. H., & Gomes, M. A. (2001). Polyaniline/lignin blends: thermal analysis and XPS. *European Polymer Journal, 37*(11), 2217–2223.

[10] Rodrigues, P. C., Cantão, M. P., Janissek, P., Scarpa, P. C., Mathias, A. L., Ramos, L. P., & Gomes, M. A. (2002). Polyaniline/lignin blends: FTIR, MEV and electrochemical characterization. *European Polymer Journal, 38*(11), 2213–2217.

[11] He, W., Zhang, X., Yu, C., Huang, D., & Li, Y. (2015). Synthesis of bamboo/polyaniline composites by in situ polymerization and their characteristics. *BioResources, 10*(2), 2969–2981.

[12] Souza, F. G., Oliveira, G. E., Rodrigues, C. H., Soares, B. G., Nele, M., & Pinto, J. C. (2009). Natural Brazilian amazonic (Curauá) fibers modified with polyaniline nanoparticles. *Macromolecular Materials and Engineering, 294*(8), 484–491.

[13] Souza, F. G., da Silva, A. M., de Oliveira, G. E., Costa, R. M., Fernandes, E. R., & Pereira, E. D. (2015). Conducting and magnetic mango fibers. *Industrial Crops and Products, 68*, 97–104.

[14] Dall'Acqua, L., Tonin, C., Peila, R., Ferrero, F., & Catellani, M. (2004). Performances and properties of intrinsic conductive cellulose–polypyrrole textiles. *Synthetic Metals, 146*(2), 213–221.

[15] Dall'Acqua, L., Tonin, C., Varesano, A., Canetti, M., Porzio, W., & Catellani, M. (2006). Vapour phase polymerisation of pyrrole on cellulose-based textile substrates. *Synthetic Metals, 156*(5), 379–386.

[16] Hosseini, S. H., & Pairovi, A. (2005). Preparation of conducting fibres from cellulose and silk by polypyrrole coating. *Iranian Polymer Journal, 14*(11), 934.

[17] Johnston, J. H., Kelly, F. M., Moraes, J., Borrmann, T., & Flynn, D. (2006). Conducting polymer composites with cellulose and protein fibres. *Current Applied Physics, 6*(3), 587–590.

[18] Montibon, E., Järnström, L., & Lestelius, M. (2009). Characterization of poly (3, 4-ethylenedioxythiophene)/poly (styrene sulfonate)(PEDOT: PSS) adsorption on cellulosic materials. *Cellulose, 16*(5), 807–815.

[19] Hu, W., Chen, S., Xu, Q., & Wang, H. (2011). Solvent-free acetylation of bacterial cellulose under moderate conditions. *Carbohydrate Polymers, 83*(4), 1575–1581.

[20] Marins, J. A., Soares, B. G., Dahmouche, K., Ribeiro, S. J., Barud, H., & Bonemer, D. (2011). Structure and properties of conducting bacterial cellulose-polyaniline nanocomposites. *Cellulose, 18*(5), 1285–1294.

[21] Müller, D., Mandelli, J. S., Marins, J. A., Soares, B. G., Porto, L. M., Rambo, C. R., & Barra, G. M. O. (2012). Electrically conducting nanocomposites: preparation and properties of polyaniline (PAni)-coated bacterial cellulose nanofibers (BC). *Cellulose, 19*(5), 1645–1654.

[22] Lee, B. H., Kim, H. J., & Yang, H. S. (2012). Polymerization of aniline on bacterial cellulose and characterization of bacterial cellulose/polyaniline nanocomposite films. *Current Applied Physics, 12*(1), 75–80.

[23] Wang, H., Zhu, E., Yang, J., Zhou, P., Sun, D., & Tang, W. (2012). Bacterial cellulose nanofiber supported polyaniline nanocomposites with flake-shaped morphology as supercapacitor electrodes. *Journal of Physical Chemistry, 116*(24), 13013–13019.

[24] Müller, D., Rambo, C. R., Recouvreux, D. O. S., Porto, L. M., & Barra, G. M. O. (2011). Chemical in situ polymerization of polypyrrole on bacterial cellulose nanofibers. *Synthetic Metals, 161*(1), 106–111.

[25] Araujo, J. R., Adamo, C. B., & De Paoli, M. A. (2011). Conductive composites of polyamide-6 with polyaniline coated vegetal fiber. *Chemical Engineering Journal, 174*(1), 425–431.

[26] Razak, S. I. A., Rahman, W. A. W. A., Sharif, N. F. A., Nayan, N. H. M., Saidi, M. A. A., & Yahya, M. Y. (2013). Polyaniline coated kenaf core and its effect on the mechanical and electrical properties of epoxy resin. *Composite Interfaces, 20*, 611–622.

[27] Merlini, C., Barra, G. M., Schmitz, D. P., Ramôa, S. D., Silveira, A., Araujo, T. M., & Pegoretti, A. (2014). Polyaniline-coated coconut fibers: structure, properties and their use as conductive additives in matrix of polyurethane derived from castor oil. *Polymer Testing, 38*, 18–25.

[28] Johnston, J. H., Moraes, J., & Borrmann, T. (2005). Conducting polymers on paper fibres. *Synthetic Metals, 153*(1), 65–68.

[29] Youssef, A. M., El-Samahy, M. A., & Rehim, M. H. A. (2012). Preparation of conductive paper composites based on natural cellulosic fibers for packaging applications. *Carbohydrate Polymers, 89*(4), 1027–1032.

[30] Ding, C., Qian, X., Shen, J., & An, X. (2010). Preparation and characterization of conductive paper via in-situ polymerization of pyrrole. *BioResources, 5*(1), 303–315.

[31] Li, J., Qian, X. R., Chen, J. H., Ding, C. Y., & An, X. H. (2010). Conductivity decay of cellulose–polypyrrole conductive paper composite prepared by in situ polymerization method. *Carbohydrate Polymers, 82*(2), 504–509.

[32] Wang, H., Hu, M., Fei, G., Wang, L., & Fan, J. (2015). Preparation and characterization of polypyrrole/cellulose fiber conductive composites doped with cationic polyacrylate of different charge density. *Cellulose, 22*(5), 3305–3319.

[33] Chen, Y., Qian, X., & An, X. (2011). Preparation and characterization of conductive paper via in situ polymerization of 3, 4-ethylenedioxythiophene. *BioResources, 6*(3), 3410–3423.

[34] Kawashima, H., Shinotsuka, M., Nakano, M., & Goto, H. (2012). Fabrication of conductive paper coated with PEDOT: preparation and characterization. *Journal Of Coatings Technology and Research, 9*(4), 467–474.

[35] Li, J., Qian, X., Wang, L., & An, X. (2010). XPS characterization and percolation behavior of polyaniline-coated conductive paper. *BioResources, 5*(2), 712–726.

[36] Goto, H. (2011). Electrically conducting paper from a polyaniline/pulp composite and paper folding art work for a 3D object. *Textile Research Journal, 81*(2), 122–127.

[37] Qian, X., Shen, J., Yu, G., & An, X. (2010). Influence of pulp fiber substrate on conductivity of polyaniline-coated conductive paper prepared by in-situ polymerization. *BioResources, 5*(2), 899–907.

[38] Yan, J., & Xu, R. (2015). Reinforced conductive polyaniline-paper composites. *BioResources, 10*(3), 4065–4076.

[39] Youssef, A. M., Kamel, S., El-Sakhawy, M., & El Samahy, M. A. (2012). Structural and electrical properties of paper–polyaniline composite. *Carbohydrate Polymers, 90*(2), 1003–1007.

[40] Razak, S. I. A., Sharif, N. F. A., & Nayan, N. H. M. (2014). Electrically conductive paper of polyaniline modified pineapple leaf fiber. *Fibers and Polymers, 15*(6), 1107–1111.

[41] Razak, S. I. A., & Sharif, N. F. A. (2015). Concurrent optimization of the mechanical and electrical properties of polyaniline modified kenaf paper. *Cellulose Chemistry And Technology, 49*(2), 195–202.

[42] Hisham, N. S. A., Razak, S. I. A., Nayan, N. H. M., & Rahman, W. A. W. A. (2015). Influence of Polyaniline coated kenaf fiber on kenaf paper sheet. *4 th International Conference on Engineering and Innovative Materials (ICEIM 2015)*. 3–4 September. Penang, Malaysia, (27), 01002.

[43] Razaq, A., Idrees, M., Malik, A., Mushtaq, N., Nadeem, M., Hussain, I., & Yar, M. (2015). Dielectric studies of composite paper reinforced with polypyrrole coated pulp fibers from wasted egg holders. *Journal of Applied Polymer Science, 132*(33), 42422.

[44] Hu, L., Zheng, G., Yao, J., Liu, N., Weil, B., Eskilsson, M., & McGehee, M. D. (2013). Transparent and conductive paper from nanocellulose fibers. *Energy & Environmental Science, 6*(2), 513–518.

[45] Luong, N. D., Korhonen, J. T., Soininen, A. J., Ruokolainen, J., Johansson, L. S., & Seppälä, J. (2013). Processable polyaniline suspensions through in situ polymerization onto nanocellulose. *European Polymer Journal, 49*(2), 335–344.

[46] Valtakari, D., Liu, J., Kumar, V., Xu, C., Toivakka, M., & Saarinen, J. J. (2015). Conductivity of PEDOT: PSS on Spin-Coated and Drop Cast Nanofibrillar Cellulose Thin Films. *Nanoscale Research Letters, 10*(1), 1–10.

[47] Montibon, E., Lestelius, M., & Järnström, L. (2010). Electroconductive paper prepared by coating with blends of poly (3, 4-ethylenedioxythiophene)/poly (4-styrenesulfonate) and organic solvents. *Journal of Applied Polymer Science, 117*(6), 3524–3532.

[48] Huang, L., Chen, K., Peng, C., & Gerhardt, R. A. (2011). Highly conductive paper fabricated with multiwalled carbon nanotubes and poly (3, 4-ethylenedioxythiophene)-poly (styrenesulfonate) by unidirectional drying. *Journal of Materials Science, 46*(20), 6648–6655.

[49] Saravanan, C., He, Z., & Ni, Y. (2014). Application of polyaniline/clay combination to cellulosic paper as an approach to conductivity development. *BioResources, 9*(2), 1886–1897.

[50] Kim, J., Deshpande, S. D., Yun, S., & Li, Q. (2006). A comparative study of conductive polypyrrole and polyaniline coatings on electro-active papers. *Polymer Journal, 38*(7), 659–668.

[51] Mahadeva, S. K., & Kim, J. (2010). Nanocoating of ionic liquid and polypyrrole for durable electro-active paper actuators working under ambient conditions. *Journal of Physics D: Applied Physics, 43*(20), 205502.

[52] John, A., Mahadeva, S. K., & Kim, J. (2010). The preparation, characterization and actuation behavior of polyaniline and cellulose blended electro-active paper. *Smart Materials and Structures, 19*(4), 045011.

[53] Mihranyan, A., Nyholm, L., Bennett, A. E. G., & Strømme, M. (2008). A novel high specific surface area conducting paper material composed of polypyrrole and Cladophora cellulose. *The Journal of Physical Chemistry B, 112*(39), 12249–12255.

[54] Nyström, G., Razaq, A., Strømme, M., Nyholm, L., & Mihranyan, A. (2009). Ultrafast all-polymer paper-based batteries. *Nano Letters, 9*(10), 3635–3639.

[55] Yuan, L., Yao, B., Hu, B., Huo, K., Chen, W., & Zhou, J. (2013). Polypyrrole-coated paper for flexible solid-state energy storage. *Energy & Environmental Science, 6*(2), 470–476.

[56] Montibon, E., Lestelius, M., & Järnström, L. (2012). Conductivity of paper containing poly (3, 4-ethylenedioxythiophene)/poly (4-styrenesulfonate) and multiwall carbon nanotubes. *Journal of Applied Polymer Science, 125*(S1), E34–E40.

[57] Liu, S., He, K., Wu, X., Luo, X., & Li, B. (2015). Surface modification of cellulose scaffold with polypyrrole for the fabrication of flexible supercapacitor electrode with enhanced capacitance. *RSC Advances, 5*(106), 87266–87276.

[58] Ansari, R., & Fahim, N. K. (2007). Application of polypyrrole coated on wood sawdust for removal of Cr (VI) ion from aqueous solutions. *Reactive and Functional Polymers, 67*(4), 367–374.

[59] Liu, X., Qian, X., Shen, J., Zhou, W., & An, X. (2012). An integrated approach for Cr (VI)-detoxification with polyaniline/cellulose fiber composite prepared using hydrogen peroxide as oxidant. *Bioresource Technology, 124*, 516–519.

[60] Richardson, M. J., Johnston, J. H., & Borrmann, T. (2006). Electronic properties of intrinsically conducting polymer-cellulose based composites. *Current Applied Physics, 6*(3), 462–465.

[61] Kelly, F. M., Johnston, J. H., Borrmann, T., & Richardson, M. J. (2007). Functionalised hybrid materials of conducting polymers with individual fibres of cellulose. *European Journal of Inorganic Chemistry, 2007*(35), 5571–5577.

[62] Ngamna, O., Morrin, A., Killard, A. J., Moulton, S. E., Smyth, M. R., & Wallace, G. G. (2007). Inkjet printable polyaniline nanoformulations. *Langmuir, 23*(16), 8569–8574.

[63] Ihalainen, P., Määttänen, A., Järnström, J., Tobjörk, D., Österbacka, R., & Peltonen, J. (2012). Influence of surface properties of coated papers on printed electronics. *Industrial & Engineering Chemistry Research, 51*(17), 6025–6036.

[64] Angelo, P. D., Cole, G. B., Sodhi, R. N., & Farnood, R. R. (2012). Conductivity of inkjet-printed PEDOT: PSS-SWCNTs on uncoated papers. *Nordic Pulp and Paper Research Journal, 27*(2), 486.

[65] Jia, H., Wang, J., Zhang, X., & Wang, Y. (2013). Pen-writing polypyrrole arrays on paper for versatile cheap sensors. *ACS Macro Letters, 3*(1), 86–90.

[66] Eriksson, M., Notley, S. M., & Wågberg, L. (2005). The influence on paper strength properties when building multilayers of weak polyelectrolytes onto wood fibres. *Journal of Colloid and Interface Science, 292*(1), 38–45.

[67] Agarwal, M., Xing, Q., Shim, B. S., Kotov, N., Varahramyan, K., & Lvov, Y. (2009). Conductive paper from lignocellulose wood microfibers coated with a nanocomposite of carbon nanotubes and conductive polymers. *Nanotechnology, 20*(21), 215602.

[68] Lee, J., Ryu, J., & Youn, H. J. (2012). Conductive paper through LbL multilayering with conductive polymer: dominant factors to increase electrical conductivity. *Cellulose, 19*(6), 2153–2164.

[69] Wistrand, I., Lingström, R., & Wågberg, L. (2007). Preparation of electrically conducting cellulose fibres utilizing polyelectrolyte multilayers of poly (3, 4-ethylenedioxythiophene): poly (styrene sulphonate) and poly (allyl amine). *European Polymer Journal, 43*(10), 4075–4091.

[70] Stejskal, J., Trchová, M., & Sapurina, I. (2005). Flame-retardant effect of polyaniline coating deposited on cellulose fibers. *Journal of Applied Polymer Science, 98*(6), 2347–2354.

[71] Stejskal, J., Trchová, M., Brodinová, J., & Sapurina, I. (2007). Flame retardancy afforded by polyaniline deposited on wood. *Journal of Applied Polymer Science, 103*(1), 24–30.

[72] Wu, X., Qian, X., & An, X. (2013). Flame retardancy of polyaniline-deposited paper composites prepared via in situ polymerization. *Carbohydrate Polymers, 92*(1), 435–440.

[73] Mao, H., Wu, X., Qian, X., & An, X. (2014). Conductivity and flame retardancy of polyaniline-deposited functional cellulosic paper doped with organic sulfonic acids. *Cellulose, 21*(1), 697–704.

[74] Zhou, Y., Ding, C., Qian, X., & An, X. (2015). Further improvement of flame retardancy of polyaniline-deposited paper composite through using phytic acid as dopant or co-dopant. *Carbohydrate Polymers*, 115, 670–676.

Chapter 5

# Space Charge–Limited Current Model for Polymers

Syed A. Moiz, Iqbal. A. Khan, Waheed A. Younis and
Khasan S. Karimov

Additional information is available at the end of the chapter

http://dx.doi.org/10.5772/63527

**Abstract**

Polymers have exceptional charge transport mechanism as a combination of delocalization and localization of charge carriers with intramolecular and intermolecular charge interaction, respectively, and most of the time, it is interpreted with Mott-Gurney space charge–limited current model. As polymers are full of traps, therefore, Mott-Gurney space charge–limited model is modified with various trap distributions as trapped space charge–limited model. The most crucial parameter affected by the nature and distribution of traps is the carrier mobility, and it is argued that space charge–limited model is an acceptable choice for the mobility measurement for polymer. Similarly, in order to account the commonly observed lowering of trap barrier height at higher electric field, the Mott-Gurney space charge–limited current is further modified with little variations, which are evaluated and discussed in detail.

**Keywords:** polymer electronics, organic semiconductor/polymer, space charge–limited current, charge transport, child's law

## 1. Introduction

From last few decades, polymer semiconductor-based electronic devices have attracted an enormous deal of interest due to their great achievements both in laboratory and as well as in commercial products [1–10]. Lightweight, flexibility, low-cost, wide-area application, deposition on various substrates, tunability, and many other advantages make these materials an excellent choice for many electronic applications such as light-emitting diode, solar cell, thin-film transistor, laser diode, etc. [11–20]. Another remarkable area of interest for conducting and semiconducting polymer is their application towards highly responsive and low-cost temperature, pressure, humidity, chemical, biochemical, and other types of sensors [21–30].

Despite all of these successes, the complete picture of charge-transport mechanism, which is highly crucial for further achievements, is still not clear [31–40]. Due to such ambiguity, different charge transport models are reported to validate the experimental results performed on various polymer devices [41–54]. Among these charge transport models, space charge–limited current model is counted as highly recited and unanimously accepted charge transport model for disorder organic/polymer semiconductor. In this chapter, we discuss the brief overview of space charge–limited current model with recent development especially for polymer semiconductors.

This chapter starts with the brief discussion of space charge–limited current model and its applications to polymer, especially for conducting and semiconducting polymer. Firstly, the concise overview of historical evaluation and charge transport mechanism of polymer will be discussed. Later on the Child's Law, Mott-Gurney space charge–limited current model, and trapped space charge–limited current model with single-trap, exponential, and Gaussian distribution of traps and then effects of shallow and deep traps on space charge–limited current are reviewed. Mobility measurement from space charge–limited current and its comparison with other mobility measuring methods such as time-of-flight and different carrier extraction by linearly increasing voltage characteristics are evaluated in the next section. Finally, the application of Poole-Frenkel on space charge–limited equation with other modifications is reviewed before conclusion.

## 2. Charge transport mechanism of conducting polymer

Before 1977, it was generally considered that polymers are electrically insulator. Most of these insulating polymers such as polyethylene, polystyrene, and polypropylene have almost

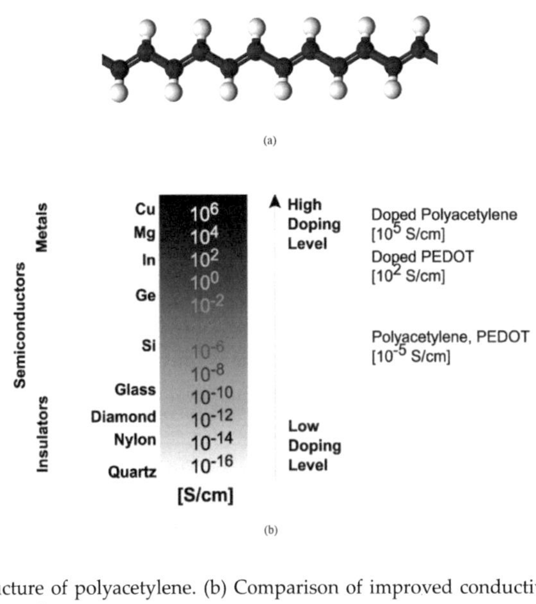


**Figure 1.** (a) Chemical structure of polyacetylene. (b) Comparison of improved conductivity of doped polyacetylene with other conventional materials.

http://dx.doi.org/10.5772/63527

negligible free charge carriers for electrical conductivity. But at the same time, three well-known scientists Heeger, MacDiarmid, and Shirakawa reported a series of great works on the doping of polymer and demonstrated that the proper doping of conjugate polymer improves its conductivity just like as usual semiconductor or metal and they referred it as semiconducting/conducting polymer [3]. They reported first time in the history that the conductivity ($10^{-5}$ S/m) of insulating polyacetylene (molecular structure shown in **Figure 1a**) thin film can be increased by doping it with iodine up to a metallic polyacetylene ($10^{5}$ S/m) just like as Cu metal as shown in **Figure 1b**. No doubt, it was a great turning point in the discipline of polymer engineering and caused to initiate a new field as "Organic Electronics" or "Polymer Electronics". On the basis of their land-mark achievements, Heeger, MacDiarmid, and Shirakawa were awarded Nobel Prize for chemistry in 2000 [2–4].

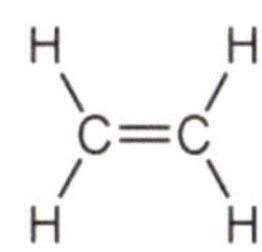

Ethylene (Ethene)

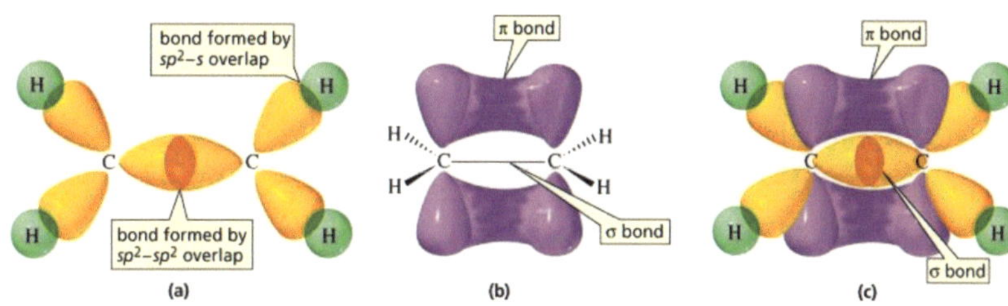


**Figure 2.** (a) Sigma (Single bond) hybridization for ethylene, (b) pi (Single bond) hybridization for ethylene and (c) both sigma and pi bond hybridization which give de-localization of pi-electron.

Broadly speaking, conjugate polymers cover the extensive portion of conducting polymers. Conjugate polymers are defined as such polymer which has a combination of alternating double bonds ($\pi$ bonds) and single bonds ($\sigma$ bonds) as their backbone structure. As compare to the $\sigma$ bond, the $\pi$ bonds are usually unstable in nature and easily ionized or removed. This special combination of single and double bond for conjugate polymer offers a unique arrangement for the de-localization of charge carriers from one end to the other end of polymer chain as extended p-orbital system [19, 55]. This p-orbital overlapping supported with contiguous single bonds allows a de-localization of $\pi$-electrons (double-bond electrons are called $\pi$-electron offers $P_x$, $P_y$, and $P_z$ hybridization for carbon atoms) throughout the double and single-bond combination just as an ethylene as shown in **Figure 2**. Such de-localization of free carriers is the main reason for the charge transport within the molecular chain (intra-molecule) of a conducting polymer [56]. Just for the discussion of $\sigma$ bond and $\pi$ bond a simple ethylene (IUPAC name ethene $H_2C = CH_2$) molecular structure is shown here in **Figure 2**. Each carbon makes $sp^2$-$sp^2$ hybridization for $\pi$ bond and the $\sigma$ bond is formed between carbon and hydrogen atom. The overall hybridization of $\sigma$ and $\pi$ bond is shown in **Figure 2c** for ethylene.

Conducting polymers offer a large number of topological defects as disorder. These defects, such as breaking of bonds in molecular chain, torsion, addition of both internal and external impurity, are produced during the synthesis of polymerization and cause to provide electrical traps for conductivity between molecular chains. Such disorder in conducting polymer presents localization of free carriers inside a molecular chain. Therefore, intermolecular chain transport requires hopping of free carries from one molecular chain to the other molecular chain. So overall, we can say that charge transport in polymer is the combination of both intermolecular and intramolecular charge transport process. Due to the hopping conductivity for intermolecular charge transport, the overall mobility of free carriers for most of polymer is very small as compared to the conventional inorganic semiconductors at normal operating conditions. From charge-transport point of view, there is no fundamental inconsistency between insulating, semiconducting, and conducting polymer except the difference between energy bandgaps, which is larger for insulating polymer semiconductor as compared to semiconducting polymer.

## 3. Metal-polymer interface

For electronic devices, metal-polymer interface plays a vital role to define the electrical response for conducing polymer. **Figure 3** demonstrates the differences between typical metal-vacuum and metal-polymer interface energy band diagrams for further discussion. Apparently both energy band diagrams look same, but actually different charge injection mechanism is observed in both cases. The Fermi-energy level ($E_f$) of metal is coincide with $E_f$ of vacuum and as well as with polymer. But the barrier height between $E_f$ of metal and vacuum energy level for metal-vacuum interface is very high as compared to the barrier height between $E_f$ of metal and polymer interface. The barrier height difference between metal $E_f$ and

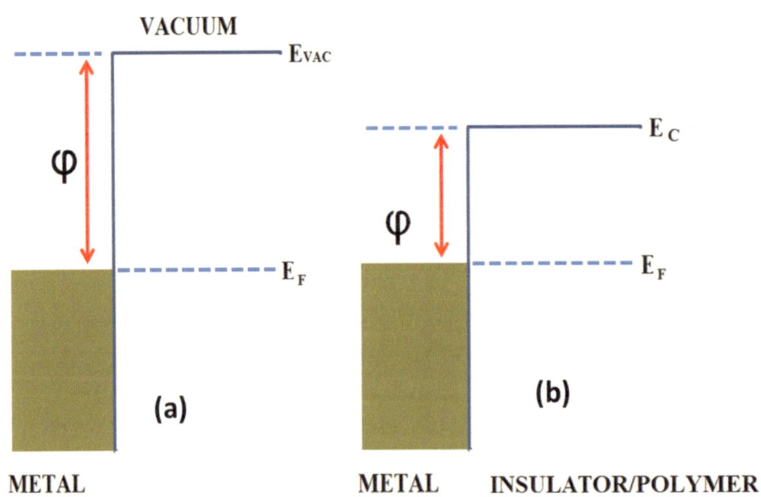


**Figure 3.** (a) Metal work function and (b) metal-polymer interface. Energy to inject electron from metal to vacuum is much higher than the required energy to inject electron from metal to polymer/insulator.

vacuum level is defined as metal work function [56]. On the basis of these facts, Mott and Gurneys proposed metal-insulator barrier height reduction theory and justified the low metal-insulator barrier height as compared to the metal work function, which causes space charge–limited current in insulating materials [57]. It is unanimously accepted that Mott and Gurney's proposed space charge–limited current is also valid for most of insulating, semiconducting, and even conducting polymers [31].

## 4. Bulk-limited and injection-limited current flow

The limitation of current by a polymer can be classified as either (i) injection-limited or (ii) bulk-limited current flow. For injection-limited, the limitation of current through polymer is imposed by non-ohmic metal-polymer interfaces, while for bulk-limited, the limitation is imposed by the bulk properties of polymer. If conducting polymer is sandwiched between two electrodes and anyone electrode offers low barrier height (~ohmic response) to the polymer-metal interface then the injected carriers from electrode forms a space charge region consisting of a large number of injected carriers and equilibrium free carriers inside polymer. As the mobility of carriers is very small, therefore, before traversing of injected carriers from one to the other electrode, more and more charges are injected. When an external electric field is applied, further charges are injected from low-barrier electrode to the polymer, and an equilibrium stage is reached when injected carriers are comparable or even higher than the free carriers concentration; at this stage, the flow of current is referred as space charge–limited current [57, 58].

## 5. Child-Langmuir space charge model

Space charge–limited current is a hot topic of research due to their great application for conducting/semiconducting polymers. The origin of space charge theory was founded by C.D. Child and I. Langmuir from 1911 to 1913, when they reported the derivation of space charge–limited current in a parallel-plane vacuum diode as [59, 60]

$$J = \frac{4\,\varepsilon_0}{9}\sqrt{\frac{2e}{m_e}}\,\frac{V^{\frac{3}{2}}}{d^2} \tag{1}$$

where J is space charge–limited current for vacuum diode, $\varepsilon_0$ is the dielectric constant for free space, e is the coulomb charge of electron, $m_e$ is the mass of electron, $V_a$ is the applied (anode) voltage, and d is vacuum spacing between two electrodes. The above equation is reported in the literature with different names such as three-halves power law, Langmuir-Child law. From the equation, it is clear that the space charge current is directly proportional to the three-halves power of the applied voltage and inversely proportional to the square of the displacement between electrodes.

## 6. Mott-Gurney space charge model

Apparently, Child-Langmuir space charge equation did not find any application for insulator or semiconducting materials due to the presence of vacuum and hence no scattering between electrodes. Therefore, Mott-Gurney proposed another space charge–limited current equation for polymer diode, which is similar to the Child-Langmuir equation with the following assumptions [57].

1. Active layer is trap-free for charge injection
2. Diffusion of carrier is negligible in active layer
3. Electric field at the injecting electrode is zero.

Generally, the assumptions 2 and 3 are still valid for most of organic/polymer semiconducting materials. But for assumption 1, space charge–limited current is further modified with new version as trapped space charge–limited current model and will be discussed in the later section.

When a voltage is applied to an active layer, sandwich between two electrodes as diode, then an electric field (E) is established inside the active layer. Such electric field forces the charges to move with velocity (v) toward another electrode, and therefore, the mobility (μ) of free carrier is defined as

$$v = \mu E \tag{2}$$

Similarly, the current density (J) passing through a semiconductor with conductivity (σ) under the influence of applied electric field E can be defined as

$$J = \sigma E \tag{3}$$

where σ is direct function of both mobility and carrier density of electron n(x) and hole p(x) and can be define as

$$\sigma = e\,n(x)\,\mu_n + e\,p(x)\,\mu_p \tag{4}$$

If we assume single-carrier devices, then Eq. (4) is simplified as

$$\sigma = e\,n(x)\,\mu \tag{5}$$

By incorporating Eq. (5), Eq. (3) becomes as

http://dx.doi.org/10.5772/63527

$$J = en(x)\mu E \tag{6}$$

The injected carriers forms space charge with the distribution of electric field inside insulator and can be defined mathematically by Poisson equation as

$$\frac{dE}{dx} = \frac{en(x)}{\varepsilon} \tag{7}$$

By solving all above equation for one-dimension path with boundary conditions V (0) = V and V (L) = 0, the pure space charge–limited current without any traps will be obtained as

$$J = \frac{9}{8}\varepsilon\varepsilon_o\mu\frac{V^2}{L^3} \tag{8}$$

It is important to note that logarithmic graph (log(J) – log (V)) of above Eq. (8) yield a straight line with slope 2, which shows trap-free space charge–limited current behavior of polymer between electrodes as sown in **Figure 4**.

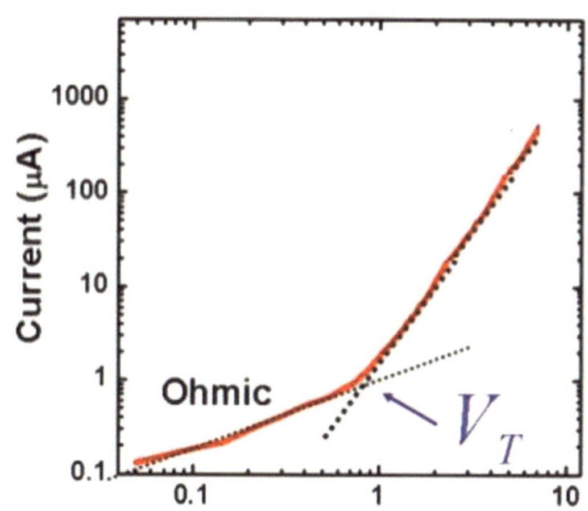


**Figure 4.** Space charge–limited current behavior for polymer semiconductor with only ohmic and trap-free space charge–limited current regions.

**Figure 4** shows the ideal current-voltage characteristics of conducting polymer diode, where there are only two regions, one is ohmic region and other is space charge region and both regions can be differentiated with the order of slope. The transition of both ohmic and space charge regions is taken place at specific voltage termed as threshold voltage $V_T$. At low voltage, polymer diode offers ohmic behavior (J α V) with slope at order 1, while at higher voltage, polymer exhibits space charge–limited current (J α $V^2$) with slope of an order 2. In other words, there is direct transition is observed from ohmic to space charge–limited region, which is not true in the presence of traps distribution inside polymer.

## 7. Trapped space charge–limited current model

Generally, an intermediate region is also observed between ohmic to space charge–limited region and this region is termed as trapped space charge region as shown in **Figure 5**. The charge transport inside polymer within this region is controlled by the trapping and de-trapping of carriers at both energetic and positional distribution. Traps are nothing just as impurity and/or structural defects which provide localized states between HOMO (highest occupied molecular orbital) and LUMO (lowest unoccupied molecular orbital) energy bandgap of polymer. These localized states trap the free carriers and avoid them to take any role for charge transport process and degraded the electrical properties of polymer and hence device [61]. When applied voltage is higher than the average energy associated with traps density, then polymer behaves trap-free space charge–limited current as shown in **Figure 5**.

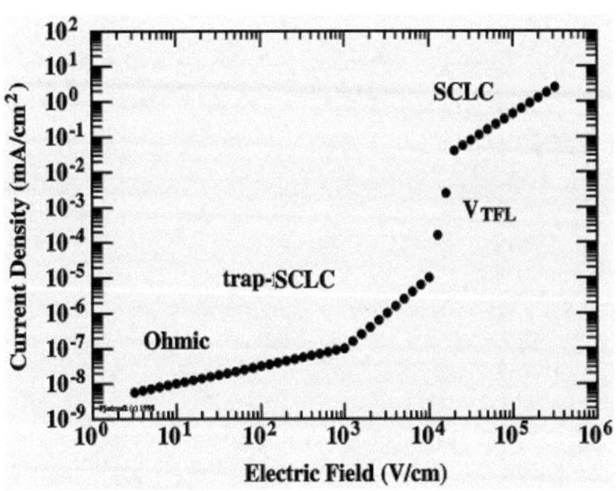


**Figure 5.** Typical space charge–limited current behavior for polymer semiconductor. Four charge-transport regions are clearly visible (i) ohmic region (J α μ V), (ii) trap-SCLC region (J α μ $V^n$) region, (iii) VTFL region (J α μ $N_t d^2$), and (iv) SCLC region (J α μ $V^2$) [31].

The trapped space charge–limited current depends on the distribution of traps state at the energy bandgap of the given polymer. Generally three distributions of electronic traps are reported for conducting polymers. These distributions are as follows:

1. Single energy level trap density.
2. Exponential distribution of trap density.
3. Gaussian distribution of trap density.

### 7.1. Single energy level trap density

For single energy level trap density ($E_T$, also called transport energy), the current for trapped space charge–limited current can be little modified as [62]

$$J = \frac{9}{8}\varepsilon\varepsilon_o\mu\theta\frac{V^2}{L^3} \tag{9}$$

where θ is defined as trap factor and can be correlated with free and trapped carrier density as [63–67]

$$\theta = \frac{n_f}{n_f + n_t} \tag{10}$$

where $n_f$ and $n_t$ are defined as free carrier and trapped carrier density, respectively. The trapped carrier density $n_t$ can be defined with single energy level $E_T$ as

$$n_T = N_T\ Exp\left(-\frac{(E_T - E_F)}{kT}\right) \tag{11}$$

where $N_T$, $E_f$, k, and T can be defined as trap density, Fermi energy level, Boltzmann constant, and ambient temperature, respectively.

### 7.2. Exponential distribution of traps

Exponential distribution of traps (g(E)) with characteristics width of trap energy distribution ($E_c$) between HOMO and LUMO energy bandgap of polymer can be identified as

$$g(E) = \frac{N_T}{E_c}\ Exp\left(-\frac{E}{E_c}\right) \tag{12}$$

$E_c$ is correlated with characteristic temperature ($T_c$) as $E_c = k\,T_c$. Now, the trapped space charge–limited current for a polymer having exponential distribution of traps can be described as [68–70]

$$J = q^{1-l}\mu_p N_v\left(\frac{2l+1}{l+1}\right)^{l+1}\left(\frac{l\varepsilon_s}{(l+1)N_t}\right)^{l}\frac{V^{l+1}}{d^{2l+1}}, \tag{13}$$

where $l = T_c / T$ and T is the absolute temperature.

### 7.3. Gaussian distribution of traps

Bassler proposed Gaussian density of states for polymer material, which are broadened by both energetic and positional disorder [41]. Therefore, many researchers apply Gaussian density of traps ($D_T$ (E)) for trapped space charge–limited current as [71].

$$D_t(E)=\frac{N_T}{\sigma\sqrt{2\pi}}\exp\left(-\frac{\left(E-\left(E_C-E_T\right)\right)^2}{2\sigma_T}\right) \tag{14}$$

where $\sigma_T$ is the width of Gaussian distribution traps and $E_c - E_T$ is the trap energy. Similarly Steiger and his workers proved that the Gaussian trapped space charge–limited current follow the same exponential distribution of traps Eq. (13), where $l$ can be redefined as [72]

$$l=\left[1+\frac{2\,\pi\,\sigma^2}{16\,k^2T^2}\right]^{\frac{1}{2}} \tag{15}$$

If $2\,\pi\sigma^2 >> k^2\,T^2$, then Eq. (15) is further simplified as

$$l=\sqrt{\frac{2\,\pi}{16}}\,\frac{\sigma}{kT} \tag{16}$$

## 8. Photocurrent space charge–limited current model

When light falls on conjugate polymer, huge number of electron and holes are uniformly generated throughout the polymer as free carriers and these carriers move to their respective electrodes as photocurrent under the influence of internal electric field generated by the difference of electrode work-function. If G is the generation rate, L is the length, and q is the electric charge of free carriers, then the photocurrent ($J_{ph}$) can be defined as [17, 34, 73]

$$J_{ph}=q\,G\,L \tag{17}$$

By incorporating the effective thickness of polymer (L) for photocurrent, Eq. (17) can be modified as

$$J_{ph}=q\,G\left(\mu_h\,\tau_h\right)^{\frac{1}{2}}V^{\frac{1}{2}} \tag{18}$$

where $\mu_h$ and $\tau_h$ are the mobility and charge carrier life time. Blom et al. observed that the photo-generated current in this region (L) follows space charge–limited current and he derived the fundamental space charge–limited photocurrent equation as [33]

$$J_{ph} = q\left(\frac{9\varepsilon_o\varepsilon_r\mu_h}{9q}\right)G^{\frac{3}{4}}V^{\frac{1}{2}} \tag{19}$$

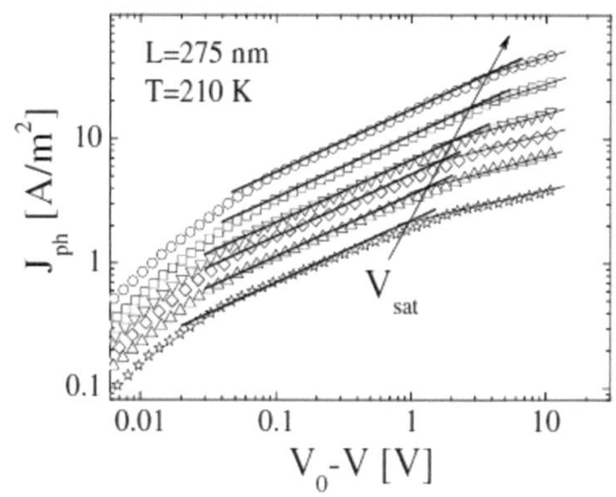


**Figure 6.** Incident light power (ILP) dependence of the photocurrent ($J_{ph}$) versus the effective voltage ($V_0$ - V) measured at T = 210 K. The solid (thick) lines represent the $J_{ph}$ using $\mu_h$ = 1.2 × $10^{-7}$ cm²/V s, $\epsilon_r$ 2.6, and G $\alpha$ ILP, where ILP was varied from 80 to 6 mW/cm². The arrow indicates the voltage ($V_{sat}$) at which $J_{ph}$ shows the transition to the saturation regime [33].

where $\epsilon_o$, $\epsilon_r$ are the dielectric constant for free air and polymer, respectively. From above equation, Blom et al. experimentally observed that the space charge–limited photocurrent is directly proportional to the three-quarter power of photo-generation rate and half power of voltage as shown in **Figure 6**.

**Figure 6** clearly shows that the space charge–limited photocurrent as calculated by Eq. (19) is in contrast with the square-root region for the given voltage range of the experimental results.

## 9. Shallow and deep traps for space charge–limited current model

Traps not only decrease the mobility for space charge–limited current but also initiate the thermal and electrical degradation for polymer electronic devices. Traps are generally classified as (i) shallow traps, and (ii) deep traps for both electron and hole. If traps are very close to the conduction band (LUMO) within energy bandgap, then traps are classified as shallow traps for electrons. Similarly if traps are in the vicinity of valence band (HOMO) within energy bandgap, then traps are identified as shallow traps for holes as shown in **Figure 8**. On the other hand, deep traps of electron and holes are exist far away (mid of the energy bandgap) from conduction (LUMO) and valence (HOMO) band, respectively, as shown in **Figure 7**.

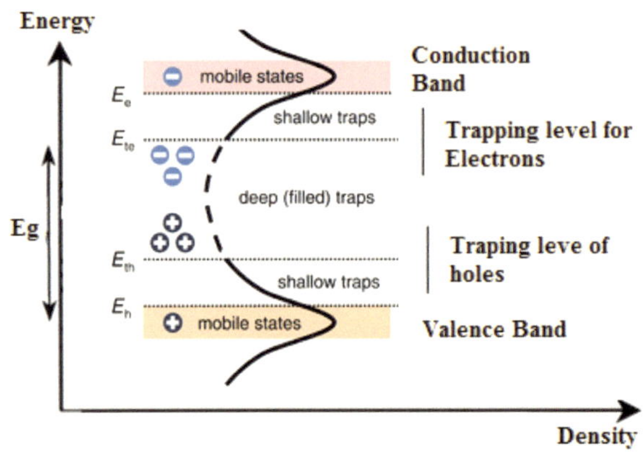


**Figure 7.** Symbolic representation of shallow and deep traps for polymer semiconductor.

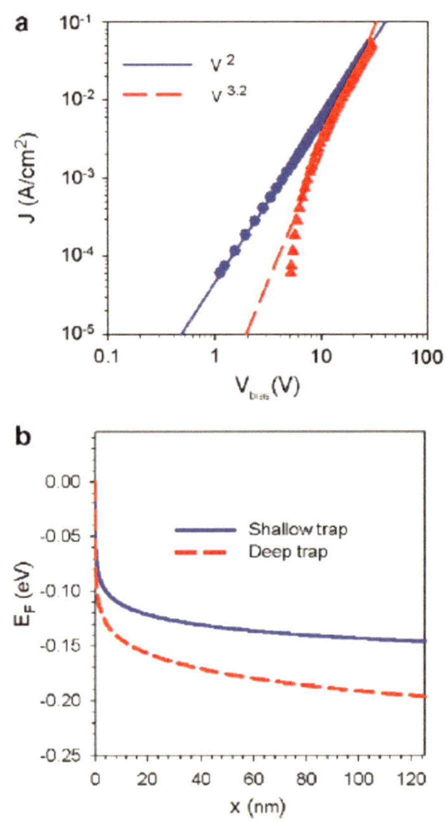


**Figure 8.** Model simulations of shallow ($E_c - E_t = 0.1$ eV) and deep ($E_c - E_t = 0.5$ eV) trap configurations represented by blue solid lines and red dashed lines. (a) Current density voltage characteristics for a shallow trap and a deep trap are plotted by blue dots and red triangles. Fittings provide the exponent of the voltage. (b) Fermi level representations at 6 V along the thickness, for a shallow trap (blue solid line) and a deep level (red dashed line) [74].

Conventional space charge and trap space charge–limited current model cannot differentiate between shallow and deep traps for both electron and hole. If space charge–limited current is modeled in frequency domain with single traps, then the dynamic picture of the traps can be obtained. By varying the single-trap energy for frequency-domain analysis of space charge–limited current model, it is possible to differential between shallow and deep traps for both electron and polymer with the condition that polymer is sandwiched between properly selected electrodes [74].

## 10. Mobility measurement and space charge–limited current

The velocity per unit electric field is defined as the mobility for polymer. Mobility is one of the most important parameter, which comprehensively defined the charge transport mechanism of polymer. The efficiency of many polymer electronic devices depends on mobility; therefore, it is very crucial to determine the exact value of mobility [43, 75]. For polymer, mobility can be measured with different methods; some of them are listed as follows:

1. Time-of-flight method [44].
2. Carrier extraction by linearly increasing voltage (CELIV) method [44] and photo-CELIV method [76, 77].
3. Space charge–limited current method [65–67].

Each of this method has some advantages and disadvantages, but these methods are the most commonly used method to determine the mobility of polymer.

### 10.1. TOF method

The TOF is the most widely used technique for the measurement of mobility. In these techniques, short pulse of light or laser is stroke over the end surface of polymer sample, which result in the generation of photo carriers from the end surface of sample.

$$\mu = \frac{L}{E.T} \tag{20}$$

These photo carriers are immediately swept away from the end surface toward the other end surface through the influence of external DC volts applied at electrodes and give rise to the transient photocurrent as shown in **Figure 9**. The time (T) taken by the photo carriers from one end to the other end surface is calculated from oscilloscope. If E is the applied electric field, L is the thickness of organic layer then the mobility can be calculated as [78].

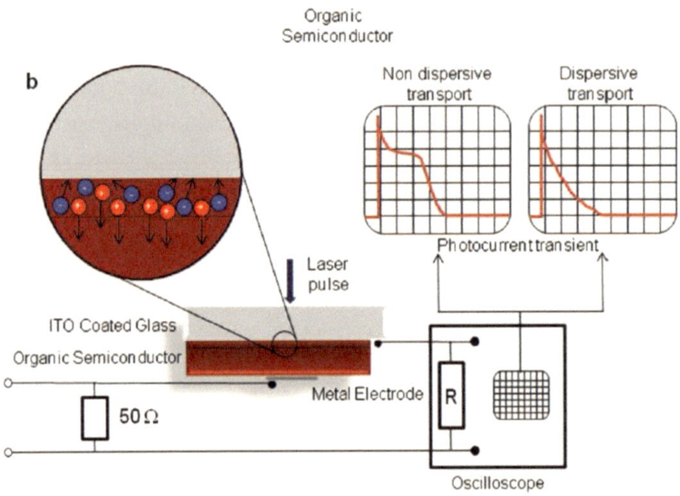


**Figure 9.** Schematic setup for typical TOF system for mobility measurements. Here the laser pulse can be illuminated through ITO electrode as well as a semitransparent metal electrode of polymer sample [43].

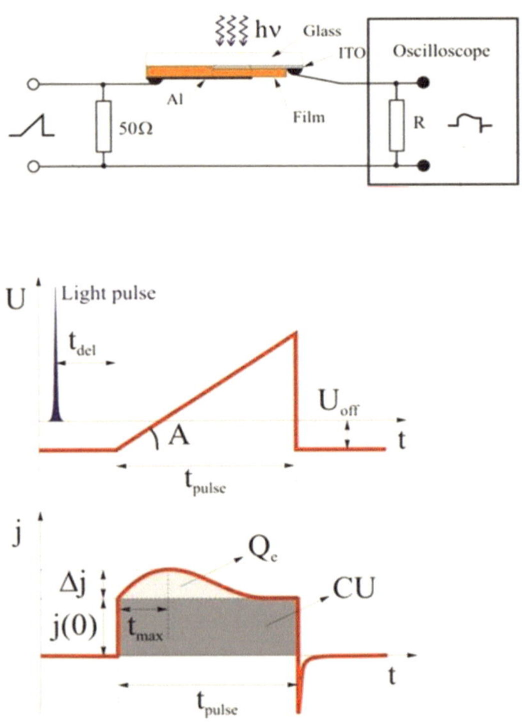


**Figure 10.** The schematic Photo CELIV measurement setup is shown. The measuring process consists of two steps. In first step a laser is shine at transparent electrode of sample which generates excess photo carriers. These carriers are extracted by linearly increasing voltage. The peak-extraction time helps to calculate the mobility.

http://dx.doi.org/10.5772/63527

### 10.2. CELIEV/photo CELIEV method

The charge extraction by increasing the linear voltage is another technique commonly uses to measure the mobility of polymer sandwiched between electrodes. The measuring process consists of two steps. In first step, a laser is shine at transparent electrode of sample which generates excess photo carriers. These carriers are extracted by linearly increasing voltage. The peak-extraction time helps to calculate the mobility of the polymer sample as shown in **Figure 10**. The detail information about these methods can be found in Ref. [79–81].

### 10.3. Dark-injection space charge–limited mobility measurement

Dark-injection space charge–limited current (DI-SCL) is another commonly used technique for the measurement of mobility for many amorphous, disorder organic/polymer materials [82–85]. To determine the mobility of hole with DI-SCL method, the polymer sample is sandwiched between electrodes and electrodes are managed in such a way that the cathode behaves as blocking electrode and a strong voltage pulse is derived at ohmic anode. The application of voltage pulse will result a transient hole current observed at oscilloscope as shown in **Figure 11**. Therefore, the hole density $\mu_h$ can be calculated as

$$\mu_h = \frac{0.787d^2}{V\tau} \tag{21}$$

where τ is the arrival time for early sheet of holes carriers to reach the respective cathode. The τ is correlated with space charge free carrier transit time ($\tau_{SCL}$) as $\tau = 0.787\ \tau_{SCL}$ [86, 87].

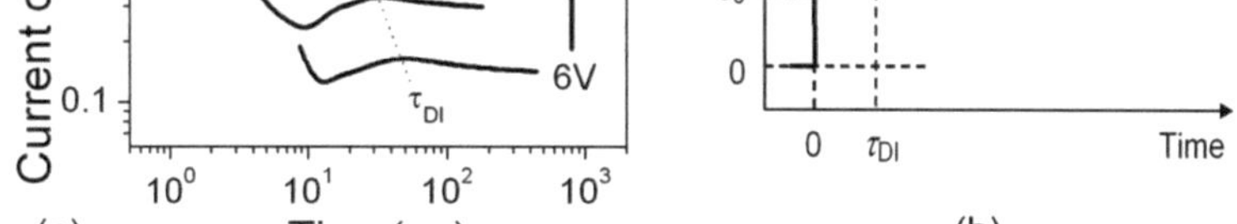


**Figure 11.** (a) A sequence of DI signals of Au (treated with UVO)/2TNATA/Au under different applied voltages. The applied voltage varied in steps of 2 V starting from 6 V. (b) An ideal DI-SCLC transient [82].

For both CELIEV and photo-CELIV mobility measurement method, it is not so simple to differential between electron and hole mobility, specially for bipolar charge-transport polymer. Similarly, the TOF method has two series drawbacks (1) the density of photo-generated carrier is not adequate (very low) to measure the mobility as compare to the usual carrier density at normal operation of many electronic devices, and another drawback is that (ii) TOF method should requires the thickness of polymer above the absorption length of polymer (>1 μm) [78]. In the same way, dark-injection space charge–limited current is required the proper selection of electrode for reliable measurements.

Recently, a comprehensive study was performed by Qiu group to compare and evaluate the performance of different mobility measurement methods reported for polymeric materials. They consider various factors such as applied electric field, injection barrier, and energetic disorder, which are crucial for the performance of polymer electronic devices and finally they concluded that the result of both space charge and TOF mobility measurement methods are very close to each other [88]. Therefore, with proper selection of electrodes, space charge–limited mobility measurement is acceptably reliable for most of the cases.

## 11. Modification of space charge–limited current model

As we discussed earlier, when the current density passing through the polymer with length (L) under the influence of applied voltage (V) shows directly proportional response with the square of voltage ($J \alpha V^2$) and inversely proportional to cube of L ($J \alpha L^{-3}$) then the polymer sample is modeled with trap-free space charge–limited current. In order to incorporate the variety of the trap distribution, the space charge–limited current model is modified as trapped space charge–limited current model. Traps are both energetically and positional distributed throughout the energy bandgap and capture the free carriers and offer some energy barrier to release them. Both space charge–limited current and trap space charge–limited currents model assume that barrier height of traps are remain constant for entire operating electric field range of the device, which is not true for some of the reported experimental results. In fact, higher electric field lowers the barrier height of traps which cause to increase the emission rate from traps and hence the current density as expected from trap space charge–limited current model as shown in **Figure 12a**. The lowering of trap barrier height or emission rate from traps at high electric field is identified as Poole-Frenkel effect [89], as shown in schematic energy band diagram (**Figure 12b**). There is some similarity between Poole-Frenkel and Schottky models [90, 91] for the lowering of traps barrier height. Schottky model represent barrier height reduction at metal-polymer interface, while Poole-Frenkel model shows the lowering of trap barrier height inside the polymer thin film. The overall current density (J) passing through the polymer due to Poole-Frenkel emission can be written as

$$J = q\mu N_C \exp\left[\frac{-q\left(\varphi_T - \sqrt{\dfrac{qE}{\pi\varepsilon_r\varepsilon_o}}\right)}{kT}\right] \tag{22}$$

where $q\,\varphi_T$ is trap energy barrier height, and other variables have already defined earlier. Poole-Frenkel behavior can be justified for any polymer device, if a linear relation is obtained between $\ln\left(\frac{J}{E}\right)$ $vs\,\sqrt{E}$ data derived from their experimental result for such device.

Murgatroyd was the first who addressed the lowering of trap barrier height by incorporating Poole-Frenkel equation into space charge–limited current equation and drive an approximate equation as [92].

$$J = \frac{9}{8}\varepsilon_r\,\varepsilon_0\,\mu_h \exp\left(0.89\gamma\sqrt{E}\right)\frac{V^2}{L^3} \tag{23}$$

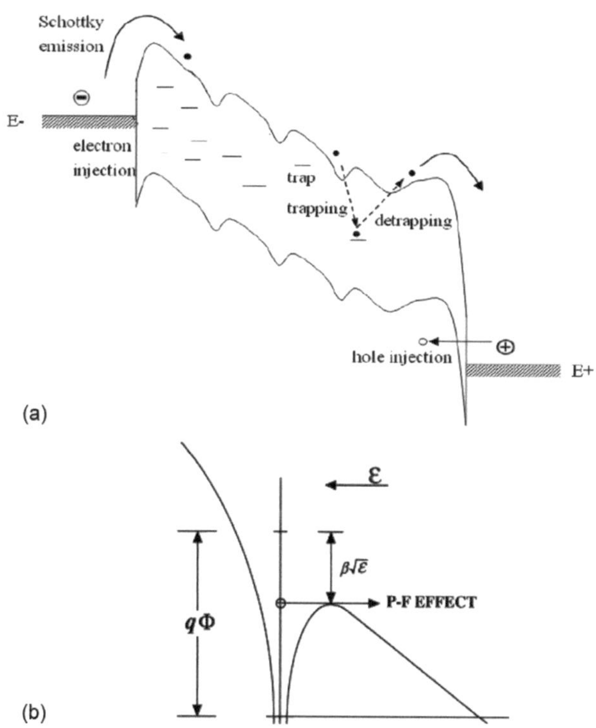


**Figure 12.** (a) Lowering of energy barrier height of traps at higher electric fields, (b) Pool-Frenkel Effect

Murgatroyd equation is simple and very handy, but it is an approximate equation. Later on, Barbe determined analytically, the effect of trap barrier height reduction on space charge–limited current model and derived an equation as [93, 94]

$$\frac{J\,x}{\mu\varepsilon\theta} = \frac{2(kT)^4}{\beta^4}\left[\exp\left(\frac{\beta\sqrt{E}}{kT}\right)\left\{\left(\frac{\beta\sqrt{E}}{kT}\right)^3 - 3\left(\frac{\beta\sqrt{E}}{kT}\right)^2 + 6\frac{\beta\sqrt{E}}{kT} - 6\right\} + 6\right] \tag{24}$$

Barbs equation is used to determine the current density (J) as a function of electric field define at position x. It is reported that both these Murgatroyd and Barbe modifications are also applicable with electrical response for many polymers devices [95–99]. As both insulating and semiconducting polymers share the same charge transport mechanism, therefore, above equations are reasonably valid for both types of polymers.

The fundamental work of Frenkel, Murgatroyd, and Barbe is to modify the space charge–limited current to accommodate the lowering of trap barrier height for only in one-dimension trap distribution. For the three-dimension trap distribution, these models not work very well. In this domain, Hartke modification is reasonably accepted for the accommodation of three-dimensional lowering of the trap barrier height for space charge–limited current model [100].

Similarly, Geurst modified the space charge–limited current model in such a way that the thickness of the semiconductor thin film is insignificant with respect to the electrode separation as 2D model [101–104]. While, Chandra derived Mott-Gurney space charge–limited current equation in two dimensions with exponentially distributed traps and other variety of polymer Schottky diodes [105–107].

## 12. Summary

Despite great recent achievements for conjugate polymer-based electronic devices, the clear picture of charge transport mechanism is still not available. Polymers have unique charge transport mechanism as a combination of delocalization and localization of charge carriers with intramolecular and intermolecular charge interaction, respectively. But it is unanimously believed that Mott-Gurney space charge–limited model is appreciably accepted for most of the polymers. As polymers are full of traps, therefore, Mott-Gurney space charge–limited model is also modified as trapped space charge–limited model according to the requirements of polymers. The nature of traps distribution inside polymer is varied and depends on many factors such as nature of materials itself, polymerization process, nature of dopants, and solvent. Generally, three types of traps distributions are reported in the literature for polymers named as single trap, exponential, and Gaussian distribution of traps. Therefore, trapped space charge–limited current model is further modified to accommodate all these distribution of traps. Just like as distribution of traps the nature of traps, such as shallow and deep traps, is also important to define the electrical response of polymer. The most crucial parameter affected by the distribution and as well as nature of traps is mobility, and space charge–limited current model with little variation is also used to measure the charge mobility of polymer. Different other methods such as time-of-flight method and charge extraction by linearly increasing voltage (CELIV) with or without light (photo—CELIV) sources are available to measure the charge mobility. By comparing these measuring methods for polymer, it is reported that the simple space charge–limited model is an acceptable choice for mobility measurements. Similarly, it is also observed the trap barrier height is significantly reduced at higher electric field and temperature due to Poole-Frenkel effect. Therefore, Murgatroyd and Barbe incorporated Poole-Frenkel effect and solved Mott-Gurney space charge–limited equation numerically

http://dx.doi.org/10.5772/63527

and analytically, respectively. It is reported that both these modifications are experimentally verified with electrical response of many polymers devices.

## Acknowledgements

The authors are thankful to Umm Al Qura University and GIK Institute of Engineering Sciences and Technology for its support to this work and the facility used. Authors are also thankful to our PhD and Master students for their cooperation, valuable information, and thoughtful suggestions. Special thanks are due for An-Zhong Lin, M.A. Turaeva, and Kh. Akhmedov for their technical support and helpful discussions.

## Author details

Syed A. Moiz[1*], Iqbal. A. Khan[1], Waheed A. Younis[1] and Khasan S. Karimov[2,3]

*Address all correspondence to: moiz_pak@yahoo.com

1 Faculty of Electrical Engineering, Umm Al Qura University, Makkah, Saudi Arabia

2 Physical Technical Institute, Dushanbe, Tajikistan

3 Faculty of Electrical Engineering, GIKI Institute, Topi, Swabi, Pakistan

## References

[1] Liao, C., Zhang, M., Yao, M.Y., Hua, T., Li, L. and Yan, F., 2015. Flexible organic electronics in biology: materials and devices. *Advanced Materials*, *27*(46), pp. 7493–7527.

[2] Forrest, S., Burrows, P. and Thompson, M., 2000. The dawn of organic electronics. *Spectrum, IEEE*, *37*(8), pp. 29–34.

[3] Inokuchi, H., 2006. The discovery of organic semiconductors. Its light and shadow. *Organic electronics*, *7*(2), pp. 62–76.

[4] Heeger, A.J., 2001. Semiconducting and metallic polymers: the fourth generation of polymeric materials (Nobel lecture). *Angewandte Chemie International Edition*, *40*(14), pp. 2591–2611.

[5] Bronstein, H. and King, F.D., 2016. Energetic tuning in spirocyclic conjugated polymers. *Polymers*, *8*(1), p. 9.

[6] Chen, J.D., Cui, C., Li, Y.Q., Zhou, L., Ou, Q.D., Li, C., Li, Y. and Tang, J.X., 2015. Single-junction polymer solar cells exceeding 10% power conversion efficiency. *Advanced Materials, 27*(6), pp. 1035–1041.

[7] Brütting, W. ed., 2006. *Physics of Organic Semiconductors*. John Wiley & Sons, Ltd., New Jersey, USA.

[8] Hush, N.S., 2003. An overview of the first half-century of molecular electronics. *Annals of the New York Academy of Sciences, 1006*(1), pp. 1–20.

[9] Chen, D., Liang, J. and Pei, Q., Flexible and stretchable electrodes for next generation polymer electronics: a review. *Science China Chemistry*, pp. 1–13.

[10] Jain, A. and George, S.J., 2015. New directions in supramolecular electronics. *Materials Today, 18*(4), pp. 206–214.

[11] Walzer, K., Maennig, B., Pfeiffer, M. and Leo, K., 2007. Highly efficient organic devices based on electrically doped transport layers. *Chemical Reviews, 107*(4), pp. 1233–1271.

[12] Petty, M.C., Bryce, M.R. and Bloor, D., 1995. *An Introduction to Molecular Electronics*. Oxford University Press, New York, USA.

[13] Ye, L. and Hou, J., 2015. Conjugated Polymer Photovoltaic Materials. In *Organic Optoelectronic Materials* (pp. 195–239). Springer International Publishing, Switzerland.

[14] Geoghegan, M. and Hadziioannou, G., 2013. *Polymer Electronics*., Oxford University Press, New York, USA.

[15] MacDiarmid, A.G., 2001."Synthetic metals": a novel role for organic polymers (Nobel lecture). *Angewandte Chemie International Edition, 40*(14), pp. 2581–2590.

[16] Shirakawa, H., 2001. The discovery of polyacetylene film: the dawning of an era of conducting polymers (Nobel lecture). *Angewandte Chemie International Edition, 40*(14), pp. 2574–2580.

[17] Brabec, C.J., Sariciftci, N.S. and Hummelen, J.C., 2001. Plastic solar cells. *Advanced Functional Materials, 11*(1), pp. 15–26.

[18] You, J., Dou, L., Yoshimura, K., Kato, T., Ohya, K., Moriarty, T., Emery, K., Chen, C.C., Gao, J., Li, G. and Yang, Y., 2013. A polymer tandem solar cell with 10.6% power conversion efficiency. *Nature Communications, 4*, p.1446.

[19] Gutman, F. and Lyons, L.E., 1981. *Organic semiconductors Part A*. Robert E. Publishing, Malabar, FL.

[20] Lewis, J., 2006. Material challenge for flexible organic devices. *Materials Today, 9*(4), pp. 38–45.

[21] Janata, J. and Josowicz, M., 2003. Conducting polymers in electronic chemical sensors. *Nature Materials, 2*(1), pp. 19–24.

[22] Moiz, S., Karimov, K.S. and Gohar, N.D., 2004. Orange dye thin film resistive hygrometers. *Eurasian Chemico-Technological Journal*, *6*(3), pp. 179–183.

[23] Karimov, K.S., IBRAHIM, I.N., Tahir, M.M., Khan, T.A. and Shafique, U., 2008. Photo organic field effect transistor's properties. *Turkish Journal of Physics*, *32*(1), pp. 13–19.

[24] Karimov, K.H., Ahmed, M.M., Karieva, Z.M., Saleem, M., Mateen, A. and Moiz, S.A., 2011. Humidity sensing properties of carbon nano-tube thin films. *Sensor Letters*, *9*(5), pp. 1649–1653.

[25] Karimov, K., Ahmed, M.M., Noshin, F., Saleem, M., Mahroof-Tahir, M., Moiz, S.A., Akhmedov, K.M., Zahid, M., Abbas, S.Z. and Rashid, A., 2015. Nickel phthalocyanine based organic photo transistor: effect of semiconductor film thickness. *The European Physical Journal Applied Physics*, *72*(2), p. 20202.

[26] Karimov, K.S., Moiz, S.A., Tahir, M.M., Ahmed, N., Tariq, R., Abbas, S.Z. and Zafar, Q., 2014. Nickel phthalocynanine-metal Schottky diode as photodetector. *Journal of Optoelectronics and Advanced Materials*, *16*(11–12), pp. 1430–1435.

[27] Karimov, K.S., Saeed, M.T., Khalid, F.A. and Moiz, S.A., 2011. Effect of displacement on resistance and capacitance of polyaniline film. *Chinese Physics B*, *20*(4), p.040601.

[28] Karimov, K.S., Sayyad, M.H., Ahmed, M.M., Khan, M.N., Karieva, Z.M., Moiz, S.A., Shah, M., Zakaullah, K. and Turaeva, M.A., 2004. Effect of temperature and humidity on electrical properties of organic orange dye complex films'. *Eurasion Chemical Technological Journal*, *6*, pp. 145–149.

[29] Khan, A., Karimov, K.S., Ahmad, Z., Sulaiman, K., Shah, M. and Moiz, S.A., 2014. Pressure sensitive organic sensor based on CNT-VO2 (3fl) Composite. *Sains Malaysiana*, *43*(6), pp. 903–908.

[30] Chani, M.T.S., Karimov, K.S., Khalid, F.A. and Moiz, S.A., 2013. Polyaniline based impedance humidity sensors. *Solid state sciences*, *18*, pp. 78–82.

[31] Peter, S., 2009. *Electrical Characterization of Organic Electronic Materials and Devices*, John Wiley and Sons, Ltd., New Jersey, USA.

[32] Moiz, S.A., Nahhas, A.M., Um, H.D., Jee, S.W., Cho, H.K., Kim, S.W. and Lee, J.H., 2012. A stamped PEDOT: PSS–silicon nanowire hybrid solar cell. *Nanotechnology*, *23*(14), p. 145401.

[33] Mihailetchi, V.D., Wildeman, J. and Blom, P.W.M., 2005. Space-charge limited photocurrent. *Physical review letters*, *94*(12), p.126602.

[34] Blom, P.W., Mihailetchi, V.D., Koster, L.J.A. and Markov, D.E., 2007. Device physics of polymer: fullerene bulk heterojunction solar cells. *Advanced Materials*, *19*(12), pp. 1551–1566.

[35] Sirringhaus, H., Brown, P.J., Friend, R.H., Nielsen, M.M., Bechgaard, K., Langeveld-Voss, B.M.W., Spiering, A.J.H., Janssen, R.A., Meijer, E.W., Herwig, P. and De Leeuw,

D.M., 1999. Two-dimensional charge transport in self-organized, high-mobility conjugated polymers. *Nature, 401*(6754), pp. 685–688.

[36] Moiz, S.A., Imran, S.M., Nahhas, A.M., Karimov, K.S. and Kim, H.T., 2014. Effect of solvent on Meyer-Neldel rule for conducting polyaniline thin film. *Journal of Optoelectronics and Advanced Materials, 16*(11–12), pp. 1405–1410.

[37] Moiz, S.A., Ahmed, M.M. and Karimov, K.S., 2005. Effects of temperature and humidity on electrical properties of organic semiconductor orange dye films deposited from solution. *Japanese Journal of Applied Physics, 44*(3R), p. 1199.

[38] Ahmed, M.M., Karimov, K.S. and Moiz, S.A., 2004. Temperature-dependent IV characteristics of organic-inorganic heterojunction diodes. *Electron Devices, IEEE Transactions on, 51*(1), pp. 121–126.

[39] Sayyad, M.H., Karimov, K.S., Ellahi, A. and Moiz, S., 2005. The photo-electrical behavior of n-Si and p-Si/orange dye/conductive glass cells. *Eurasian Chemico-Technological Journal, 7*(2), pp. 133–137.

[40] Karimov, K.S., Akhmedov, K.M., Dzhuraev, A.A., Khan, M.N., Abrarov, S.M. and Fiodorov, M.I., 2000. Organic-on-inorganic Ag/n-GaAs/p-CuPc/Ag photoelectric sensor. *Eurasian Chemico-Technological Journal, 2*(3–4), pp. 251–256.

[41] Bässler, H., 1993. Charge transport in disordered organic photoconductors a Monte Carlo simulation study. *Physica Status Solidi(b), 175*(1), pp. 15–56.

[42] Taylor, D.M., 2006. Space charges and traps in polymer electronics. *IEEE Transactions on Dielectrics and Electrical Insulation, 5*(13), pp. 1063–1073.

[43] Kokil, A., Yang, K. and Kumar, J., 2012. Techniques for characterization of charge carrier mobility in organic semiconductors. *Journal of Polymer Science Part B: Polymer Physics, 50*(15), pp. 1130–1144.

[44] Campbell, A.J., Bradley, D.D. and Antoniadis, H., 2001. Dispersive electron transport in an electroluminescent polyfluorene copolymer measured by the current integration time-of-flight method. *Applied Physics Letters, 79*(14), pp. 2133–2135.

[45] Wang, Yan-Ping, Jin-Ying Huang, Jiang-Shan Chen, Xian-Feng Qiao, De-Zhi Yang, Dong-Ge Ma, and Li-Song Dong, 2016. "The Hole Transport Characteristics of 1, 4, 5, 8, 9 and 11-Hexaazatriphenylene-Hexacarbonitrile by Blending. *Chinese Physics Letters 33*(2), p. 027302.

[46] Coropceanu, V., Cornil, J., da Silva Filho, D.A., Olivier, Y., Silbey, R. and Bréédas, J.L., 2007. Charge transport in organic semiconductors. *Chemical Reviews, 107*(4), pp. 926–952.

[47] Brédas, J.L., Calbert, J.P., da Silva Filho, D.A. and Cornil, J., 2002. Organic semiconductors: a theoretical characterization of the basic parameters governing charge transport. *Proceedings of the National Academy of Sciences, 99*(9), pp. 5804–5809.

[48] Pivrikas, A., Sariciftci, N.S., Juška, G. and Österbacka, R., 2007. A review of charge transport and recombination in polymer/fullerene organic solar cells. *Progress in Photovoltaics: Research and Applications*, *15*(8), pp. 677–696.

[49] Karimov, K.S., Sayyad, M.H., Ali, M., Khan, M.N., Moiz, S.A., Khan, K.B., Farah, H. and Karieva, Z.M., 2006. Electrochemical properties of Zn/orange dye aqueous solution/carbon cell. *Journal of Power Sources*, *155*(2), pp. 475–477.

[50] Karimov, K.S., Ahmed, M.M., Moiz, S.A., Babadzhanov, P., Marupov, R. and Turaeva, M.A., 2003. Electrical properties of organic semiconductor orange nitrogen dye thin films deposited from solution at high gravity. *Eurasian ChemTech J*, *5*, pp. 109–113.

[51] Karimov, K.S., Ahmed, M.M., Moiz, S.A. and Fedorov, M.I., 2005. Temperature-dependent properties of organic-on-inorganic Ag/p-CuPc/n-GaAs/Ag photoelectric cell. *Solar Energy Materials and Solar Cells*, *87*(1), pp. 61–75.

[52] Takeshita, S., Modeling of space-charge limited current injection incorporating an advanced model of the Poole–Frenkel effect, MS thesis, Electrical Engineering Clemson University, 2008, Clemson, USA.

[53] Laquai, F., Andrienko, D., Mauer, R. and Blom, P.W., 2015. Charge carrier transport and photo generation in P3HT: PCBM photovoltaic blends. *Macromolecular Rapid Communications*, *36*(11), pp. 1001–1025.

[54] Ahmed, M.M., Karimov, K.S. and Moiz, S.A., 2008. Photoelectric behavior of n-GaAs/orange dye, vinyl-ethynyl-trimethyl-piperidole/conductive glass sensor. *Thin Solid Films*, *516*(21), pp. 7822–7827.

[55] Molapo, K.M., Ndangili, P.M., Ajayi, R.F., Mbambisa, G., Mailu, S.M., Njomo, N., Masikini, M., Baker, P. and Iwuoha, E.I., 2012. Electronics of conjugated polymers (I): polyaniline. *International Journal of Electrochemical Science*, *7*(12), pp. 11859–11875.

[56] Salaneck, W.R., Friend, R.H. and Brédas, J.L., 1999. Electronic structure of conjugated polymers: consequences of electron–lattice coupling. *Physics Reports*, *319*(6), pp. 231–251.

[57] Mott, N.F. and Gurney, R.W., 1940. *Electronic Processes in Ionic Crystals*. Oxford University Press, New York.

[58] Ng, K.K., *Complete Guide to Semiconductor Devices*, 2nd ed., John Wiley & Sons, New York, 2002.

[59] Child, C.D., 1911. Discharge from hot CaO. *Physical Review (Series I)*, *32*(5), p. 492.

[60] Lau, Y.Y., 2001. Simple theory for the two-dimensional Child–Langmuir law. *Physical Review Letters*, *87*(27), p. 278301.

[61] Schmechel, R. and Von Seggern, H., 2004. Electronic traps in organic transport layers. *Physica Status Solidi (a)*, *201*(6), pp. 1215–1235.

[62] Lampert, M.A., 1964. Volume-controlled current injection in insulators. *Reports on Progress in Physics*, *27*(1), p. 329.

[63] Moiz, S.A., Ahmed, M.M. and Karimov, K.S., 2005. Estimation of electrical parameters of OD organic semiconductor diode from measured IV characteristics. *ETRI Journal*, *27*(3), pp. 319–325.

[64] Moiz, S.A., Ahmed, M.M., Karimov, K. and Mehmood, M., 2007. Temperature-dependent current–voltage characteristics of poly-N-epoxypropylcarbazole complex. *Thin Solid Films*, *516*(1), pp. 72–77.

[65] Moiz, S.A., Ahmed, M.M., Karimov, K.S., Rehman, F. and Lee, J.H., 2009. Space charge limited current–voltage characteristics of organic semiconductor diode fabricated at various gravity conditions. *Synthetic Metals*, *159*(13), pp. 1336–1339.

[66] KARIMOV, K., Qazi, I., Moiz, S.A. and Murtaza, I., 2008. Electrical properties of organic semiconductor copper phthalocyanine thin films deposited from solution at high gravity. *Optoelectronics and Advanced Materials – Rapid Communications*, *2*(4), pp. 219–223.

[67] Moiz, S.A. and Nahhas, A.M., 2013. Temperature Dependent Electrical Response of Orange-Dye Complex Based Schottky Diode. *International Journal of Electronics and Communication Engineering & Technology (IJECET)*, *4*(2), pp. 269–279.

[68] Mandoc, M.M., De Boer, B. and Blom, P.W.M., 2006. Electron-only diodes of poly (dialkoxy-p-phenylene vinylene) using hole-blocking bottom electrodes. *Physical Review B*, *73*(15), p. 155205.

[69] Van Mensfoort, S.L.M., Billen, J., Vulto, S.I.E., Janssen, R.A.J. and Coehoorn, R., 2009. Electron transport in polyfluorene-based sandwich-type devices: quantitative analysis of the effects of disorder and electron traps. *Physical Review B*, *80*(3), p. 033202.

[70] Mark, P. and Helfrich, W., 1962. Space-charge-limited currents in organic crystals. *Journal of Applied Physics*, *33*(1), pp. 205–215.

[71] Nicolai, H.T., Mandoc, M.M. and Blom, P.W.M., 2011. Electron traps in semiconducting polymers: exponential versus Gaussian trap distribution. *Physical Review B*, *83*(19), p. 195204.

[72] Steiger, J., Schmechel, R. and Von Seggern, H., 2002. Energetic trap distributions in organic semiconductors. *Synthetic Metals*, *129*(1), pp. 1–7.

[73] Goodman, A.M. and Rose, A., 1971. Double extraction of uniformly generated electron-hole pairs from insulators with non injecting contacts. *Journal of Applied Physics*, *42*(7), pp. 2823–2830.

[74] Montero, J.M., Bisquert, J., Garcia-Belmonte, G., Barea, E.M. and Bolink, H.J., 2009. Trap-limited mobility in space-charge limited current in organic layers. *Organic Electronics*, *10*(2), pp. 305–312.

[75] Koo, Y.M., Choi, S.J., Chu, T.Y., Song, O.K., Shin, W.J., Lee, J.Y., Kim, J.C. and Yoon, T.H., 2008. Ohmic contact probed by dark injection space-charge-limited current measurements. *Journal of Applied Physics, 104*(12), p. 123707.

[76] Juška, G., Arlauskas, K., Viliūnas, M. and Kočka, J., 2000. Extraction current transients: new method of study of charge transport in microcrystalline silicon. *Physical Review Letters, 84*(21), p. 4946.

[77] Mozer, A.J., Sariciftci, N.S., Lutsen, L., Vanderzande, D., Österbacka, R., Westerling, M. and Juška, G., 2005. Charge transport and recombination in bulk heterojunction solar cells studied by the photoinduced charge extraction in linearly increasing voltage technique. *Applied Physics Letters, 86*(11), p. 112104.

[78] Markham, J.P., Anthopoulos, T.D., Samuel, I.D., Richards, G.J., Burn, P.L., Im, C. and Bässler, H., 2002. Nondispersive hole transport in a spin-coated dendrimer film measured by the charge-generation-layer time-of-flight method. *Applied Physics Letters, 81*(17), pp. 3266–3268.

[79] Juška, G., Arlauskas, K., Viliūnas, M., Genevičius, K., Österbacka, R. and Stubb, H., 2000. Charge transport in $\pi$-conjugated polymers from extraction current transients. *Physical Review B, 62*(24), p. R16235.

[80] Hanfland, R., Fischer, M.A., Brütting, W., Würfel, U. and MacKenzie, R.C., 2013. The physical meaning of charge extraction by linearly increasing voltage transients from organic solar cells. *Applied Physics Letters, 103*(6), p.063904.

[81] Neukom, M.T., Reinke, N.A. and Ruhstaller, B., 2011. Charge extraction with linearly increasing voltage: a numerical model for parameter extraction. *Solar Energy, 85*(6), pp. 1250–1256.

[82] Cheung, C.H., Kwok, K.C., Tse, S.C. and So, S.K., 2008. Determination of carrier mobility in phenylamine by time-of-flight, dark-injection, and thin film transistor techniques. *Journal of Applied Physics, 103*(9), p.093705.

[83] Abkowitz, M., Facci, J.S. and Stolka, M., 1993. Behavior of an ideal injecting contact on a trap-free polymer. *Applied Physics Letters, 63*(14), pp. 1892–1894.

[84] Campbell, A.J., Bradley, D.D., Antoniadis, H., Inbasekaran, M., Wu, W.W. and Woo, E.P., 2000. Transient and steady-state space-charge-limited currents in polyfluorene copolymer diode structures with ohmic hole injecting contacts. *Applied Physics Letters, 76*(13), pp. 1734–1736.

[85] Campbell, I.H., Smith, D.L., Neef, C.J. and Ferraris, J.P., 1999. Consistent time-of-flight mobility measurements and polymer light-emitting diode current-voltage characteristics. *Applied Physics Letters, 74*(19).

[86] Lampert, M.A. and Mark, P., 1970. *Current Injection in Solids*. Academic Press, New York, USA.

[87] Kao, K.C. and Hwang, W., 1981. *Electrical Transport in Solids Semiconductors*, Pergamon Press, Oxford, New York, USA.

[88] Li, H., Duan, L., Zhang, D. and Qiu, Y., 2014. Electric field inside a hole-only device and insights into space-charge-limited current measurement for organic semiconductors. *The Journal of Physical Chemistry C, 118*(19), pp. 9990–9995.

[89] Frenkel, J., 1938. On pre-breakdown phenomena in insulators and electronic semiconductors. *Physical Review, 54*(8), p.647.

[90] Moiz, S.A., Jee, S.W., Um, H.D. and Lee, J.H., 2010. Electrical characterization of metal–silicon microwire interface using conductive atomic force microscope. *Japanese Journal of Applied Physics, 49*(4R), p.045003.

[91] Jee, S.W., Kim, J., Jung, J.Y., Um, H.D., Moiz, S.A., Yoo, B., Cho, H.K., Park, Y.C. and Lee, J.H., 2010. Ni-catalyzed growth of silicon wire arrays for a Schottky diode. *Applied Physics Letters, 97*(4), p.042103.

[92] Murgatroyd, P.N., 1970. Theory of space-charge-limited current enhanced by Frenkel effect. *Journal of Physics D: Applied Physics, 3*(2), p.151.

[93] Barbe, D.F., 1971. Space-charge-limited current enhanced by Frenkel effect. *Journal of Physics D: Applied Physics, 4*(11), p.1812.

[94] Barbe, D.F. and Westgate, C.R., 1970. Bulk trapping states in β-phthalocyanine single crystals. *The Journal of Chemical Physics, 52*(8), pp. 4046–4054.

[95] Mihailetchi, V.D., Xie, H.X., de Boer, B., Koster, L.A. and Blom, P.W., 2006. Charge transport and photocurrent generation in poly (3-hexylthiophene): methanofullerene bulk heterojunction solar cells. *Advanced Functional Materials, 16*(5), pp. 699–708.

[96] Malliaras, G.G., Salem, J.R., Brock, P.J. and Scott, C., 1998. Electrical characteristics and efficiency of single-layer organic light-emitting diodes. *Physical Review B, 58*(20), p. R13411.

[97] Goh, C., Kline, R.J., McGehee, M.D., Kadnikova, E.N. and Fréchet, J.M., 2005. Molecular-weight-dependent mobilities in regioregular poly (3-hexyl-thiophene) diodes. *Applied Physics Letters*, 86(12), p. 122110.

[98] Schwartz, G., Pfeiffer, M., Reineke, S., Walzer, K. and Leo, K., 2007. Harvesting triplet excitons from fluorescent blue emitters in white organic light-emitting diodes. *Advanced Materials*, 19(21), pp. 3672–3676.

[99] Gao, W. and Kahn, A., 2002. Electronic structure and current injection in zinc phthalocyanine doped with tetrafluoro tetracyano quino di methane: interface versus bulk effects. *Organic Electronics*, 3(2), pp. 53–63.

[100] Hartke, J.L., 1968. The three-dimensional Poole-Frenkel effect. *Journal of Applied Physics*, 39(10), pp. 4871–4873.

http://dx.doi.org/10.5772/63527

[101] Geurst, J.A., 1966. Theory of space-charge-limited currents in thin semiconductor layers. *Physica Status Solidi*(b), 15(1), pp. 107–118.

[102] Grinberg, A.A., Luryi, S., Pinto, M.R. and Schryer, N.L., 1989. Space-charge-limited current in a film. Electron Devices, *IEEE Transactions on*, 36(6), pp. 1162–1170.

[103] Jurchescu, O.D. and Palstra, T.T., 2006. Crossover from one-to two-dimensional space-charge-limited conduction in pentacene single crystals. *Applied Physics Letters, 88*(12), p. 122101.

[104] Fraboni, B., Fraleoni-Morgera, A. and Cavallini, A., 2010. Three-dimensional anisotropic density of states distribution and intrinsic-like mobility in organic single crystals. *Organic Electronics, 11*(1), pp. 10–15.

[105] Chandra, W., Ang, L.K., Pey, K.L. and Ng, C.M., 2007. Two-dimensional analytical Mott–Gurney law for a trap-filled solid. *Applied Physics Letters, 90*(15), p. 153505.

[106] Chandra, W., Ang, L.K. and Koh, W.S., 2009. Two-dimensional model of space charge limited electron injection into a diode with Schottky contact. *Journal of Physics D: Applied Physics, 42*(5), p. 055504.

[107] Chandra, W. and Ang, L.K., 2010. Space charge limited current in a gap combined of free space and solid. *Applied Physics Letters, 96*(18), p. 183501.

Chapter 6

# Perspectives of Conductive Polymers Toward Smart Biomaterials for Tissue Engineering

Anca Filimon

Additional information is available at the end of the chapter

http://dx.doi.org/10.5772/63555

**Abstract**

Developing the stimuli-responsive biomaterials with tailor properties represents an important goal of the tissue-engineering community. Such biomaterial promises to become the conductive polymers (CPs), as a novel generation of organic materials that have both electrical and optical properties similar to those of metals and inorganic semiconductors but which also exhibit the attractive properties associated with conventional polymers, that is, easy synthesis and flexibility in processing. The fact that several tissues are responsive to electrical fields and stimuli has made conductive polymers attractive for various biological and medical applications. In this context, the chapter provides information on the basic properties of the conductive polymers and how these polymers can be optimized to generate specific properties for biomedical applications. The synthesis routes of novel materials and specific design techniques, as well as the mechanisms by which electrical conduction affects cells/tissues, are examined, and the significant impact of the conductive polymers in the biomedical field, that is, biosensors, tissue engineering, and neural probes, is demonstrated.

**Keywords:** conductive polymers, functionalization strategies, smart materials, biocompatibility, tissue engineering

## 1. Introduction

Electroactive biomaterials represent a part of a new generation of "smart biomaterials" that allow the direct delivery of electrical, electrochemical, and electromechanical stimulation to cells and/or tissues [1, 2]. The electroactive biomaterials' class includes conductive polymers

(CPs), electrets, and piezoelectric and photovoltaic materials [3]. Electrets and piezoelectric materials make possible the delivery of an electrical stimulus without being necessary an external power source, but the control over the stimulus is limited. On the other hand, the conductive polymers (CPs), allow excellent control of the electrical stimulus and possess both the electrical and optical properties, exhibiting a high conductivity/weight ratio, and also can be made biocompatible, biodegradable, and porous [1].

Produced for the first time few decades ago, the polypyrrole, PPy, dates since 1960 and only in 1977, the polyacetylene (PA) doped with iodine was recognized as the first inherent conductive polymer [4], and today there are several tens of conductive polymer systems (e.g., polyacetylene (PA), polythiophene (PT), poly(3,4-ethylenedioxythiophene) (PEDOT), and polyaniline (PANI)) [5]. Therefore, the conducting polymers, known as "synthetic metals," are polymers with a highly conjugated polymeric chain.

It can be presumed that typically organic polymers, considered to be insulators, become conductors, when they adopt different electronic structures. In conducting polymers delocalization of the electrons is generated by the carbon atoms engaged in $\pi$ bonding, with $sp^2p_z$ configuration, and by overlapping the orbitals of successive atoms, providing the mobility of polymer chains and implicitly the specific properties (e.g., the electrical conductivity, low ionization potential and energy optical transitions, and also high electron affinity). Therefore, the specific properties abovementioned make these polymers considerable for a wide range of applications, from the electronic to medical field [6]. Additionally, the conducting polymers can be designed for the transport of the small electronic signals in the body, acting as artificial nerves) [5].

Consequently, the conductive polymers represent dynamic structures with complex properties which have gained a great interest in various scientific fields because of their vast versatility. On the other hand, they possess very good electrical and optical properties and allow an excellent control of the electrical stimulus, and furthermore, a great advantage of conductive polymers is that the physicochemical properties can be tailored to the specific needs required by different applications, by incorporating antibodies, enzymes, and other biological moieties. Additionally, their properties can be altered and controlled through stimulation (e.g., electricity, light, pH) even after synthesis.

These complex properties can be controlled only if the nature of the processes that regulate them during the synthesis of the conducting polymers and the extent to which these properties are changed by the application of an electrical stimulus are known and understood.

Considering the vast possibilities to use the new electroactive materials in biomedicine, this chapter gathers all of the available information concerning the most commonly conductive polymers and their mechanism of conduction. The implications of the classical and modern synthesis methods in designing the new electroactive systems are analyzed in accordance with the phenomena underlying their conductivity and the ways to tailor their biocompatibility, biomolecule doping, and drug release for tissue-engineering applications.

## 2. Conductivity source and doping of the conductive polymers

Conductive polymers can conduct charges as a result of ease with which electrons jump within and between the polymer chains [2]. These polymers are considered unique, because they possess a conjugated backbone, meaning that it is formed by a series of alternating double and single bonds with $sp^2$-hybridized atoms [4]. Both types of linkages are chemically strong, single ones containing localized σ-bond, while those double a less strongly localized π-bond [7] (see **Scheme 1**).

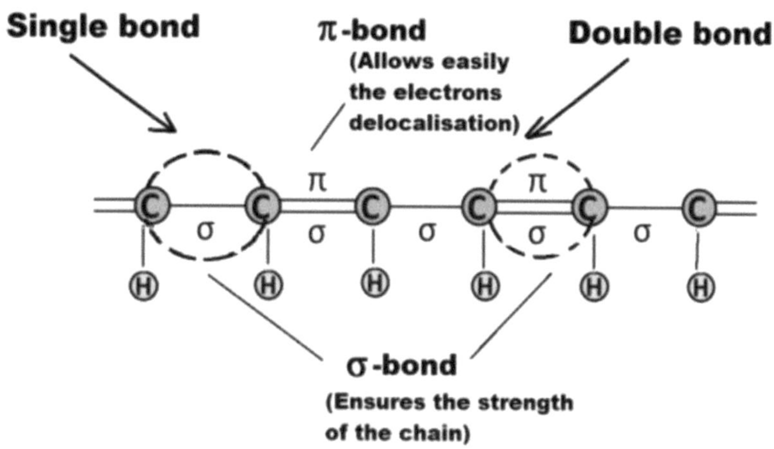


**Scheme 1.** Schematic representation of a conjugated chain, containing alternating single and double bonds.

Such an alternating double bond system exhibits a delocalization over the entire chain, as result of the electronic wave functions, allowing charge mobility along the polymer backbone and between the adjacent chains, mobility which is limited by both disorder and Coulombic interactions between the electrons and holes [8]. Consequently, the conductivity is a measure of electrical conduction and thus represents the capacity of a material to conduct current. Generally, the materials with conductivities less than $10^{-8}$ S/cm are considered insulators, materials with conductivities between $10^{-8}$ and $10^3$ S/cm are considered semiconductors, and materials with conductivities greater than $10^3$ S/cm are considered conductors [8]. Therefore, dopants exhibit a main role in improving of the polymers' conductivity [2, 7, 9]. Thus, the polymers in their pure (undoped) state are described as electronic insulators; when these polymers are doped, the conductivity changes from insulators to metals.

Doping is the process of oxidation (p-doping) or reduction (n-doping) of a neutral polymer, delivering a counter anion or cation (i.e., dopant), respectively. Thereby, the dopant introduces a charge carrier into the system by removing or adding electrons from/to the polymer chain, relocalizing them as polarons (i.e., radical ions) or bipolarons (i.e., dications or dianions) into the polymer (**Figure 1**) [2, 7, 9, 10].

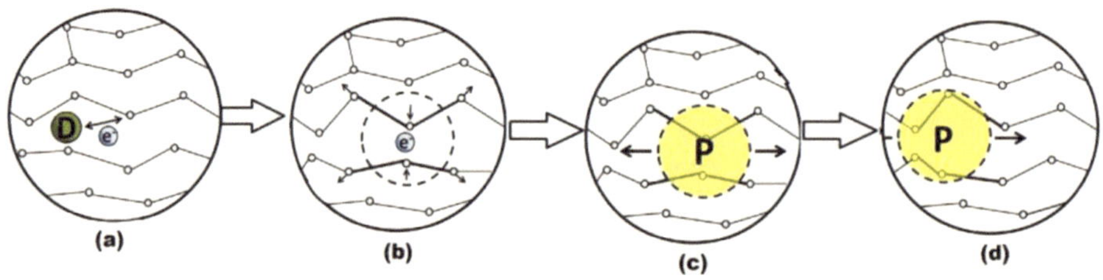


**Figure 1.** Illustration of the charge transport mechanism in conductive polymers: (a) the dopant removes or adds an electron from/to the polymer chain, generating a charge delocalized; (b) the chain is energetically favorable to localize this charge and surround it with a local distortion of the crystal lattice; (c) the charge surrounded by a distortion appears as a radical ion associated with a lattice distortion (known as a polaron); and (d) the polaron can move along the polymer chain, allowing it to conduct electricity.

Therefore, by doping, new electronic states are introduced within the band gap of material, which strongly interact between them at high concentrations. As a result of these interactions, an overlapping of the electronic wave functions occurs and generates a band of electronic state within the band gap, in the place of the discrete levels.

Upon doping, a conducting polymer system is produced as a result of close association of the counterions with the charged CP backbone. The attraction of the electrons from one repeat unit to the nuclei neighboring units determines charge mobility along the chains or between the chains, process known as "electron hopping." The ordered movement of these charge carriers along the conjugated CP backbone produces electrical conductivity. A conducting polymer can be considered more conductive, when the band gap energy is smaller (i.e., distance between conducting band and valence band).

As abovementioned, the doping process produces a number of charged carriers in polymer, and these carriers must be mobile in order to contribute to conductivity. This parameter, $\sigma$, is proportional with the carrier concentration, $n$, and mobility, $\mu$.

The transport mechanism of the carriers in a conducting polymer is manifested more likely as a "hopping transport," similarly with that of amorphous semiconductors and less similar with that of crystalline semiconductors—band transport. Therefore, can be concluded that the doping creates the active sites—polarons—which allow carriers (electronic and holes) to move from one site to another, by hopping mechanism.

There are many factors that influence the charge transport mechanisms and implicitly conductivity, namely, the dopant, oxidation level/doping percentage, synthesis method, and temperature. These will affect not only electroactivity but also the surface and bulk structural properties of the polymer, for example, the color, porosity, and volume [2, 8, 11]. Doping can be performed chemically, electrochemically, or via photodoping [8, 11, 12]. Therefore, small (e.g., $Cl^-$) and large (e.g., sodium polystyrenesulfonate, PSS) dopants can modulate both the electrical conductivities and surface structural properties. On the other hand, larger dopants, such as hyaluronic acid (HA), can affect the material properties more dramatically (surface topography and physical handling properties [13]) and can increase the polymer density. Large dopants are better integrated into the polymer and will not be leached out with time or by the application of an electrical stimulus, thereby giving the polymer a greater electrochemical stability [9, 14, 15].

Surface properties, as the roughness, morphology, wettability, and stiffness, are known to affect the adhesion and proliferation of multiple cell types [16]. It is important to note that the nature of the dopants has a strongly influence on the bulk and surface properties of conductive polymers [17]. Indeed, by use of different dopants, different modifications of these properties have been observed. For example, hyaluronic acid (HA)-doped PPy is rougher and more brittle than PSS-doped PPy [16]. Chloride anions are frequently used to dope CPs, due to their biological compatibility, for example, PPyCl.

To incorporate molecules that are not capable of redox chemistry, such as biological dopants, it is necessary to synthesize/dope CPs electrochemically. In literature [18], the relevant biological effects of dopants—dextran sulfate (DS), poly(2-methoxyaniline-5-sulfonic acid) (PMAS), para-toluenesulfonic (pTS) acid, HA, and chitosan (CS)—were compared, and it was found that the PMAS- and CS-doped films possessed a much lower surface roughness and Young's modulus than the films prepared with the other dopants. In the same study, it was also evidenced that the PMAS- and CS-doped films have sustained the adhesion and differentiation of skeletal myoblasts much better than their counterparts. These observations show the importance of relationship between dopant—material property—cellular behavior and highlight the preservation of this correlation when choosing the doping molecule.

It can be concluded that dopants with large molecule are physically trapped and implicitly more stable in the CP, without being easily excluded and/or exchanged by the application of an electrical potential. Instead, small dopants can be easily diffused from CP and/or exchanged with other ions within the surrounding environment. Therefore, properties of conductive polymers can be adjusted by the doping process.

## 3. Synthesis routes and processing

Currently, the conductive polymers can be synthesized by two main methods, chemically or electrochemically, each having advantages and disadvantages [19]. The chemical synthesis provides not only many routes to synthesize a wide variety of CPs but also permits to obtain a scale-up of such materials, which, currently, by electrochemical synthesis is not possible. Thus, the chemical synthesis methods include either condensation polymerization, that is, step-growth polymerization, or addition polymerization, that is, chain-growth polymerization.

During the chemical synthesis the monomer solution is mixed with an oxidizing agent (e.g., ferric chloride, ammonium persulfate), obtaining polymers in the various forms, which makes the method to be selected for commercial applications [20]. The conductivity of the created polymer is highly susceptible to the choice and purity of the solvent, oxidant, relative concentration of the reagents, reaction time, temperature, stirring rate, etc. Unfortunately, the conductivity of the polymers synthesized by the chemical method is always lower than that of their electrochemically synthesized counterparts [20].

The electrochemical synthesis is a common alternative method for CPs obtaining, because this synthetic procedure is relatively straightforward [19]. This technique dates from 1968 when "pyrrole black" was obtained as a precipitate, using a platinum electrode, by exposing an aqueous solution of pyrrole and sulfuric acid to an oxidative potential [8, 21]. Nowadays, the electrochemical polymerization is performed using a three-electrode system (working, counter, and reference electrodes) placed into a solution containing the monomer of the polymer, appropriate solvent, and electrolyte (dopant) (**Figure 2**) [8, 22].

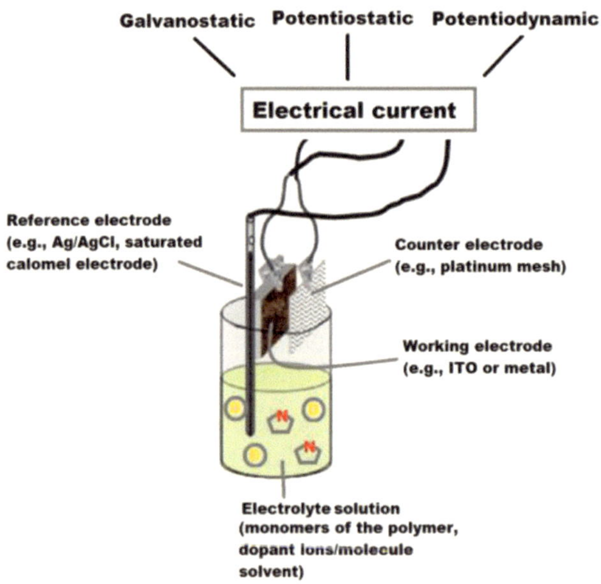


**Figure 2.** Schematic representation of the equipment for electrochemical synthesis.

This method allows the deposition of a thin film of the polymer with a well-controlled thickness and morphology. The electrical current is passed through the solution and the electrodeposition occurs at the positively charged working electrode or anode, forming insoluble polymer chains on the electrode surface.

Generally, the electrochemical polymerization can be achieved using three techniques: potentiostatic (to obtain thin films), galvanostatic (to obtain thick films), and potentiodynamic [23, 24]. In the potentiostatic polymerization, the potential of the electrodes is controlled, while the current varies [23]. This protects the integrity of the component to be coated, making this method ideal for the manufacture of biosensors. The electrical current can vary depending on a number of factors (e.g., the electrode material, the plating conditions), thus a coulometer being necessary to control the amount of polymer that is deposited. Instead, in galvanostatic polymerization, the electrical current is controlled; this means that the rate at which the polymer is deposited remains constant and can be controlled accurately [23]. In the course of the potentiodynamic deposition, the polymerizing potential is swept between a low and high potential limit in cycles. This determines the deposition of polymer in layers, each layer becoming electrically active before the next to be synthesized [25]. Important variables, including the deposition time and temperature, solvent system, electrolyte, electrode system, and deposition charge, must be considered, because each of these parameters exhibits an effect on film morphology (thickness and topography), conductivity, and mechanical properties, these having a direct impact on the utility of the material for biomedical applications. Consequently, the electrochemical polymerization represents a new way to achieve a range of composite materials with the special properties. Therefore, the doping and processing take place simultaneously in the electrochemical polymerization.

One way to compensate the shortcomings of a conductive polymer is to use it together with another polymer, combining the positive qualities of both materials. For example, functionalized polysulfone with quaternary ammonium groups (PSFQ) was combined with polyvinyl alcohol (PVA) and/or cellulose acetate phthalate (CAP) into its matrix, yielding flexible, biocompatible, and biodegradable composites with improved conductivity compared to the PSFQ [26]. Interactions between blood and a polymer surface depend on blood composition, blood flow, and physicochemical properties of the polymer surface, such as hydrophobicity/hydrophilicity, roughness, and flexibility, or on the toxicological and electrical properties [27, 28]. Consequently, it is assumed that the quaternization effect and choosing of an appropriate additive (PVA and/or CAP) significantly improve the ionic conductivity and also could optimize electric properties required by ionic exchange membrane [29]. Additionally, PSFQ/CAP and PSFQ/PVA composites were able to maintain its electroactivity and established compatibility with some blood compound (e.g., red blood cells (RBC) and platelets) and also with plasma protein (e.g., albumin, immunoglobulin G (IgG), and fibrinogen) [26, 27, 30]. These results seem to be applicable for evaluating bacterial cell adhesion on polymer surfaces and could be employed for studying possible induced infections or for obtaining biomembranes.

Electrospinning is a versatile process that allows to process some conductive polymers into nano- and microfibers [31]. In this context, solutions processable of cationic ionomers (PSFQ) have received widespread attention for their promising roles as exchange membranes and antibacterial coatings. Therefore, the solutions of the polysulfone ionomers were processed by electrospinning to create new fibrous materials that can modulate biomembrane properties. In particular, their biological activity (investigated against Gram-positive *Staphylococcus aureus* and the Gram-negative *Escherichia coli* bacteria) depends on their structure and physicochemical properties which affects the interaction with the cytoplasmic membrane of bacteria and influences their cell metabolism [32]. Consequently, the electrospinning proved to extend the possible applications of quaternized polysulfones by preparing continuous fine fibers with characteristics which recommend them for applications as biomaterials and membranes in various fields.

Additionally, the conductive polymers have also been successfully polymerized inside of the hydrogel networks [33]. This allows the creation of electroactive hydrogels, which combine the redox switching capabilities of conductive polymers with the fast ion mobility and biocompatibility of the hydrogels. These electroactive hydrogels can be of various sizes, binding the bioactive molecules and nanotemplated in order to mimic the extracellular matrix

[34]. These properties make them ideal for implantable biosensors, drug release devices, and deep-brain stimulators.

## 4. General functionalization strategies of conductive polymers for specific applications

As already mentioned, CPs have electrical and optical properties similar to those of metals and inorganic semiconductors and also are ease of synthesized and processed, which makes them suitable for a wide range of applications in the microelectronics industry [35] and, more recently, in the biological field. Good cellular response to the biomaterial is essential for many biomedical applications [36]. Therefore, many types of conductive polymers (e.g., PPy, PANI, and polyethyleneimine (PEI)) have been shown to support the growth of a large variety of cell [1, 2]. Additionally, the improvement of the conductive polymer biocompatibility can be easily accomplished by bonding the biocompatible molecules, segments, and side chains on the polymer [37]. However, there are few studies of the reduced biocompatibility [5, 38]. It was found that this variation in biocompatibility is due to the dopants and different preparation protocols used in the experiments; for example, it was shown that rinsing, extraction, and aging have a significant effect on the biocompatibility [39], generating changes of the polymer surface topography, which can have as a result an altered cell behavior [5].

Studies of the conductive polymers for applications in biomedical field were expanded greatly; since 1980s these materials were compatible with many biological molecules such as those used in biosensors. Most of CPs present significant characteristics for biomedical applications, namely, the biocompatibility, ability to entrap and/or release the biological molecules (i.e., reversible doping), or ability to transfer the charge by biochemical reaction. Additionally, they have the capacity to easily modify the electrical, chemical, and physical properties for a better match to specific applications [36, 40].

Although a wide range of CPs has been explored for a number of applications, as described in more detail below, considerable research remains to be performed, contributing thus to extending possible applications of these "smart" materials.

### 4.1. Overview on the conductive polymers functionalization

There is always the desire to optimize the material properties when it targeted a specific application. Thus, the optimization of the roughness, porosity, hydrophobicity, conductivity, degradability properties of the conductive polymers, and binding of the biological molecules (capable to make them promising candidates for biomedical applications) can be achieved through different techniques (see **Figure 3**) [8].

**(1)** Physical adsorption represents the simplest method. The resulted product is sensitive to pH, and adsorbed molecule dissociates and thus compromises the conductivity of the polymer, making the material to become "inactive." In this method, a solution containing functionalizing agent is placed in contact with the polymer, after it was synthesized. The biomolecules are physically absorbed due the static interactions between the polymer matrix and charge of the molecule [8, 41, 42].

**(2)** Another non-covalent method is entrapment [43]. This occurs by mixing the desired molecule with the monomer of the polymer, dopant, and solvent, prior to the synthesis. After the electrochemical polymerization, the molecules of the functionalizing agent from the proximity of the electrode are incorporated into the growing polymer. This technique was firstly applied to link large molecules (e.g., enzymes, DNA), which once entrapped are unable to leave the polymer due to their size [44].

**(3)** The process of doping CPs, necessary to induce conductivity, can also be exploited to modify CPs non-covalently and to introduce new properties for a desired application. This allows the bonding of a wide range of biomolecules (dopants), as long as they are charged [45, 46]. For example, by doping the collagen, heparin, chitosan, and ATP have been successfully bound in conductive polymers as growth factors [2, 12].

**(4)** Introducing of the bioactive molecules through doping exhibits a higher negative effect on the polymer's conductivity than the covalent bonding [2, 11, 45]. Thus, alternatively, covalent methods can be used to permanently functionalize CPs. By this method, the biological molecules will be strongly bound and will not be released, thereby enhancing the long-term stability of the polymer [45].

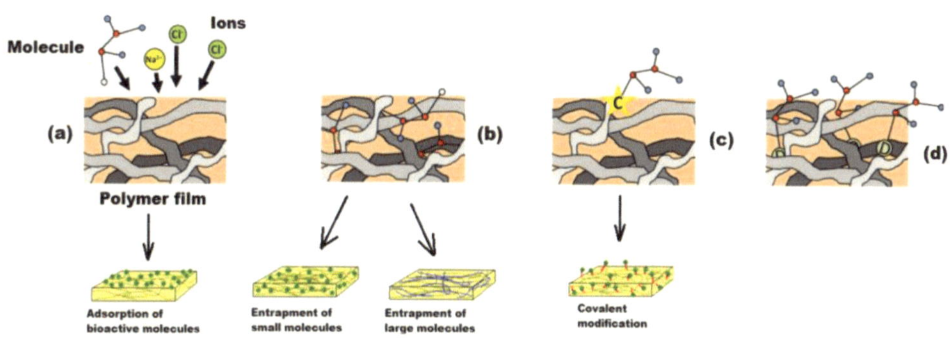


**Figure 3.** Modification strategies applied to conductive polymers: (a) physical adsorption, (b) entrapping, (c) covalent bonding, and (d) exploiting the doping.

It is important to note that the steric effects of any incorporated functional group may disrupt the planarity of the conjugated system and implicitly decrease the conductivity. Functionalization of CPs with different biomolecules has allowed medical engineers to modify CPs with sensible biological elements. Thus, through different ways, CPs with enhanced adhesion and proliferation of a cell variety with improved biocompatibility have been developed.

## 4.2. Specific applications: from concept to biomedical perspectives

Conducting polymers have attracted great interest as suitable matrices for biomolecules, because they improve the stability, speed, and sensitivity [47, 48] and, therefore, can be widely used in medical diagnostics [49, 50]. The conductive and semiconductive properties of these polymers make them an important materials' class used in a wide range of applications, from the electronic and biotechnological ones (such as rechargeable batteries, electronic displays, solar cells, ion-exchange membrane in fuel cells, chemical sensors, and biosensors) to drug release systems and tissue applications (used to transport small electronic signals in the body acting as an artificial nerves or used as films in a neurotransmitter as a drug release system into the brain).

### *4.2.1. Biosensor applications*

Biosensors are of great interest, because these interesting bio-devices have been shown to have applications in clinical diagnostics, environmental monitoring, food freshness, and bioprocess monitoring [47, 51–53]. The first biosensor device was created by integrating an enzyme into an electrode [54], but subsequently, for clinical purposes a great progress in monitoring and diagnosing metabolites (e.g., glucose, hormones, antibodies, antigens) has been made. A biosensor is a chemical sensing device in which a biologically entity is coupled to a transducer to allow the quantitative development of some complex biochemical parameter [55]. The interaction of the detection element with the interest analyte produces a chemical signal that is transmitted to the transducer, which ultimately transforms the input into an electrical signal (**Figure 4**). The major challenge when using CPs in the design of the biosensor with electrochemical transducer is to understand the mechanism of the electrons transfer. Depending on the way in which the chemical signal is transmitted, the biosensors can be divided in several categories: amperometric, potentiometric, conductometric, optical, calorimetric, and piezoelectric [56].

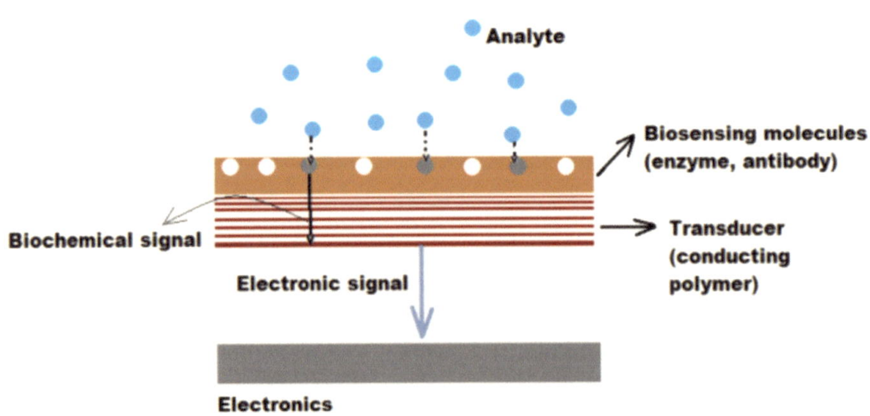


**Figure 4.** Scheme of a conductive polymer-based biosensor.

There are three so-called generations of biosensors:

**(1)** First generation of biosensors where the normal product of the reaction diffuses to the transducer and causes the electrical response.

**(2)** Second generation of biosensors involves the specific "mediators" between the reaction and the transducer, generating an improved response in two steps. Firstly, a redox reaction between the enzyme and substrate, reoxidized by the mediator, occurs, and secondly, the mediator is oxidized by the electrode.

**(3)** Third generation of biosensors is autonomous and arises from the nature of the sensor; the reaction itself causes the response and no product or mediator is not directly involved. The third generation of sensors is accompanied by co-immobilization of the enzyme and mediator at the electrode surface, that is, the direct electrical contact of enzyme to electrode occurs.

An important aspect in the biosensor applications is represented by the integration of conductive polymer in biological recognition components (such as glucose, cholesterol, urea, triglycerides, creatinines, and pesticides) [8]. Rahman et al. [41, 57], using the covalent immobilization method of pyruvate oxidase on the nanoparticle-comprised poly-(terthiophene carboxylic acid), poly-TTCA, have obtained a biosensor for amperometric detection of the phosphate ions. For example, literature reported that for the fabrication of the DNA sensors, copolymers of PPy and oligonucleotides bearing a pyrrole group were used [8]. Additionally, literature has been shown as example the immunosensors for detecting *Listeria monocytogenes*, created by covalent binding of the *Listeria* monoclonal antibody to a copolymer of carboxylic acid-functionalized PPy and regular PPy [8]. Trojanowicz and Miernik [58] have developed avidin–biotin (N-(biotinoyl)-1,2-dihexadecanoyl-sn-glycero-3-phosphoethanolamine, triethylammonium salt) interactions for the immobilization of the glucose oxidase on bilayer lipid membrane formed on polypyrrole and poly-o-phenylylenediamine electrodeposited onto platinum wire, respectively. These types of membrane-based glucose biosensors cause a significant reduction from the electroactive species (e.g., ascorbic acid, cholesterol, and uric acid), giving a stable response [59].

The recent progress in the molecular biology/recombinant DNA technologies has opened enormous possibilities for the microorganisms' adjustment to improve the activity of an existing enzyme or to emit a foreign enzyme/protein in a host cell to be used for enhancing the specific activity. There have been various strategies to modify the microbes for application to microbial biosensors. For example, Lei et al. [59, 60] have reported a cell biosensor using PNP-degrader *Pseudomonas putida* JS444 for the estimation of organophosphorus nerve agent with p-nitrophenyl constituent. This biosensor was found to be stable for only 5 days.

Recent advances in carbon nanotubes (CNTs) include their incorporation in CP-based biosensors. For example, preliminary studies have been performed on the exploration of the properties of both PPy/CNT and PANI/CNT devices as pH sensors [61]. Moreover, for label-free detection of DNA, DNA-doped PPy in conjunction with CNTs was used. Similarly, DNA sensors have been created from a composite of PPy and CNTs functionalized with carboxylic groups by covalently immobilization of DNA on CNTs [62]. In general, the presence of CNTs generates an increase of the biosensors overall sensitivity and selectivity.

### 4.2.2. *Tissue-engineering applications*

The development of biosensors has created a basis for using the conductive polymers in many biological environments; in fact it was the first step to explore CPs as biomaterials for different cells. Literature has shown that the cells (e.g., fibroblasts, neurons, and osteoblasts) responded to the electrical fields created by electrets [63] or between electrodes, *in vitro* and *in vivo* studies [64, 65]. The general properties of CPs, desired for the tissue-engineering applications, include the conductivity, reversible oxidation, redox stability, biocompatibility, hydrophobicity (contact angle with water is 40–70°, promoting cell adhesion), three-dimensional geometry, and surface topography.

Literature states that PPy was one of the first CPs studied for its effect on the mammalian cells [66]. Moreover the adhesion and growth of various cells (endothelial cells [66, 67], rat pheochromocytoma cells (PC12) [68], neurons, and support cells), associated with dorsal root ganglia (DRG) [69], primary neurons [70], keratinocytes [71], and mesenchymal stem cells [72], were reported. Because the advantage of using CPs in tissue-engineering applications consists in ability to subject the cells to an electrical field, studies concerning the cell compatibility, when a current or voltage is applied to PPy, have been made. In addition to the biocompatibility of PPy, several studies have shown that the electrical stimulation using PPy doped with p-toluene sulfonate (PPyTS) can modulate the cellular response. Most of these studies were based on the passive adsorption of the biomolecules from serum or on the coating of purified protein solutions at PPy surface. In one of the first studies [66] performed to assess the suitability of CPs to support the cell growth and control the cell function, aortic endothelial cells were cultured on fibronectin (FN)-coated PPy films and exposed to the oxidizing potentials (see **Figure 5**). Oxidized PPy leads to the cell spreading, because the reduction of PPy to its neutral state causes the cell rounding and a concomitant drop in DNA synthesis (≈98%). Despite the changes in cellular morphology, the cell viability and adhesion were very good on both oxidized and neutral PPy.

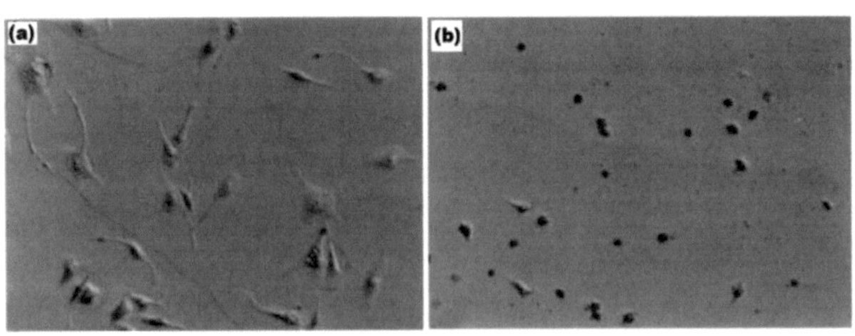


**Figure 5.** Photomicrograph of endothelial cells cultured on fibronectin-coated PPyTS: (a) PPy in native oxidized state and (b) PPyTs reduced by the application of −0.5 V for 4 h.

Also, in literature many studies on the entrapped biomolecules, for example, adenosine 50-triphosphate (ATP) [73] and nerve growth factor (NGF) [68] in PPy, as well as other CPs for both drug delivery and tissue-engineering applications, exist. The dopant confers to the material new properties; thus, function of dopant nature, some specific properties of PPy, needed in tissue applications (e.g., the growth of different cell or specific aspects of wound healing), can be optimized. For example, PPy doped with HA has been synthesized and investigated for tissue-engineering applications, because of HA's inherent role in the wound healing and angiogenesis [8, 13]. According to **Figure 6**, *in vivo* studies realized for PPyPSS films and PPyHA bilayer films showed that the PPyHA bilayer films were biocompatible and promoted vascularization as a result of the HA presence. Both films were implanted in subcutaneous pouches at rats. Tissues surrounding the material were harvested after 2 weeks, fixed, imbedded, and stained with hematoxylin and eosin.

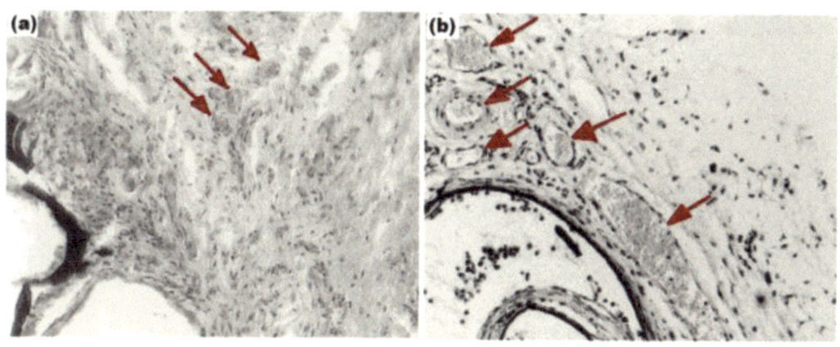


**Figure 6.** *In vivo* tissue response to PPyPSS films (a) and PPyHA bilayer films (b). The black lines in the images represent the films and the blood vessels are denoted by arrows.

It is important to note that both the surface topography and conductivity are significantly altered when the biomolecules are exclusively used as dopants (see examples in **Figure 7**) [13, 71].

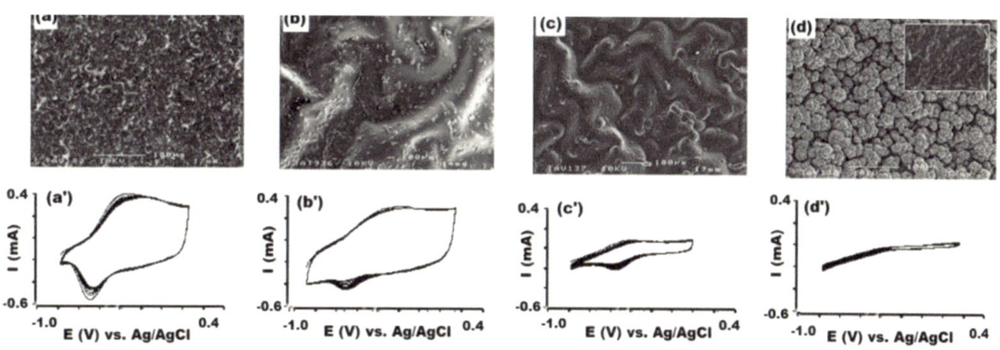


**Figure 7.** Topography using scanning electron microscopy (SEM) (a–d) and corresponding cyclic voltammograms (CVs indicating electrical activity of the materials) (a′–d′) for thick films of PPyCl (a,a′), PPy doped with poly(vinyl sulfate) (b,b′), PPy doped with dermatan sulfate (c,c′), and PPy doped with collagen (d,d′) (inset is for thin film of collagen-doped PPy). A narrow CV spectrum correlates to decreased electroactivity.

Researchers have shown that the exploration of PANI for tissue-engineering applications has progressed more slowly than the development of PPy for similar applications. However, recently, there were more evidences concerning the ability and variants of PANI to support the cell growth and PANI's biocompatibility *in vivo* arousing interest to be used in tissue-engineering applications. For this reason, other methods have been sought to modify PANI to render it biocompatible while maintaining the desirable electrical properties of the material. For example, wettability properties of PANI have been modified by the entrapment of a triblock copolymer, [poly(ethylene oxide)–poly(propylene oxide)–poly(ethylene oxide) (PEO–PPO–PEO)] (see **Figure 8**) [8, 74].

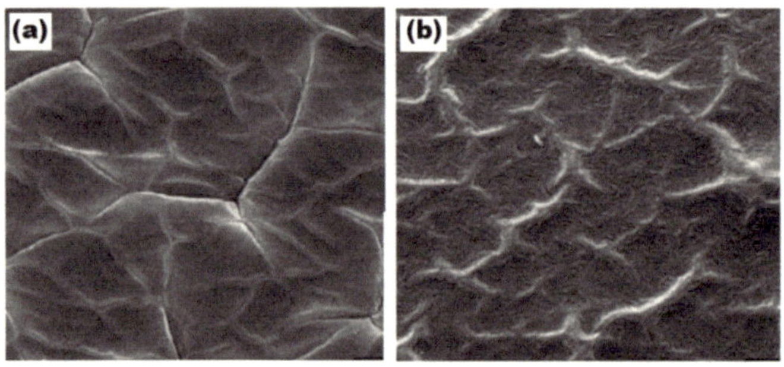


**Figure 8.** SEM image of PANI films immersed in N-methylpyrrolidinone (NMP) for 300 s followed by dipping in water (a) and immersed in 0.03 g/mL pluronic polymer/NMP solution followed by dipping in water (b) for 20 μm scan area and magnification of 1000x.

Moreover, literature presents that the cross-linked composites of PANI–gelatin exhibit good biocompatibility *in vivo* [8]. Due to the optimal mechanical properties, the gelatins allow it to be electrically spun into fibers, generating three-dimensional scaffolds as a function of PANI:gelatin ratio (**Figure 9**) [75].

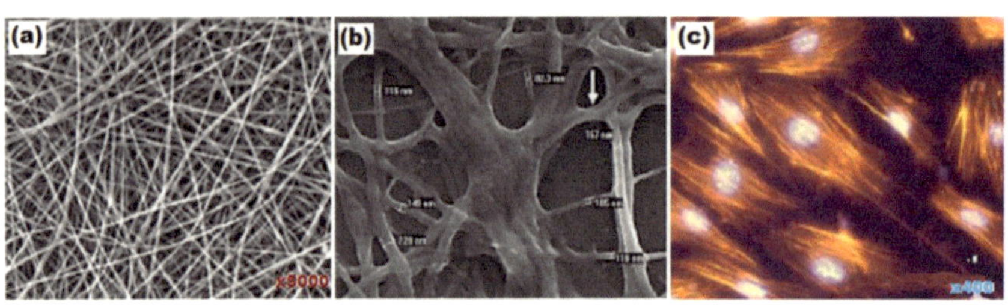


**Figure 9.** SEM images of PANI–gelatin blend fibers at 45:55 (w/w) ratio (a) and of myoblast cells cultured on 45:55 PANI–gelatin blend fibers (b) and (c) morphology of myoblast cells at 20 h after post-seeding on 45:55 PANI–gelatin blend fibers (staining is for nuclei-bisbenzimide and actin cytoskeleton phalloidin—fibers autofluoresce).

*4.2.3. Neural probe applications*

Many of the advances made in tissue-engineering applications, especially regarding the neurons, are relevant for the optimized neural electrodes' development. Significant in this research area is the necessity of an interface between the electrodes and neural tissue and

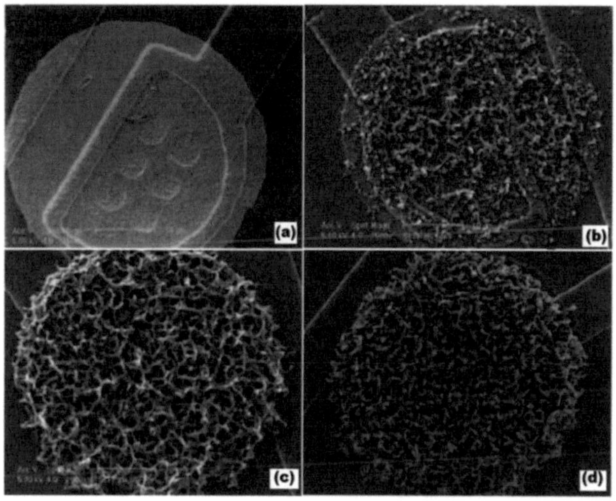


**Figure 10.** SEM images (a–d) of PPySLPF-coated electrode sites function of the deposition time which increased, corresponding to a total charge of 0 mC (a), 1 mC (b), 4 mC (c), and 10 mC (d). The area of the uncoated electrode is 1250 μm$^2$ (scan area for all images is 10 μm).

efficient transmission of the signal between the cells and electrode, thus integrating the device seamlessly with the native neuronal signaling network [8]. Therefore, conductive polymers (e.g., PPy commonly explored, PEDOT recently studied) can be considered as potential candidates for neuronal probe applications, due to their special properties, namely, the high surface area and ability to promote an effective ion exchange between the recording sites and surrounding tissue. In this context, George et al. [76] have designed a neural probe, which could also be used as a neural scaffold, starting from PPy doped with PSS or sodium dodecylbenzenesulfonate (NaDBS). The aim of this study was to investigate the effectiveness of an implant specially manufactured by the electrochemical deposition (the dopant and temperature were varied) of PPy on a patterned gold template. It was found that PPy doped with NaDBS is more conductive than PPyPSS, which could increase the signal transportation from the cells to electrode. Moreover, PPy augmented with biological moieties may offer advantages for the neural probe applications. The careful selection of the bioactive molecule is essential to enhance the neuron adhesion, which will increase the signal received from neurons and in the same time, to minimize the astrocyte adhesion, which would interfere with the neuronal signal. For example, a silicon-based 4-pronged neural probe was micropatterned with a layer of gold. Thus, PPy doped with either SLPF (the silk-like polymer having fibronectin fragments) or laminin-derived nonapeptide p31 (CDPGYIGSR) was deposited on the gold sites [77]. The entrapment of these peptide sequences has enhanced the cell attachment, growth, and migration, but at a particular deposition time, the surface area and the conductivity of PPy were maximized. The integration of the probe into the neural tissue and increasing the conductivity lead to an increase in surface area (see **Figure 10**), improving the signal transportation. Finally, it was demonstrated that glial cells preferentially attached to PPySLPF and neuroblastoma cells adhered significantly better to PPyCDPGYIGSR compared to

$PPyCH_3COO^-$, further suggesting the enhancement of electrode–neuron interactions in the presence of PPy doped with bioactive peptides.

In addition to CP modification with peptides, the deposition of the CP within a hydrogel network represents another attractive strategy to better integration of CP with target cells [78]. Hydrogels are attractive, because they are biocompatible and porous and can be tailored to possess the mechanical properties of the surrounding tissue (e.g., brain tissue), thus creating a better electrode–cell interface.

As previously mentioned, PEDOT has recently explored as an alternative to PPy, because it is more stable to oxidation and more conductive. In literature comparative studies concerning the electrochemical deposition of PEDOT and PEDOT–MeOH for the neural probes were presented [79]. Both materials lead to an improvement of the electrochemical stability and surface area increasing, compared to the controls' sample, and implicitly to a decrease of the impedance. The two forms of PEDOT were successfully doped with the laminin-derived DCDPGYIGSR peptide and the rat glial cells preferentially grown on PEDOT–DCDPGYIGSR. PEDOT–MeOH has not been tested *in vivo*, but the advantage of using this CP over PEDOT is that its monomer is more soluble in water, which would permit the polymer synthesis in the aqueous systems that are necessary for the incorporation of many biomolecules. Additionally, future studies will be focused on the depositing agents in order to minimize the immune response, to reduce the encapsulation, and to induce the nerve growth toward the electrode [80].

#### *4.2.4. Drug delivery*

Besides biosensors, tissue engineering, and neural probes, there are other important investigations of CPs for biological applications. These include the using of CPs as drug-delivery mediators, actuators, and antioxidants. Many disciplines of science, including the medical, pharmaceutical, and agricultural fields, require the controlled delivery of the chemical compounds [33]. This has been a great challenge, but in present, the use of the conductive polymers as a substrate material for controllable drug-delivery devices promises to overcome this [33]. The molecules linked in such polymers through doping can be expelled through the application of an electrical potential. The fact that they can be porous and show delocalized charge carriers helps to the diffusion of the linked molecules [33]. This represents another reason for conductive polymers to be considered suitable for drug release applications.

In experiments, concerning drug-delivery applications, many therapeutic drugs including 2-ethylhexyl phosphate [81], dopamine [82], naproxen [83], heparin [84], and dexamethasone have already been bound and successfully released from these polymers. By electrical stimulation [84], the release of the heparin from PVA hydrogels covalently immobilized on PPy films has been triggered. There are a few factors that limit the application of the conductive polymers for drug release, namely, the molecules loaded in polymer tend to leach out through diffusion being replaced by other molecules from the polymer's environment [85]. This passive loss of charge becomes worse when a relatively small amount of drug is bounded in polymer. Additionally, both charge and molecular weight restrict the molecules which can be bound and released. This can be easily overcome through the use of biotin–streptavidin coupling; the biotin acts as the dopant, while the bioactive molecule is covalently bound to the biotin and then released by electrical stimulation. Another advantage is that the biotin provides more uniform release kinetics. A problem that persists is the tiredness of the conductive polymer with repeated cycles of electrical stimulation: repeated cycles can cause irreversible oxidation in the polymer, which coincide with the dedoping and the reduced conductivity, which ultimately limits its useful lifetime [33].

## 5. Conclusions

Unlike many other materials, CPs have uses in a diverse array of applications ranging from photovoltaic devices to nerve regeneration. The unique property that ties all these applications is their conductivity and, in addition, ease of preparation and functionalization. This fact is especially true in biomedicine, whether for biosensors or for control over cell proliferation and differentiation. Tissue engineering is a new concept in which cells are seeded on material scaffolds and then implanted in defected part of body. The ability of conductive scaffolds to accept and modulate the growth of a few different cells, including endothelial, nerve, and chromaffin cells, has shown a bright future of this "smart" material in the field of tissue engineering and regenerative medicine.

## Acknowledgements

This work was supported by a grant of the Romanian National Authority for Scientific Research, CNCS—UEFISCDI, project number PN-II-RU-TE-2012-3-143.

## Author details

Anca Filimon

Address all correspondence to: capataanca@yahoo.com

"Petru Poni" Institute of Macromolecular Chemistry, Iasi, Romania

## References

[1] Lakard B, Ploux L, Anselme K, Lallemand F, Lakard S, Nardin M, Hihn JY. Effect of ultrasounds on the electrochemical synthesis of polypyrrole, application to the adhesion and growth of biological cells. Bioelectrochemistry. 2009;75:148–157. DOI: 10.1016/j.bioelechem.2009.03.010.

[2] Ghasemi-Mobarakeh L, Prabhakaran MP, Morshed M, Nasr-Esfahani MH, Baharvand H, Kiani S, Al-Deyab SS, Ramakrishna S. Application of conductive polymers, scaffolds and electrical stimulation for nerve tissue engineering. Journal of Tissue Engineering and Regenerative Medicine. 2011;5:17–35. DOI: 10.1002/term.383.

[3] Rivers TJ, Hudson TW, Schmidt CE. Synthesis of a novel, biodegradable electrically conducting polymer for biomedical applications. Advanced Functional Materials. 2002;12:33–37. DOI: 10.1002/1616 3028(20020101)12:1<33::AID-ADFM33>3.0.CO;2-E.

[4] Shirakawa H, Louis EJ, MacDiarmid AG, Chiang CK, Heeger AJ. Synthesis of electrically conducting organic polymers: halogen derivatives of polyacetylene, (CH)x. Journal of the Chemical Society, Chemical Communications. 1977:578–580. DOI: 10.1039/C39770000578.

[5] Ateh DD, Navsaria HA, Vadgama P. Polypyrrole-based conducting polymers and interactions with biological tissues. Journal of the Royal Society Interface. 2006;3:741–752. DOI: 10.1098/rsif.2006.0141.

[6] Weetall HH, Druzhko A, de Lera AR, Alvarez R, Robertson B. Measurement of proton release and uptake by analogs of bacteriorhodopsin. Bioelectrochemistry. 2000;51:27–33. DOI: 10.1016/S0302-4598(99)00072-0.

[7] Ravichandran R, Sundarrajan S, Venugopal JR, Mukherjee S, Ramakrishna S. Applications of conducting polymers and their issues in biomedical engineering. Journal of the Royal Society Interface. 2010;7:S559–S579. DOI: 10.1098/rsif.2010.0120.focus.

[8] Guimard NK, Gomez N, Schmidt CE. Conducting polymers in biomedical engineering. Progress in Polymer Science. 2007;32:876–921. DOI: 10.1016/j.progpolymsci.2007.05.012.

[9] Liu X, Gilmore KJ, Moulton SE, Wallace GG. Electrical stimulation promotes nerve cell differentiation on polypyrrole/poly(2-methoxy-5 aniline sulfonic acid) composites. Journal of Neural Engineering. 2009;6:1–10. http://dx.doi.org/10.1088/1741-2560/6/6/065002.

[10] Bredas JL, Streer GB. Polarons, bipolarons and solitons in conductive polymers. Accounts of Chemical Research. 1985;18:309–315. DOI: 10.1021/ar00118a005.

[11] Cortés MT, Moreno JC. Artificial muscles based on conducting polymers. e-Polymers. 2003;4:1–42.

[12] Gilmore KJ, Kita M, Han Y, Gelmi A, Higgins MJ, Moulton SE, Clark MG, Kapsa R, Wallace GG. Skeletal muscle cell proliferation and differentiation on polypyrrole substrates doped with extracellular matrix component. Biomaterials. 2009;30:5292–5304. DOI: 10.1016/j.biomaterials.2009.06.059.

[13] Collier JH, Camp JP, Hudson TW, Schmidt CE. Synthesis and characterization of polypyrrole–hyaluronic acid composite biomaterials for tissue engineering applications. Journal of Biomedical Materials Research, Part B: Applied Biomaterials.

2000;50:574–584. DOI: 10.1002/(SICI)1097-4636(20000615)50:4<574::AID-JBM13>3.0.CO;2-I.

[14] Wallace G, Spinks G. Conducting polymers—bridging the bionic interface. Soft Matter. 2007;3:665–671. DOI: 10.1039/B618204F.

[15] Shi G, Rouabhia M, Wang Z, Dao LH, Zhang Z. A novel electrically conductive and biodegradable composite made of polypyrrole nanoparticles and polylactide. Biomaterials. 2004;25:2477–2488. DOI: 10.1016/j.biomaterials.2003.09.032.

[16] Rosales-Leal JI, Rodríguez-Valverde MA, Mazzaglia G, Ramón-Torregrosa PJ, Díaz-Rodríguez L, García-Martínez O, Vallecillo-Capilla M, Ruiz C, Cabrerizo-Vílchez MA. Effect of roughness, wettability and morphology of engineered titanium surfaces on osteoblast-like cell adhesion. Colloids and Surfaces A: Physicochemical and Engineering Aspects. 2010;365:222–229. DOI: 10.1016/j.colsurfa.2009.12.017.

[17] Pelto J, Haimi S, Puukilainen E, Whitten PG, Spinks GM, Bahrami-Samani M, Ritala M, Vuorinen T. Electroactivity and biocompatibility of polypyrrole–hyaluronic acid multi-walled carbon nanotube composite. Journal of Biomedical Materials Research, Part A. 2010;93:1056–1067. DOI: 10.1002/jbm.a.32603.

[18] Gelmi A, Higgins MJ, Wallace GG. Physical surface and electromechanical properties of doped polypyrrole biomaterials. Biomaterials. 2010;31:1974–1983. DOI: 10.1016/j.biomaterials.2009.11.040.

[19] Diaz AF, Bargon J. Electrochemical synthesis of conducting polymers. In: Skotheim TA, editor. Handbook of conducting polymers. Vol. I. New York: Marcel Dekker; 1986. p. 81–115.

[20] Calvo PA, Rodriguez J, Grande H, Mecerreyes D, Pomposo JA. Chemical oxidative polymerization of pyrrole in the presence of m-hydroxybenzoic acid- and m-hydroxycinnamic acid-related compounds. Synthetic Metals. 2002;126:111–116. DOI: 10.1016/S0379-6779(01)00560-4.

[21] Dall'Olio A, Dascola G, Varacco V, Bocchi V. Electron paramagnetic resonance and conductivity of an electrolytic oxypyrrole [(pyrrole polymer)] black. Comptes Rendus de l'Académie des Sciences—Series IIC—Chemistry. 1968;267:433–435.

[22] Martins NCT, Moura e Silva T, Montemora MF, Fernandes JCS, Ferreira MGS. Electrodeposition and characterization of polypyrrole films on aluminium alloy 6061-T6. Electrochimica Acta. 2008;53:4754–4763. DOI: 10.1016/j.electacta.2008.01.059.

[23] Wallace GG, Smyth M, Zhao H. Conducting electroactive polymer-based biosensors. Trends in Analytical Chemistry. 1999;18:245–251. DOI: 10.1016/S0165-9936(98)00113-7.

[24] Kim DH, Richardson-Burns SM, Hendricks JL, Sequera C, Martin DC. Effect of immobilized nerve growth factor on conductive polymers: electrical properties and cellular response. Advanced Functional Materials. 2007;17:79–86. DOI: 10.1002/adfm.200500594.

[25] Patra S, Barai K, Munichandraiah N. Scanning electron microscopy studies of PEDOT prepared by various electrochemical routes. Synthetic Metals. 2008;158:430–435. DOI: 10.1016/j.synthmet.2008.03.002.

[26] Filimon A. High performance polymeric biomaterials based on functionalized polysulfones with medical applications [research project]. Iassy, Romania: Romanian National Authority for Scientific Research; 2012–2016.

[27] Filimon A, Avram E, Dunca S. Surface and interface properties of functionalized polysulfones: cell–material interaction and antimicrobial activity. Polymer Engineering and Science. 2015;55:2184–2194. DOI: 10.1002/pen.24103.

[28] Labarre D. Improving blood-compatibility of polymeric surfaces. Trends in Biomaterials and Artificial Organs. 2001;15:1–3.

[29] Buruiana LI, Avram E, Musteata VE, Filimon A. Optical and electronic properties of quaternized polysulfone/polyvinyl alcohol blends in relation to structure of the polymers. Materials Chemistry and Physics. 26 April 2016 Online. DOI:10.1016/j.matchemphys.2016.04.051.

[30] Dobos AM, Filimon A, Avram E, Ioanid GE. Impact of surface properties of blends based on quaternized polysulfones on modeling and interpretation the interactions with blood plasma. In: Proceedings of the International Conference on e-Health and Bioengineering (EHB-2015); 19–21 November 2015; Iaşi: EHB; 2015. p. 1–4.

[31] Subramanian A, Krishnan UM, Sethuraman S. Axially aligned electrically conducting biodegradable nanofibers for neural regeneration. Journal of Materials Science: Materials in Medicine. 2012;23:1797–1809. DOI: 10.1007/s10856-012-4654-y.

[32] Filimon A, Avram E, Olaru N, Doroftei F, Dunca S. Electrospun quaternized polysulfone fibers with antimicrobial activity. In: Proceedings of the International Conference on e-Health and Bioengineering (EHB-2015); 19–21 November 2015; Iaşi: EHB; 2015. p. 1–4

[33] Guiseppi-Elie A. Electroconductive hydrogels: synthesis, characterization and biomedical applications. Biomaterials. 2010;31:2701–2716. DOI: 10.1016/j.biomaterials.2009.12.052.

[34] Justin G, Guiseppi-Elie A. Electroconductive blends of poly(HEMA-co-PEGMA-co-HMMAco–SPMA) and poly(Py-co-PyBA): in vitro biocompatibility. Journal of Bioactive and Compatible Polymers. 2010;25:121–140. DOI: 10.1177/0883911509350660.

[35] Gurunathan K, Murugan AV, Marimuthu R, Mulik UP, Amalnerkar DP. Electrochemically synthesized conducting polymeric materials for applications towards technology in electronics, optoelectronics and energy storage devices. Materials Chemistry and Physics. 1999;61:173–191. DOI: 10.1016/S0254-0584(99)00081-4.

[36] Zhang Q, Yan Y, Li S, Feng T. The synthesis and characterization of a novel biodegradable and electroactive polyphosphazene for nerve regeneration. Materials Science and

Engineering: C. Materials for Biological Applications. 2010;30:160–166. DOI: 10.1016/j.msec.2009.09.013.

[37] Williams RL, Doherty PJ. A preliminary assessment of poly(pyrrole) in nerve guide studies. Journal of Materials Science: Materials in Medicine. 1994;5:429–433.

[38] Kim S, Oh WK, Jeong YS, Hong JY, Cho BR, Hahn JS, Jang J. Cytotoxicity of, and innate immune response to, size-controlled polypyrrole nanoparticles in mammalian cells. Biomaterials. 2011;32:2342–2350. DOI: 10.1016/j.biomaterials.2010.11.080.

[39] Ferraz N, Strømme M, Fellstrom B, Pradhan S, Nyholm L, Mihranyan A. In vitro and in vivo toxicity of rinsed and aged nanocellulose–polypyrrole composites. Journal of Biomedical Materials Research, Part A. 2012;100:2128–2138. DOI: 10.1002/jbm.a.34070.

[40] Crompton TR. Applications of reinforced plastics. In: Crompton TR, editor. Plastics reinforcement and industrial applications. Boca Raton, London, New York: Taylor and Francis; 2015. p. 238.

[41] Ahuja T, Mir IA, Kumar D, Rajesh. Biomolecular immobilization on conducting polymers for biosensing applications. Biomaterials. 2007;28:791–805. DOI: 10.1016/j.biomaterials.2006.09.046.

[42] Arora K, Chaubey A, Singhal R, Singh RP, Pandey MK, Samanta SB, Malhotra BD, Chand S. Application of electrochemically prepared polypyrrole–polyvinyl sulphonate films to DNA biosensor. Biosensors and Bioelectronics. 2006;21:1777–1783. DOI: 10.1016/j.bios.2005.09.002.

[43] Cosnier S. Biomolecule immobilization on electrode surfaces by entrapment or attachment to electrochemically polymerized films. A review. Biosensors and Bioelectronics. 1999;14:443–456. DOI: 10.1016/S0956-5663(99)00024-X.

[44] Prabhakar N, Arora K, Singh SP, Singh H, Malhotra BD. DNA entrapped polypyrrole–polyvinyl sulfonate film for application to electrochemical biosensor. Analytical Biochemistry. 2007;366:71–79. DOI: 10.1016/j.ab.2007.03.009.

[45] Lee JY, Lee JW, Schmidt CE. Neuroactive conducting scaffolds: nerve growth factor conjugation on active ester-functionalized polypyrrole. Journal of the Royal Society Interface. 2009;6:801–810. DOI: 10.1098/rsif.2008.0403.

[46] Thompson BC, Richardson RT, Moulton SE, Evans AJ, O'Leary S, Clark GM, Wallace GG. Conducting polymers, dual neurotrophins and pulsed electrical stimulation—dramatic effects on neurite outgrowth. Journal of Controlled Release. 2010;141:161–167. DOI: 10.1016/j.jconrel.2009.09.016.

[47] Arora K, Chand S, Malhotra BD. Recent developments in bio-molecular electronics techniques for food pathogens. Analytica Chimica Acta. 2006;568:259–274. DOI: 10.1016/j.aca.2006.03.078.

[48] Brown JQ, McShane J. Modeling of spherical fluorescent glucose microsensor systems: design of enzymatic smart tattoos. Biosensors and Bioelectronics. 2006;21:1760–1769. DOI: 10.1016/j.bios.2005.08.013.

[49] Collis GE, Burrell AK, Scott SM, Officer DL. Toward functionalized conducting polymers: synthesis and characterization of novel β-(styryl)terthiophenes. The Journal of Organic Chemistry. 2003;68:8974–8983. DOI: 10.1021/jo034855g.

[50] Adhikari B, Majumdar S. Polymers in sensor applications. Progress in Polymer Science. 2004;29:699–766. DOI: 10.1016/j.progpolymsci.2004.03.002.

[51] Yu B, Moussy Y, Moussy F. Lifetime improvement of glucose biosensor by epoxy-enhanced PVC membrane. Electroanalysis. 2005;17:1771–1779. DOI: 10.1002/elan.200503301.

[52] Geetha S., Rao CRK, Vijyan M, Trivedi DC. Biosensing and drug delivery by polypyrrole. Analytica Chimica Acta. 2006;568:119–125. DOI: 10.1016/j.aca.2005.10.011.

[53] Terry LA, White SF, Tigwell LJ. The application of biosensors to fresh produce and the wider food industry. Journal of Agricultural and Food Chemistry. 2005;53:1309–1316. DOI: 10.1021/jf040319t.

[54] Clark LC, Lyons C. Electrode systems for continuous monitoring in cardiovascular surgery. Annals of the New York Academy of Sciences. 1962;102:29–45. DOI: 10.1111/j.1749-6632.1962.tb13623.x.

[55] Deshpande MV, Amalnerkar DP. Biosensors prepared from electrochemically synthesized conducting polymers. Progress in Polymer Science. 1993;18:623–649. DOI: 10.1016/0079-6700(93)90005-W.

[56] Gerard M, Chaubey A, Malhotra BD. Application of conducting polymers to biosensors. Biosensors and Bioelectronics. 2002;17:345–359. DOI: 10.1016/S0956-5663(01)00312-8.

[57] Rahman MA, Park D, Chang S, McNeil CJ, Shim Y. The biosensor based on the pyruvate oxidase modified conducting polymer for phosphate ion determinations. Biosensors and Bioelectronics. 2006;21:1116–1124. DOI: 10.1016/j.bios.2005.04.008.

[58] Trojanowicz M, Miernik A. Bilayer lipid membrane glucose biosensors with improved stability and sensitivity. Electrochimica Acta. 2001;46:1053–1061. DOI: 10.1016/S0013-4686(00)00684-8.

[59] Malhotra BD, Chaubey A, Singh S. Prospects of conducting polymers in biosensors. Analytica Chimica Acta. 2006;578:59–74. DOI: 10.1016/j.aca.2006.04.055.

[60] Lei Y, Mulchandani P, Chen W, Mulchandani A. Direct determination of p-nitrophenyl substituent organophosphorus nerve agents using a recombinant *Pseudomonas putida* JS444-modified clark oxygen electrode. Journal of Agricultural and Food Chemistry. 2005;53:524–527. DOI: 10.1021/jf048943t.

[61] Ferrer-Anglada N, Kaempgen M, Roth S. Transparent and flexible carbon nanotube/polypyrrole and carbon nanotube/polyaniline pH sensors. Physica Status Solidi B. 2006;243:3519–3523. DOI: 10.1002/pssb.200669220.

[62] Xu Y, Ye X, Yang L, He P, Fang Y. Impedance DNA biosensor using electropolymerized polypyrrole/mulitwalled carbon nanotubes modified electrode. Electroanalysis. 2006;18:1471–1478. DOI: 10.1002/elan.200603544.

[63] Fine G, Valentini RF, Bellamkonda R, Aebischer P. Improved nerve regeneration through piezoelectric vinylidenefluoride–trifluoroethylene copolymer guidance channels. Biomaterials. 1991;12:775–780. DOI: 10.1016/0142-9612(91)90029-A.

[64] Giaever I, Keese CR. Monitoring fibroblast behavior in tissue culture with an applied electric field. Proceedings of the National Academy of Sciences of the United States of America. 1984;81:3761–3764

[65] Kerns JM, Pavkovic IM, Fakhouri AJ, Wickersham KL, Freeman JA. An experimental implant for applying a DC electrical field to peripheral nerve. Journal of Neuroscience Methods. 1987;19:217–223. DOI: 10.1016/S0165-0270(87)80005-5.

[66] Wong JY, Langer R, Ingber DE. Electrically conducting polymers can noninvasively control the shape and growth of mammalian cells. Proceedings of the National Academy of Sciences of the United States of America. 1994;91:3201–3204. DOI: 10.1073/pnas.91.8.3201.

[67] Garner B, Georgevich A, Hodgson AJ, Liu L, Wallace GG. Polypyrrole–heparin composites as stimulus-responsive substrates for endothelial cell growth. Journal of Biomedical Materials Research, Part A. 1999;44:121–129. DOI: 10.1002/(SICI)1097-4636(199902)44:2<121::AID-JBM1>3.0.CO;2-A.

[68] Hodgson AJ, John MJ, Campbell T, Georgevich A, Woodhouse S, Aoki T, et al. Integration of biocomponents with synthetic structures—use of conducting polymer polyelectrolyte composites. Proceedings of SPIE—The International Society for Optical Engineering. 1996;2716:164–176.

[69] Song H-K, Toste B, Ahmann K, Hoffman-Kim D, Palmore GT R. Micropatterns of positive guidance cues anchored to polypyrrole doped with polyglutamic acid: a new platform for characterizing neurite extension in complex environments. Biomaterials. 2006;27:473–484. DOI: 10.1016/j.biomaterials.2005.06.030.

[70] Gomez N, Lee JY, Nickels JD, Schmidt CE. Micropatterned polypyrrole: a combination of electrical and topographical characteristics for the stimulation of cells. Advanced Functional Materials. 2007;17:1645–1653. DOI: 10.1002/adfm.200600669.

[71] Ateh DD, Vadgama P, Navsaria HA. Culture of human keratinocytes on polypyrrole-based conducting polymers. Tissue Engineering. 2006;12:645–655. DOI: 10.1089/ten.2006.12.645.

[72] Castano H, O'Rear EA, McFetridge PS, Sikavitsas VI. Polypyrrole thin films formed by admicellar polymerization support the osteogenic differentiation of mesenchymal stem cells. Macromolecular Bioscience. 2004;4:785–794. DOI: 10.1002/mabi.200300123.

[73] Pyo M, Reynolds JR. Electrochemically stimulated adenosine 50-triphosphate (ATP) release through redox switching of conducting polypyrrole films and bilayers. Journal of Materials Chemistry. 1996;8:128–133

[74] Li ZF, Ruckenstein E. Two liquid adsorptive entrapment of a pluronic polymer into this surface of polyaniline films. Journal of Colloid and Interface Science. 2003;264:370–377. DOI: 10.1016/S0021-9797(03)00442-9.

[75] Li M, Guo Y, Wei Y, MacDiarmid AG, Lelkes PI. Electrospinning polyaniline-containing gelatin nanofibers for tissue engineering applications. Biomaterials. 2006;27:2705–2715. DOI: 10.1016/j.biomaterials.2005.11.037.

[76] George PM, Lyckman AW, LaVan DA, Hegde A, Leung Y, Avasare R, Testa C, Alexander MA, Langer R, Sur M. Fabrication and biocompatibility of polypyrrole implants suitable for neural prosthetics. Biomaterials. 2005;26:3511–3519. DOI: 10.1016/j.biomaterials.2004.09.037.

[77] Cui X, Lee VA, Raphael Y, Wiler JA, Hetke JF, Anderson DJ, Martin DC. Surface modification of neural recording electrodes with conducting polymer/biomolecule blends. Journal of Biomedical Materials Research, Part B. 2001;56:261–272. DOI: 10.1002/1097-4636(200108)56:2<261::AID-JBM1094>3.0.CO;2-I.

[78] Kim D-H, Abidian M, Martin DC. Conducting polymers grown in hydrogel scaffolds coated on neural prosthetic devices. Journal of Biomedical Materials Research, Part A. 2004;71:577–585. DOI: 10.1002/jbm.a.30124.

[79] Xiao YH, Cui XY, Hancock JM, Bouguettaya M, Reynolds JR, Martin DC. Electrochemical polymerization of poly(hydroxymethylated-3,4-ethylenedioxythiophene) (PEDOT–MeOH) on multichannel neural probes. Sensors and Actuators B. 2004;99:437–443. DOI: 10.1016/j.snb.2003.12.067.

[80] Ludwig KA, Uram JD, Yang J, Martin DC, Kipke DR. Chronic neural recordings using silicon microelectrode arrays electrochemically deposited with a poly(3,4-ethylenedioxythiophen) (PEDOT) film. Journal of Neural Engineering. 2006;3:59–70.

[81] Massoumi B, Entezami AA. Electrochemically stimulated 2-ethylhexyl phosphate (EHP) release through redox switching of conducting polypyrrole film and polypyrrole/poly(N-methylpyrrole) or self-doped polyaniline bilayers. Polymer International. 2002;51:555–560. DOI: 10.1002/pi.885.

[82] Miller LL, Zhou XU. Poly(N-methylpyrrolylium) poly(styrenesulfonate). A conductive, electrically switchable cation exchanger that cathodically binds and anodically releases dopamine. Macromolecules. 1987;20:1594–1597. DOI: 10.1021/ma00173a027.

[83] Kontturi K, Pentti P, Sundholm G. Polypyrrole as a model membrane for drug delivery. Journal of Electroanalytical Chemistry. 1998;453:231–238. DOI: 10.1016/S0022-0728(98)00246-0.

[84] Li Y, Neoh KG, Kang ET. Controlled release of heparin from polypyrrole–poly(vinyl alcohol) assembly by electrical stimulation. Journal of Biomedical Materials Research, Part A. 2005;73:171–181. DOI: 10.1002/jbm.a.30286.

[85] Gomez N, Schmidt CE. Nerve growth factor-immobilized polypyrrole: bioactive electrically conducting polymer for enhanced neurite extension. Journal of Biomedical Materials Research, Part A. 2007;81:135–149. DOI: 10.1002/jbm.a.31047.

# Chapter 7

# Electrical Properties of Polymer Light-Emitting Devices

Lucas Fugikawa Santos and Giovani Gozzi

Additional information is available at the end of the chapter

http://dx.doi.org/10.5772/64358

**Abstract**

In this chapter, we present a brief introduction to semiconducting properties of conjugated polymers and the motivation to apply this class of materials in electronic/optoelectronic devices such as polymer light-emitting diodes (PLEDs). We describe, in detail, the operating mechanisms of PLEDs, with particular focus on the effects of charge injection and transport and their dependence on the external electric field and temperature. The mechanisms of current injection from the electrodes into the organic semiconductor are initially treated using traditional models for thermionic emission and tunnelling injection. More recent models considering the influence of metal/semiconductor interface recombination and of energetic and spatial disorder in the injection currents are also introduced and discussed. In addition, models considering space-charge-limited currents and trap-filling-limited currents are employed to describe the charge transport characteristics in the bulk. Furthermore, we present a brief discussion on ideas concerning the effects of the disorder on the charge-carrier transport behaviour.

**Keywords:** polymer light-emitting diodes, conjugated polymers, electrical properties, space-charge-limited currents, organic semiconductors

## 1. Introduction

In the past three decades, we have witnessed an impressive fast-growing number of scientific publications concerning the electronic and optoelectronic properties of semiconducting polymers, from fundamental theoretical studies to reports of cutting-edge applications. In the same period, the device performance, quantified by device parameters such as charge-carrier mobility, electroluminescence efficiency or photovoltaic energy conversion efficiency, has presented a remarkable improvement permitting the achievement of devices with performance compactible to commercial needs. The great advantage of semiconducting polymers compared to other electronic materials is that they are low-cost, lightweight, flexible and can be solution-processed, allowing the deposition over large areas and facilitating upscaling production.

A common characteristic of all semiconducting polymers is the chain conjugation, i.e., the alternation of single and double bonds of the carbon atoms in the backbone chain. When two carbon atoms form a double bond, three equivalent orbitals are formed due to the $sp^2$ hybridization and are responsible for three coplanar σ-bonds. The remaining orbital, $p_z$, is perpendicular to $sp^2$ orbitals plane and the interatomic spacing is such that an overlapping of the wave function of neighbouring electrons occurs, giving rise to a π-bond. In a conjugated structure, however, the π-electrons are delocalized over all carbon atoms along the conjugation extension, forming a set of bonding molecular orbitals (MO). The π-electrons in the highest energy occupied molecular orbital (HOMO) are more susceptible to be excited and to suffer a transition to the lowest energy unoccupied molecular orbital (LUMO) in a process commonly referred as a $\pi$-$\pi^*$ transition. The minimum energy necessary for the electron to undergo this transition is the energetic bandgap of the semiconducting polymer. For a more detailed study concerning the formation of molecular orbitals and electronic transitions, we recommend the studies of Kao and Hwang [1] and Pope and Swenberg [2].

An electronic $\pi$-$\pi^*$ transition introduces an electron in the LUMO, leaving a vacant state (hole) in the HOMO. Such transition can be thermally or optically induced, due to the absorption of a sufficiently energetic photon, in the latter case. When such transition originates an excited state characterized by an electrostatically bound electron-hole pair, it is designated as an exciton. Excitons can migrate, diffuse, dissociate or simply decay (which can be radiative or non-radiative), playing an important role in the physics of optoelectronic devices such as polymer light-emitting diodes (PLEDs) and organic photovoltaic solar cells (OPVs). Moreover, charge transport in conjugated polymers takes places via charged excited states (electrons in the LUMO or holes in the HOMO) which may be originated from exciton dissociation, charge injection from electrodes, thermal generation (intrinsic carriers), electrochemical doping, etc. In conjugated polymers with degenerate ground state, charged excited states are associated with conformational defects named solitons, whereas for polymers with non-degenerate ground state such defects are called polarons [3–5]. Polarons are half-spin semi-particles which can be negatively or positively charged, playing a similar role in conjugated polymers as electrons and holes in inorganic semiconductors or in organic molecular crystals. In the context of this chapter, we will interchange the terms for negatively (positively) charged polarons and electrons (holes) with no loss of rigor.

The schemes of the chemical structures of some conjugated polymers, frequently used as conducting/semiconducting layer of polymeric electronic/optoelectronic devices, are presented in **Table 1**.

From the previous exposition, we can assume that all characteristics of electronic processes of semiconducting materials such as charge injection, transport, photo-generation and recombination are originated in the conjugation of the polymer backbone chain. In the following sections, we will show how these properties can be applied in electronic and optoelectronic devices, with special focus on polymer light-emitting diodes, and discuss their basic operating mechanisms.

| Polymer | Acronym | Scheme |
|---|---|---|
| *trans-polyacetylene* | *t*-PA | 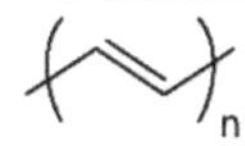 |
| Polyaniline (emeraldine base) | PANI | 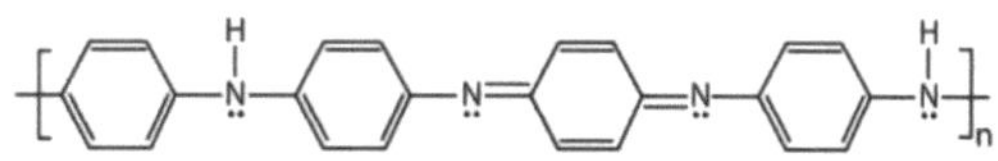 |
| Poly(3,4-ethylenedioxythiophene) | PEDOT | 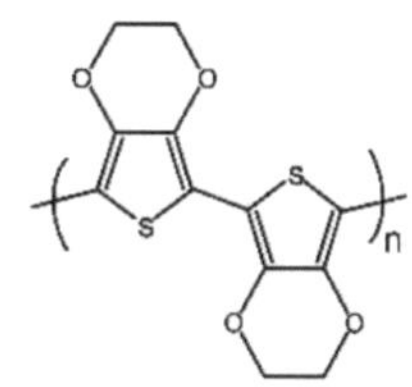 |
| Poly(*p*-phenylene vinylene) | PPV | 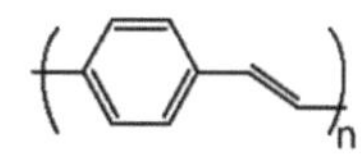 |
| Poly[2-methoxy-5-(2-ethylhexyloxy)-1,4-phenylenevinylene] | MEH-PPV | 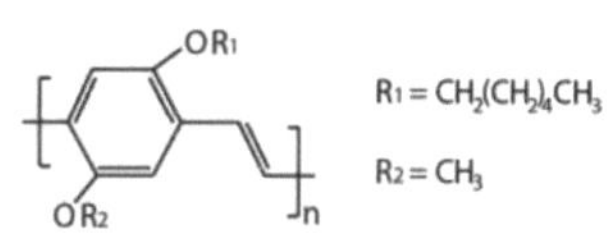 |
| Poly(9,9-di-n-octylfluorenyl-2,7-diyl) | PFO | 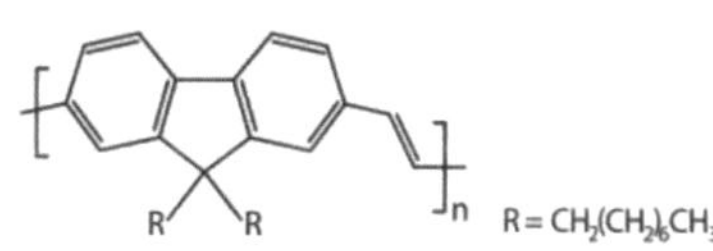 |
| Poly(3-hexylthiophene) | P3HT | 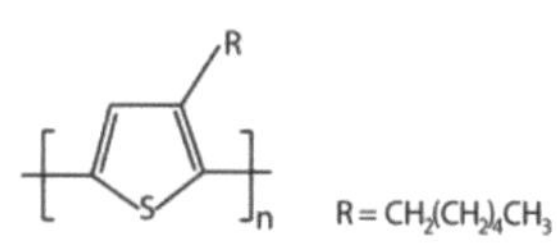 |

**Table 1.** Scheme of chemical structures of some conjugated polymers used is polymer electronic devices.

## 2. Polymer light-emitting diodes (PLEDs)

The first electroluminescent device with active layer based on a conjugated polymer was reported in 1990 by researchers from Cambridge University [6]. Although organic light-emitting diodes (OLEDs) with a high efficiency light output were reported 3 years before by researchers from Eastman Kodak [7], the great novelty of this work was the remarkably easy manufacturing process, based on the deposition of a thin-film, by spin-coating from a solution of a polymer precursor onto indium-tin oxide (ITO)/glass substrates. After thermal conversion of the polymer precursor into poly(*p*-phenylene vinylene) (PPV), metallic electrodes (Al) were thermally evaporated on top of the polymeric film, forming a diode that, when forward polarized (ITO electrode biased positively), emitted a yellowish-green light (thanks to a bandgap energy of about 2.7 eV). Polymer light-emitting diodes (PLEDs) are, in fact, OLEDs; however, a different acronym is frequently employed to emphasize the difference whether the active layer comprises a conjugated polymer (solution processed) or a molecular solid (processed via thermal evaporation). OLEDs based on molecular solids present higher crystallinity, better thickness control, can be manufactured in multiple layers configuration and, in most of the cases, achieve higher performance than PLEDs. Nevertheless, polymers can be solution processed, allowing low-cost deposition techniques, compatibility to flexible substrates and easy expansion to large-area mass production (by means of spray or roll-to-roll deposition).

**Figure 1a** schematically represents the cross section of a simple, single-layer PLED structure. The 'straight-band' diagrams of **Figure 1b–e** are just a simplified vision of the formation of the metal/semiconductor junctions (which are, actually, Schottky type) at the interfaces. To obtain a PLED, it is necessary to use a high work-function ($\Phi_A$) electrode (close to the HOMO level) as an anode and a low work-function ($\Phi_C$) metal (close to the LUMO level) as a cathode. **Figure 1b** shows the energy diagrams for the isolated semiconductor and metallic electrodes before the contacts are made. When the device is reverse biased (ITO electrode biased negatively) or at a positive voltage lower than the difference between the electrodes' work functions divided by the elementary charge ($[\Phi_A - \Phi_C]/q$), commonly known as *built-in* voltage, $V_{bi}$, the barriers for the injection of electrons into the LUMO and holes into the HOMO of the conjugated polymer are very high (**Figure 1d**) and the charge-carrier transport is dominated by the thermally generated intrinsic carriers whose density is usually low, resulting in a relatively low-current device. At an external bias equal to $V_{bi}$, the band diagram becomes 'flat' (**Figure 1c**) and the situation is equivalent to the isolated energy levels before the formation of the junctions (**Figure 1b**). For voltages higher than $V_{bi}$, the band diagrams are reversed compared to the equilibrium configuration (**Figure 1d**) and the energy barriers for injection of electrons into the LUMO and holes into the HOMO are much lower, resulting in a considerably higher current device. In this situation (forward bias), electrons in the LUMO flow towards the anode and holes in the HOMO flow towards the cathode. A fraction of these injected charge carriers recombines inside the device active layer, giving origin to excitons that can undergo radiative decay, resulting in electroluminescence (EL).

A high-performance PLED is characterized by a high EL efficiency, which is mainly limited by the exciton recombination efficiency, the optical coupling device geometry and the balance in the electrode injection and bulk transport of both types (electrons and holes) of charge carriers. Each of these limiting factors in the device EL efficiency depends on several parameters demanding a thorough study. For brevity, we will focus, in the present chapter, only on the charge injection and transport processes, leaving the former two cases for further reading [8, 9].

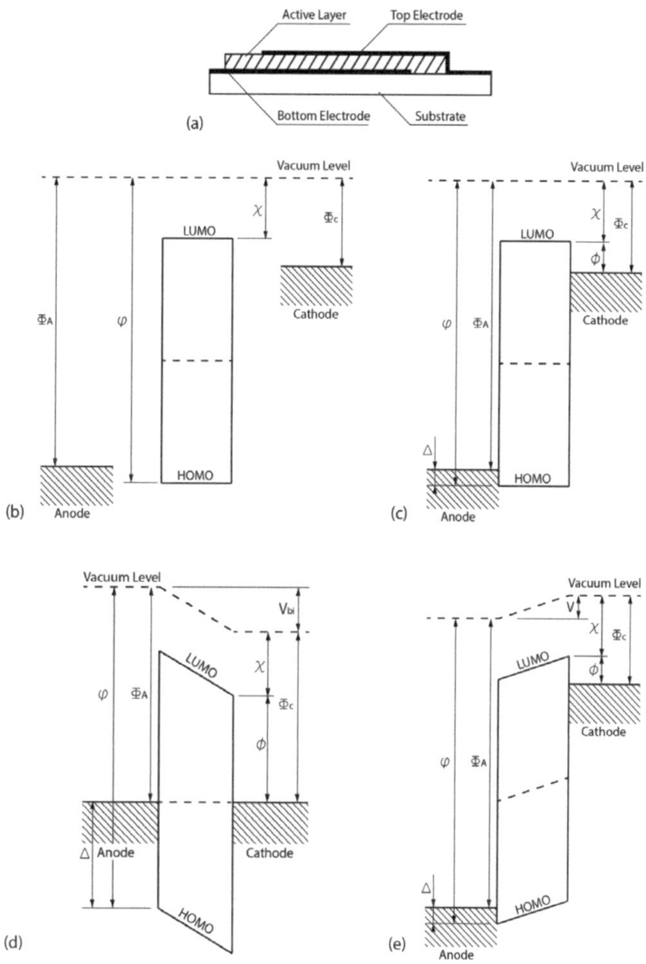


**Figure 1.** (a) Basic structure of a PLED; (b) energy band diagram before the contact formation; (c) 'flat-band' condition; (d) band diagram for the device in equilibrium and (e) band diagram for the device in forward bias. In these figures, $\Phi_A$ and $\Phi_C$ represents, respectively, the anode and the cathode work functions, $\chi$, the electronic affinity, $\phi$, the ionization potential of the conjugated polymer, $\varphi$ and $\Delta$, the energy barrier for injection of, respectively, electrons and holes at forward bias and $V$, the voltage between the device electrodes.

In an ideal PLED, electrons and holes would not have any restriction for injection from the electrodes and the density of injected charge carriers would be balanced, as well as the charge transport across the polymer layer would have equal constraints for both types of charge carriers. This picture is, though, far from being realistic. Energetic barriers need to be surpassed for the injection of electrons and holes from the electrodes into the polymeric layer, and it is virtually impossible to have equivalent injection for both electrons and holes. Although the charge injection represents a bottleneck in the current flow process, after the charge carriers are injected into the semiconducting polymer, the charge-carrier mobility becomes the decisive factor for the charge balance inside the device active layer. The charge-carrier mobility of electrons in conjugated polymers is frequently found to be several times smaller than the hole mobility [10], which results in an asymmetric charge density distribution, strongly affecting the EL efficiency.

In a general way, the total current in a PLED can be described by

$$J \propto P_{inj} \times M_{bulk} \tag{1}$$

where $P_{inj}$ is a function dependent on the device injection current and $M_{bulk}$ is a function dependent on the transport current in the bulk is the transport current in the bulk. Eq. (1) means that, if the charge injection is restricted, the total device current will be limited, no matter how efficient is the charge-carrier transport in the bulk. Equivalently, if constraints in the charge-carrier transport exist, they will limit the total current in spite of how good are the electrodes for charge injection. In the following two subsections, we will present and discuss the main mechanisms that rule charge injection and transport in semiconducting polymers.

### 2.1. Charge-injection mechanisms in PLEDs

The ideal condition for charge injection from an electrode into a semiconducting polymer is when the work function of the cathode ($\Phi_C$) and of the anode ($\Phi_A$) matches, respectively, the LUMO and HOMO levels of the organic semiconductor. In such situation, when the difference between the electron affinity ($\chi$) and the cathode work function and the ionization potential ($\phi$) and the anode work function is in the order of magnitude or smaller than the thermal fluctuation energy, $k_BT$, we say that the device has ohmic contacts and that, consequently, the device current will be mainly limited by the transport in the bulk. A device with ohmic contacts does not necessarily present an ohmic behaviour, i.e., a linear dependence between current and voltage. Ohmic contacts simply mean that the barriers for charge injection from the electrodes are small enough to be neglected in the description of the total net current flowing across the device. The general case, however, is when these energy barriers for injection cannot be neglected and barriers $\varphi = \Phi_C - \chi$ for electron injection and $\Delta = \phi - \Phi_A$ for hole injection exist.

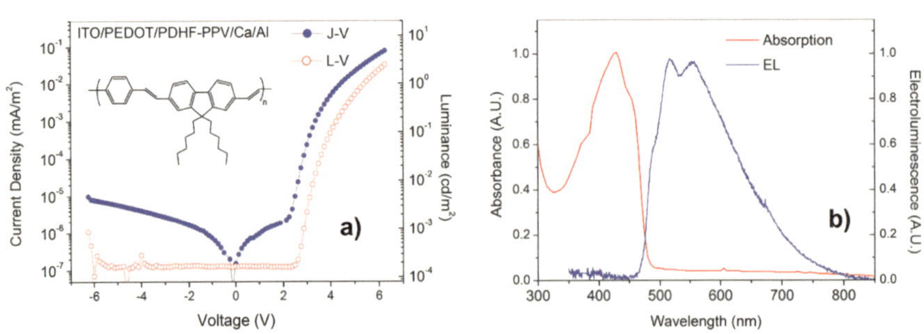


**Figure 2.** (a) Current-voltage (*I-V*) and luminance-voltage (*L-V*) characteristics of a typical PLED in structure ITO/PEDOT/PDHF-PPV/Ca/Al. (b) Absorption and emission (EL) spectra for the conjugated polymer used as active layer (PDHF-PPV).

**Figure 2a** presents the current-voltage (*I-V*) and luminance-voltage (*L-V*) characteristics of a typical PLED fabricated using a structure ITO/PEDOT/PDHF-PPV/Ca/Al. The poly(3,4-ethylenedioxythiophene), PEDOT, layer serves as a hole injection layer between the active emissive layer (made of poly (9,9–dihexyl fluorene diyl phenylene-alt-1,4-phenylene vinylene), PDHF-PPV) and the ITO electrode, providing almost ohmic contacts for hole injection into the polymer HOMO. On the other hand, the Ca electrode offers an almost ohmic contact for electron injection into the organic semiconductor LUMO. As a consequence, the onset for the forward current and for the EL device (**Figure 2a**) occurs almost simultaneously (due to balanced injection of both charge carriers) at a voltage that is nearly equal to $\Phi_A - \Phi_C/q$, and slightly above 2.0 *V*. The absorption and emission spectra of PDHF-PPV are shown in **Figure 2b**.

In a real PLED, the height of these barriers may be very different, leading to a highly unbalanced injection of electrons and holes. In such a case, the total current density will be dominated by the majority carriers, which are the most injected charge carrier type. On the other hand, the light emission will be limited by the density of injected minority charge carriers, since, for electroluminescence, both types of charge carriers are needed.

**Table 2** presents some values for the ionization potential, electron affinity and energy bandgap of some conjugated polymers used in PLEDs, as well as the work function of conductive materials that can be used as electrodes. Although PEDOT is a conjugated polymer, it is highly conductive and only the value of the work function is computed **Table 2**. These data allow one to evaluate the barrier height for carrier injection in the correspondent molecular orbital (HOMO or LUMO), being essential for the design of the PLED with the desired characteristics.

| | Bandgap $E_g$ (*eV*) | Electron affinity $\chi$ (*eV*) | Ionization potential $\phi$ (*eV*) | Work function $\Phi$ (*eV*) |
|---|---|---|---|---|
| PPV | 2.7 | 2.8 | 5.5 | – |
| MEH-PPV | 2.2 | 3.0 | 5.2 | – |
| PFO | 3.0 | 2.9 | 5.9 | – |
| PEDOT | – | – | – | 5.0–5.2 |
| ITO | – | – | – | 4.7–4.8 |
| Al | – | – | – | 4.06–4.41 |
| Au | – | – | – | 5.1–5.47 |
| Ag | – | – | – | 4.26–4.74 |
| Mg | – | – | – | 3.6–3.7 |
| Ca | – | – | – | 2.87–3.00 |

**Table 2.** Some values for the ionization potential, electron affinity and energy bandgap of some conjugated polymers used as active layer of PLEDs and the work function of materials commonly used as electrodes.

The device total injection current behaviour can be appropriately described by the injection properties of the lowest height energy barrier contact, in the case of highly unbalanced electron and hole injection or, when the barrier heights at both metal/semiconductor interface are comparable to each other, by a function proportional to the sum of each separate injection current contribution.

The charge injection from an electrode into an organic semiconductor can be treated by a variety of different approaches. The first models used to explain the charge injection characteristics in PLEDs were imported and adapted from the traditional models used in inorganic semiconductor devices and insulators. However, the experimental data obtained from PLEDs can hardly be totally described by a simple model and, frequently, a combination of different processes, occurring concomitantly, is used. Another methodology, specifically developed for organic/amorphous semiconductors, includes the effects of energetic and positional disorder in the injection mechanisms. Next, we will present and discuss the most commonly used models to describe charge injection in PLEDs.

### 2.1.1. *Injection by tunnelling*

When a PLED is at forward bias (**Figure 1e**) and the energy barrier height $\varphi$ (or $\Delta$) at the metal/polymer interface cannot be neglected, one possibility for the injection of an electron (hole) into the LUMO (HOMO) is the tunnelling across this barrier. Due to the band bending caused by the external applied field, such barrier can be approximated to a triangular barrier whose width depends on the electric field. The tunnelling current density can be evaluated by considering the Wentzek-Kramers-Brillouin (WKB) approximation [1] for tunnelling through a potential given by the image force model. The analytic dependence of the current density dominated by tunnelling injection is then given by

$$J_{tu}(F) = J_0 \frac{F^2}{\phi} \exp\left(-\frac{2\alpha\phi^{3/2}}{3qF}\right) \tag{2}$$

where $F$ is the electric field strength, $\varphi$ is the energy barrier height at the contact, $J_0$ is a free fitting parameter for the current density, $q$ is the elementary charge and $\alpha = \frac{4\pi(2m^*)^{1/2}}{h}$, where $m^*$ is the effective mass of the charge carrier and $h$ is the Planck constant. For a quick test if the experimental data can be fitted by Eq. (2) (and, consequently, that the major contribution to the device current is injection by tunnelling) is to make a plot of $\ln(J/F^2)$ vs. $1/F$ and expect for a linear behaviour.

An important feature of the tunnelling injection model is that it does not consider the temperature dependence on the device current, which is frequently present in PLEDs. Despite this aspect, the tunnelling injection model was successfully used to demonstrate the dependence on the barrier height and on the electric field for a complete set of experimental data obtained from devices that the metals used as electrodes, and the active layer thicknesses were systematically varied [11].

### *2.1.2. Current injection via thermionic emission*

Another possible contribution to the injection current across an energy barrier at the semiconductor/metal interface comes from the thermionic emission of charge carriers from the electrodes. In this model, the charge carriers from the electrodes, which have enough thermal energy to overcome the energy barrier at the contact, can be injected into the LUMO/HOMO of the semiconducting polymer. The density of charge carriers which satisfies this condition follows a Boltzmann distribution $n = n_0 \exp(-\phi/k_B T)$, where $n_0$ is the charge-carrier density in the electrode, $\varphi$, is the energy barrier height and $T$, is the absolute temperature. A simple expression for the current density controlled by thermionic emission is given by

$$J_{th}(T) = AT^2 exp(-\phi / k_B T) \tag{3}$$

where $A$ is the Richardson constant. Obviously, the effective barrier height is not constant and it depends on the electric field. Such dependence appears with the inclusion of the image force barrier lowering term, which can be considered only when the device is forward biased. The inclusion of this term results in the following effective potential in the vicinity of the injecting contact:

$$U(x,F) = \phi - \frac{q^2}{4\pi\varepsilon}\frac{1}{x} - qFx \tag{4}$$

where $x$ represents the position from the metal/semiconductor interface and $\varepsilon$, is the electrical permittivity of the semiconductor. **Figure 3** shows a schematic representation of effective potential due to barrier lowering in a semiconductor/metal interface. Such potential results in an effective energy barrier height

$$\phi^* = \phi - 2\left(\frac{q^3}{4\pi\varepsilon}\right)^{\frac{1}{2}} F^{\frac{1}{2}} \tag{5}$$

which can replace the fixed barrier height in Eq. (3) to yield the thermionic emission injection current as a function dependent on temperature and electric field (or external applied voltage).

Such a barrier lowering term can also be achieved by considering a Poole-Frenkel-type effect represented by an effective barrier height

$$\phi^*(F) = \phi - aF^{\frac{1}{2}} \tag{6}$$

which gives the same functional dependence of Eq. (3) by considering a barrier height given by Eq. (5). Experimental data from PLED current-voltage curves (*I-V*) [12] were quite well fitted by considering such effective barrier height, but considering that the pre-factor *a* in Eq. (6) has a thermal activation energy, which is not initially predicted by Eq. (5).

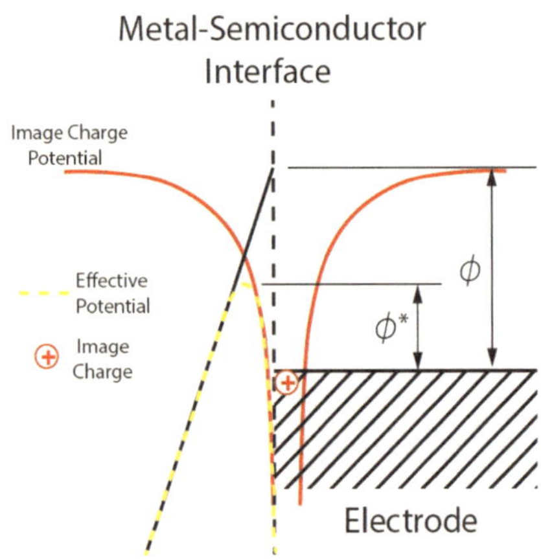


**Figure 3.** Schematic representation of the effective potential at a metal/semiconductor interface considering barrier lowering term due to image force effect.

### 2.1.3. *Interface recombination effects*

The charge injection problem can also be approached by using a more complete model which includes both contributions from tunnelling injection and thermionic emission, with the additional inclusion of a back flowing current contribution due to interface recombination. If one contact has a considerably higher effective energy barrier than the other, the injection current from that contact can be neglected and practically only one type of charge carrier is injected from the contacts (single-carrier device). In this situation, the charge transport in the device can be described as a combination of the continuity equation

$$\frac{\partial p(x,t)}{\partial t} + \frac{1}{q}\frac{\partial J_p(x)}{\partial x} = \Upsilon - \Omega \tag{7}$$

and a drift-diffusion form for the current density for injected holes

$$J_p = q\mu_p\left[p(x)F(x) - \frac{k_B T}{q}\frac{\partial p(x)}{\partial x}\right] \tag{8}$$

where $p(x)$ is the carrier density of holes, $F(x)$, the electric field, $\Upsilon$ and $\Omega$ are, respectively, the carrier generation and recombination rates and $\mu_p$, the effective hole mobility in the organic semiconductor. The carrier mobility can also implicitly depend on temperature and electric field. For a single-carrier device, the recombination rate term as well as the generation rate term can be evidently neglected, since we are considering organic semiconductors in the dark and with high enough bandgap energies. In Eqs. (7) and (8), the total charge inside the active layer is given by the injected charges from the electrodes and its distribution permits the evaluation of the electric field from the Poisson equation. To solve this set of equations, it is necessary to establish the boundary conditions, which are determined by the carrier currents at the interfaces. As stated in the two previous subsections, charge carriers (holes) can be injected by thermionic emission and tunnelling. Once injected in the polymer, however, holes can flow back into the metal, in a process known as interface recombination. Considering these three contributions to the current, the injection current density for holes at the contact ($x = 0$) can be represented in the form

$$J_p(0) = J_{th} + J_{tu} - J_{IR} \tag{9}$$

where $J_{IR}$ is the interface recombination current density contribution, which is proportional to the hole density at the interface, $J_{IR} = \nu p(0)$. The kinetic recombination coefficient $\nu$ can be determined by the detailed balance between thermionic emission and interface recombination [13–15], resulting in

$$J_p(0) = \nu\left\{p_{qe}\left[F(0)\right] - p(0)\right\} + J_{tu} \tag{10}$$

where $\nu p(0)$ is the interface recombination current and $p_{qe}[F(0)]$ is the quasi-equilibrium carrier density at the contact, considering the electric field influence ($F(0)$) on the barrier energy lowering. Therefore, $\nu p_{qe}[F(0)]$ is equal to the thermionic emission current density. This equation can be solved for $p(0)$

$$p(0) = p_{qe}(F(0)) + \frac{(J_{tu} - J_d)}{\nu} \tag{11}$$

where $J_d$ is the device's net current density.

The device current is commonly very small compared to the sum of the tunnelling and the thermionic emission current densities and cannot change significantly the carrier density at the semiconductor/metal interface. As a consequence, the interface recombination current density can be considered as a competition process between the tunnelling injection current density and the thermionic emission injection current density. An important feature of this model is that the magnitude of the device current density can be considerably smaller than that obtained by the direct application of thermionic emission and tunnelling injection expressions, resulting in a more realistic description of the device behaviour. A detailed discussion of this model can be found in Refs. [13–15].

Another model which considers carrier recombination in the injection current was proposed by Scott and Malliaras [16, 17], where the current injection is dependent explicitly on the charge-carrier mobility in the organic semiconductor. This model also considers that the net injection current is the result of the balance between the inflowing charge carriers and the surface recombination rate at the contact interface. Assuming that the surface recombination in a contact depends on the drift/diffusion rate of charge carriers away from the region where the image charge potential has a strong influence on the injected carriers, it is reasonable to consider that the carrier mobility plays an important role in the recombination process. The resulting injection current is

$$J_{inj} = 4\psi^2 N_0 q \mu F \exp(-\phi / k_B T) \exp(f^{1/2}) \quad (12)$$

where $\psi$ is a function which depends smoothly on the electric field [17], $N_0$ is the density of states in the organic semiconductor, $\mu$ is the carrier mobility of the injected carrier at the contact, $F$, is the electric field strength, $q$, is the elementary charge and $\varphi$ is the height of the Schottky barrier at the interface. The Schottky barrier lowering effect is included by the second exponential term, $f = q^3 F / \left[4\pi\epsilon (k_B T)^2\right]$.

The advantage of this model is that it includes thermionic emission injection, image force lowering and surface recombination in a single analytic function instead of a set of differential equations coupled to balanced equilibrium equations, which is considerably convenient for fitting experimental data. This model, however, assumes that the considered carrier mobility is independent on the electric field, or that this dependence is implicit in $\psi$, which also has a weak dependence on $F$.

*2.1.4. Effects of disorder in the current injection*

The models for current injection presented so far were adapted from traditional models of charge injection into inorganic semiconductors. The energy barrier concepts used were, therefore, formulated by assuming the injection from the Fermi level of a metal to a continuum of delocalized states represented by the energy bands of the semiconductor (valence band for holes and conduction band for electrons). In amorphous inorganic semiconductors and organic semiconductors, due to the disorder characteristic of these materials, the electronic states available to be occupied by a charge carrier injected from the contact are better described as a distribution of localized states with a dispersion in energy. Moreover, in organic semiconductors (conjugated polymers or molecular solids), the localization of these states and their energy dispersion can be even more pronounced than in amorphous inorganic semiconductors, highlighting the influence of disorder in the charge injection and transport processes [18]. In this sense, it is reasonable to consider the use of models which include the effects of disorder to have a more realistic picture of the mechanism of charge injection into organic semiconductors.

Models considering thermally assisted tunnelling injection into localized states in a semiconducting polymer [19] and thermally limited injection current into a disordered molecular insulator via Monte Carlo simulations [20] were successfully developed to introduce the effects of disorder in the charge injection problem in organic semiconductors. A particularly interesting model, based on an analytic approach, was formulated by considering that the charge carriers are injected from the electrode into a distribution of localized hopping states in the organic semiconductor, followed by either electrode recapture or diffusion away from the attractive image potential [21].

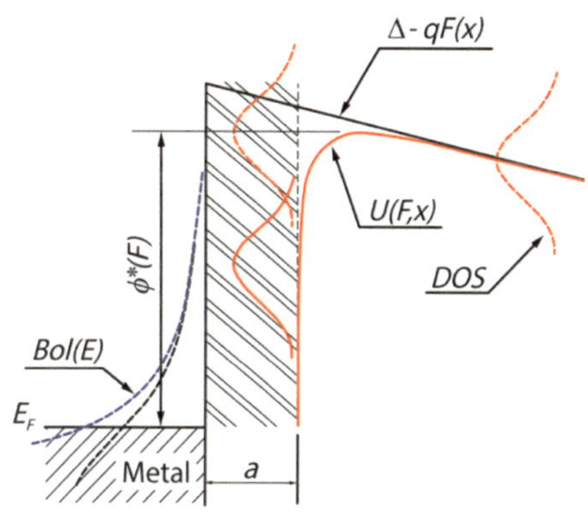


**Figure 4.** Representation of a metal/polymer contact in a model considering charge-carrier injection into a DOS. Dashed line Gaussian curves represent the distribution of available energy states in the semiconductor. The solid line Gaussian curve represents the first DOS after populated by an injected charge carrier, given by the convolution of the first unoccupied DOS and the edge of the Fermi distribution in the metal. The figure is adapted from Refs. [20, 21].

In this model, the initial step of the carrier injection process is the hoping from the Fermi level of an electrode, which is below the centre of the density of states (DOS) of the conjugated polymer by an energy $\Delta$, to a localized state at a distance $x_0 \geq a$ from the interface, where $a$ is the minimum distance between two neighbour hopping states. The potential, relative to the Fermi level of a metal, in a state of energy E, situated at a distance $x$ from the contact, is given by

$$U(E,x) = \Delta - \frac{q^2}{16\pi\varepsilon x} - qFx + E \quad (13)$$

Once injected into a state of energy $E_i$, the subsequent jump of the charge carrier is towards the easiest neighbour target state of energy $E_f$. If $E_f > E_i$, the probability for the jump follows a Boltzmann distribution and is independent on the energy of the previous state. Such process is also valid for the first jump from the electrode to the semiconductor. For sufficiently high energy barriers, the first step is the most difficult and, therefore, limits the injection rate. The subsequent drift-diffusion process inside the polymer is considered an equilibrium process, which determines the probability of the carriers to migrate into the bulk. **Figure 4** shows a schematic representation of the injection process from the Fermi level of a metal to an energetically and spatially disordered semiconductor.

The analytic dependence of this injection current is given by

$$J_{inj} = q\nu_0 \int_a^{\infty} dx_0 \exp(-2\gamma x_0) w_{esc}(x_0) \times \int_{-\infty}^{+\infty} dE' Bol(E') g(U_0(x) - E') \tag{14}$$

where $w_{esc}(x_0)$ is the probability of the carrier to avoid recombination at the interface, $a$, is the distance from the electrode to the first localized state, $\nu_0$, is the hopping frequency, $\gamma$, is the inverse of the carrier localization radius and the $Bol(E)$ function is defined as

$$Bol(E) = \begin{cases} \exp\left(-\dfrac{E}{k_B T}\right), E > 0 \\ 1, E \leq 0 \end{cases} \tag{15}$$

which means that the probability of the charge carrier to jump to a higher energy state has an Arrhenius-type dependence and to jump to a lower energy state is equal to 1.

The electrostatic potential energy at a distance $x$ from the electrode is

$$U_0(x) = \Delta - \frac{q^2}{16\pi\varepsilon x} - qFx \tag{16}$$

The escape probability $w_{esc}(x_0)$ at the interface can be evaluated by solving the unidimensional Fokker-Planck equation for the energy distribution given by Eq. (16). The result is

$$w_{esc}(x_0) = \frac{\int_a^{x_0} dx \exp\left[-\dfrac{q}{k_B T}\left(F_0 x + \dfrac{q}{16\pi\varepsilon}\dfrac{1}{x}\right)\right]}{\int_{-\infty}^{-\infty} dx \exp\left[-\dfrac{q}{k_B T}\left(F_0 x + \dfrac{q}{16\pi\varepsilon}\dfrac{1}{x}\right)\right]} \tag{17}$$

The advantage of this method as compared to the other models of charge injection into a disordered system is that it gives an analytic expression for the injection current, without the need of solving numerically a set of differential equations or performing complicated numerical simulations. **Figure 5** shows some results for the current density injected into a disordered organic semiconductor by considering Eqs. (14)–(17).

The current-density injection curves (arbitrary units) presented in **Figure 5a** were calculated at 300 K with a localization radius of $10^{-10}$ m, a first step distance of 0.6 nm, a dispersion of the Gaussian DOS of 80 meV and the energy barrier height Δ varying from 0.2 to 0.7 eV. In **Figure 5b**, Δ is fixed at 0.3 eV and the temperature is varied from 150 to 300 K, with all other parameters kept constant. An important feature from these results is that the current density does not have an Arrhenius dependence on the temperature, as expected from thermionic emission injection. A more detailed discussion of the results produced from this model is found in Ref. [21].

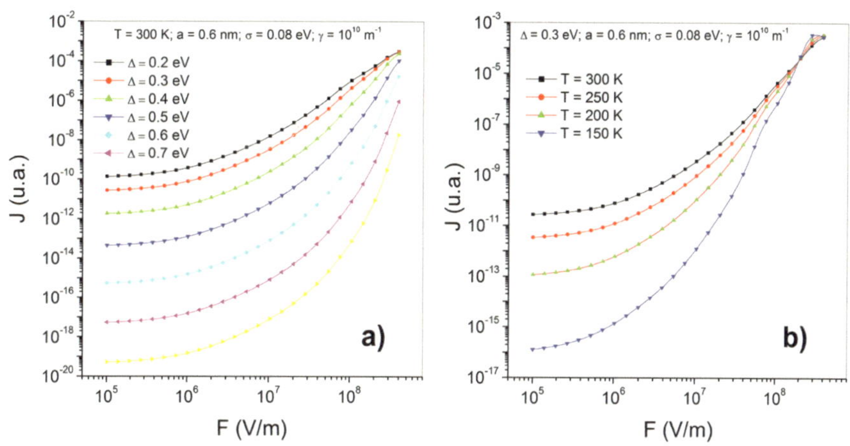


**Figure 5.** (a) Injection current density (arbitrary units) versus electric field simulated for different values of the zero-field energy height barrier (Δ) at 300 K. (b) Temperature dependence of the injection current density for a zero-field energy height barrier of 0.3 eV.

The model described above presents a very interesting view of the inclusion of disorder effects in the current injection from a metal into a semiconducting polymer, with evident influence on the temperature, especially in the low electric field regime. At high electric fields, the current density becomes almost independent on the temperature (**Figure 5b**), resembling a tunnelling injection dependency with, however, a very distinct functional dependence. To circumvent this problem, it is possible to conceive a model based on an electronic hopping-type injection into a DOS with the inclusion of the tunnelling probability across a triangular barrier represented by the potential at the contact interface [22]. This model considers that the carrier injection from the electrode is determined by the sum of the probabilities of each separate process to occur (hopping-type injection or tunnelling), resulting in better fitting to experimental results in a broader range of electric field and temperature. Ref. [22] is recommended for a more detailed view of this model.

### 2.2. Influence of the bulk transport on the current

In the previous section, we presented the most common models used to explain the current injection mechanisms in PLEDs and made a brief introduction to models which include the effects of disorder in the charge injection process. The total device current, however, depends on Eq. (1), which means that each contribution (injection or transport in the bulk) is individually a limiting process. If the amount of charges injected from the contacts does not change significantly the density of intrinsic charge carriers, the polymer film can easily transport these charges and the injection current determines the net device current. On the other hand, if the contacts are ohmic, the polymer bulk properties will dominate the electrical behaviour of the device. Therefore, in a real device, both contributions exist and the electrical characteristics of the device are determined by which process is dominant in each regime or, if they are comparable, to a combination of both. In this section, we will focus on the main characteristics of charge transport in the bulk and on the theoretical approaches used to fit experimental data.

### 2.2.1. Space-charge-limited currents (SCLC)

In PLEDs, the conjugated polymer which comprises the device active emissive layer is usually in its intrinsic, non-doped form. This means that the amount of intrinsic charge carriers is considerably low (due to the relatively high bandgap) and the organic semiconductor is poorly conductive or almost an insulator. In order to achieve enough current density to promote a sufficiently visible electroluminescence, a high external electric field must be provided (above $10^6$ V/m) to induce the injection of excess charge carriers into the organic semiconductor. Therefore, the operating characteristics of a PLED depend on the injection and transport of this excess charge injected in the polymer bulk. In the present subsection, we will focus on the transport characteristics of excess charge in PLEDs considering that the contacts are ohmic, i.e., they can provide as much charge as the dielectric volume can support.

For a semiconductor (or insulator) with ohmic contacts, the current due to the intrinsic charge carriers is determined by Ohm's law $J = qn\mu F$, where $F$ is the electric field, $n$ is the charge-carrier density in the correspondent band/molecular orbital and $\mu$ is the carrier mobility. This equation stands for single-carrier devices. For double-carrier devices, the carrier density is the sum of the densities of both charge carriers (electrons and holes), $n = n_e + n_h$, as well as the mobility is the sum of both mobilities, $\mu = \mu_e + \mu_h$. Additionally, since intrinsic charge carriers are thermally generated (geminate generation) and considering that the charge distribution is nearly uniform all over the polymer bulk, as a result from Poisson's equation, the local electric field is approximately constant, $F = V/L$, where $V$ is the external applied voltage and $L$ is the film thickness. Same situation occurs when the excess injected charge-carrier density is small as compared to the intrinsic charge-carrier density, resulting in a linear dependence of the device current on the applied voltage (and inverse dependence on the film thickness).

When the excess injected charge-carrier density exceeds considerably the intrinsic charge-carrier density, the local electric field is distorted and the dependence of the current on the applied voltage becomes non-linear. For a device that has an ohmic electrode for one type of charge carrier and a blocking electrode for the other, it is a good approximation to consider it a single-carrier device. If there are no charge-carrier traps in the semiconductor/insulator, the device net current is assumed to follow Mott-Gurney law [1, 23, 24]:

$$J = \frac{9}{8}\varepsilon\mu\frac{V^2}{L^3} \tag{18}$$

where $\varepsilon$ is the electric permittivity of the semiconductor. Eq. (18) is frequently known as trap-free (SCLC) [1, 24, 25]. In this simple picture, the device current is expected to be initially linear (when the charge injection is low and the current is dominated by the intrinsic charge carriers) and becomes quadratic as space-charge effects from excess injected charge carriers surpass the amount of intrinsic charge carriers. Typical analysis uses a log-log plot of the device current (or current density) versus voltage, where the current is expected to be represented by a straight line with slope equal to 1 (linear behaviour), changing to a straight line with slope equal to 2 (quadratic dependence) when excess injected charges become dominant. Additional confirmation of the dependence denoted by Eq. (18) is the analysis of the device current against the device thickness (where an inverse cubic dependence is expected).

The quadratic dependence from Eq. (18) happens only by considering that the carrier mobility is field independent. The charge-carrier mobility in polymers, however, due to the intrinsic disorder, is generally field dependent. A commonly found field dependence [26] is

$$\mu(F) = \mu_0 \exp(\gamma\sqrt{F}) \tag{19}$$

where $\mu_0$ is the zero-field carrier mobility and $\gamma$ is a parameter which provides the temperature dependence. The device current can then be evaluated by solving the following set of equations [27]:

$$J = n(x)q\mu(F(x))F(x) \tag{20}$$

$$\frac{\varepsilon}{q}\frac{dF(x)}{dx} = n(x) \tag{21}$$

where $n(x)$is defined as the carrier density at position $x$. Assuming ohmic contacts, $n(0)$ can be estimated from the effective carrier density of states in the HOMO (for holes) or LUMO (for electrons) of the polymer and can be used as boundary condition for Eqs. (20) and (21).

Empirically, it is also observed that the zero-field carrier mobility is thermally activated [27, 28]:

$$\mu_0(T) = \mu_{T_0} \exp\left(-\frac{\Delta}{k_B T}\right) \tag{22}$$

Fitting the experimental results for each temperature by using Eqs. (20) and (21), it is also possible to find the temperature dependence of the $\gamma$ parameter in Eq. (19), which is usually of the type [27, 28]:

$$\gamma(T) = \beta\left(\frac{1}{k_B T} - \frac{1}{k_B T_0}\right) \tag{23}$$

where $\beta$ and $T_0$ are empirical parameters evaluated from the fittings.

Eqs. (19)–(23) describe the electric field and temperature dependence on the charge-carrier transport in a single-carrier, trap-free semiconductor with field-dependent carrier mobility. Such model can be quite successfully used to describe the current due to holes in PLEDs, by considering that the barrier for hole injection from ITO or ITO/PEDOT electrodes to the HOMO of the polymer is usually small (almost ohmic) and that, in conjugated polymers, holes are not strongly trapped in the bulk. In a system where the charge-carrier trap distribution is in equilibrium with the correspondent band/molecular orbital, the traps are known as shallow traps. Shallow traps do not change significantly the electric field distribution and the net effect in the charge transport can be approximated by considering an effective mobility proportional to the trap-free mobility $\mu_{\mathrm{eff}} = \mu\theta$, where $\theta$ is independent or weakly dependent on the electric field, but dependent on the energy trap distribution [1, 24, 30–32].

Electrons, in contrast, are usually trapped in significantly deeper energy states within the energy bandgap, resulting in a much lower effective carrier mobility due to dispersive transport [29], and in a significantly non-uniform charge-carrier distribution, strongly influencing the local electric field distribution. The resulting transport current in an 'electron-only' single-carrier device can be evaluated from theories of space-charge-limited currents in an insulator with a distribution of trap states [1, 24, 30–32]. For low applied voltages, where the electron injection current is low, the intrinsic electrons dominate the charge transport, and the device current follows Ohm's law. When the excess-charge-injected carriers dominate the charge transport, the free-carrier density depends on the trap density of states that can be considered to depend exponentially on the energy:

$$n_t(E) = \left(\frac{N_t}{k_B T_t}\right)\exp\left(\frac{E - E_{\mathrm{LUMO}}}{k_B T_t}\right) \tag{24}$$

where $n_t(E)$ is the trap density of states at energy $E$, $E_{\mathrm{LUMO}}$ is the energy of the LUMO of the polymer, $N_t$, is the total density of traps and $k_B T_t$ is a temperature characteristic of the trap distribution. The trap distribution of Eq. (24) results in a trap-filling-limited (TFL) current dependence that can be expressed in the form [24, 33]:

http://dx.doi.org/10.5772/64358

$$J = N_{LUMO} q \mu_e \left( \frac{\varepsilon}{qN_t} \right)^r \frac{V^{r+1}}{L^{2r+1}} C(r) \tag{25}$$

where $N_{\mathrm{LUMO}}$ is the effective density of states in the LUMO, $r$ is a fitting parameter defined by $r = T_t/T$ and $C(r) = r^r(2r+1)^{r+1}(r+1)^{-(r+2)}$. Whilst Eq. (18) gives a quadratic dependence of the current on the applied voltage and the Eqs. (19)–(23), a slight deviation from quadratic behaviour, Eq. (25) can give much steeper current dependences on the electric field. For system with deep distributions of carrier traps, the log-log plots of the $J$ versus $V$ curves can result in straight lines with slopes much higher than 2. Experimental results from devices, which could be approximately considered as hole-only (ITO/PPV/Au) and electron-only (Ca/PPV derivative/Ca) devices (see **Table 2** for values of work function and HOMO and LUMO energies), were fitted by these equations to corroborate the validity in the study of separate electrical transport of electrons and holes in conjugated polymers [28, 33].

For an operating PLED, however, both charge carriers (electrons and holes) must be present in the polymer bulk. Such a double-carrier device can be manufactured by using good hole- and good electron-injecting electrodes sandwiching the organic semiconductor, like a ITO/PPV/Ca structure. To model the whole device current, we cannot simply consider the independent contributions for hole (expected to be SCLC) and for electron (considered to be TFL) transport because of two additional processes which have extreme importance in double-carrier devices: charge-carrier recombination and charge neutralization. The recombination term can be considered as Langevin type, or bimolecular, where the recombination rate is proportional to the product of both electron and hole densities. The charge neutralization effect results in a net charge within the polymer bulk much lower than the actual amount of charge injected from the electrodes, permitting higher charge injection currents than in a single-carrier device. Considering field-independent mobility for electrons and holes in a trap-free semiconductor/insulator, the SCLC in the 'plasma limit' [24] is expressed by:

$$J = \left( \frac{9\pi}{8} \right)^{1/2} \varepsilon \left( \frac{2q\mu_h\mu_e(\mu_h + \mu_e)}{\varepsilon B} \right)^{1/2} \frac{V^2}{L^3} \tag{26}$$

where $B$ is the bimolecular recombination constant. One important feature of Eq. (26) is that the higher the recombination, the lower the device current. Such behaviour occurs because, by increasing recombination, the amount of charge neutralization decreases. The expected dependence on the applied voltage and film thickness is, though, exactly the same for single-carrier devices in the trap-free SCLC regime. As described above, hole transport in a conjugated polymer can be considered not highly dependent on traps, but a field dependence on the carrier mobility is at least expected. Moreover, electrons are usually strongly trapped in such a way that Eq. (26) may hardly be directly applied in experimental results for PLEDs.

The inclusion of a field-dependent carrier mobility and of trapping effects in the transport mechanism of a double-carrier device complicates significantly the charge transport problem, making it impossible to find a simple analytic solution as in Eq. (26). The problem is then solved by considering that the contributions of the free charge-carrier distributions ($n_h$ $(x)$ for holes and $n_e(x)$ for electrons) to the total device current result in [34, 35]:

$$J = q\left[\mu_h\left(F\left(x\right)\right)n_h\left(x\right) + \mu_e\left(F\left(x\right)\right)n_e\left(x\right)\right]F(x). \tag{27}$$

These free-carrier distributions are related by Poisson equation:

$$\frac{\varepsilon}{q}\frac{dF(x)}{dx} = n_h\left(x\right) - n_e\left(x\right) - n_{t,e}(x) \tag{28}$$

where $n_{t,e}(x)$ is the density of trapped electrons. The continuity equation (considering steady-state regime) is determined by the bimolecular recombination

$$\frac{1}{q}\frac{dJ_e}{dx} = -\frac{1}{q}\frac{dJ_h}{dx} = Bn_h\left(x\right)n_e(x) \tag{29}$$

In this model, holes are not considered to be significantly trapped. Moreover, Eq. (29) says that only mobile electrons contribute to charge-carrier recombination. If the contacts are ohmic, the carrier density at the interfaces can be considered as the density of states in the correspondent molecular orbitals, $n_h(x = 0) = N_{\text{HOMO}}$ and $n_e(x = L) = N_{\text{LUMO}}$ (the hole-injecting electrode is at $x = 0$). These conditions are used as boundary conditions to find the solution of Eqs. (27)–(29). The field and temperature dependence of the carrier mobility is provided by Eqs. (19), (22) and (23). By considering the exponential distribution of traps for electrons from Eq. (24), it is possible to obtain the density of trapped electrons

$$n_{t,e}\left(x\right) = N_t\left[\frac{n_e(x)}{N_{\text{LUMO}}}\right]^{T/T_t} \tag{30}$$

To properly determine the current characteristics of a double-carrier device using the above equations, one must first produce and characterize hole-only [27] and electron-only [33] devices, varying the temperature and device thickness in order to obtain the needed parameters from Eqs. (22) and (23) (for holes) and (24) and (25) (for electrons). Therefore, Eqs. (27)–(29) can be solved numerically by considering Eq. (30), with the bimolecular constant $B$ as the unknown parameter to be determined. This procedure was applied in the fitting of experimental data of double-carrier devices, in order to find optimum mobility values for PLED performance [35, 36].

### 2.2.2. *Effects of disorder in the charge-carrier transport*

The previous subsection presented a discussion on how the effects of space-charges in the bulk affect the charge-carrier transport and, consequently, the device current in dc current-voltage measurements. The models considered, though, a field-independent mobility (which is hardly found in organic semiconductors) or an empirical temperature and local electric field dependence resembling a Poole-Frenkel behaviour, represented by the combination of Eqs. (19), (22) and (23),

$$\mu(T,F) = \mu_{T_0} \exp\left(-\frac{\Delta}{k_B T}\right) \exp\left[\beta F^{1/2}\left(\frac{1}{k_B T} - \frac{1}{k_B T_0}\right)\right] \tag{31}$$

Despite the relatively good consistency with experimental results, the arguments used to justify such dependence are still controversial. The Poole-Frenkel behaviour is considered as a consequence of the electric field lowering of the potential barrier that a carrier needs to overcome to leave a charged centre. An argument against the attribution of the mobility dependence to Poole-Frenkel effect is that such behaviour is observed in a large variety of materials which, in most of the cases, the amount of charged traps in the bulk can be considered negligible.

A more reasonable consideration is that such dependence comes from the disorder inherent of organic semiconductors. The charge transport mechanism in these materials is then associated with the hopping of carriers within a randomly positioned and energetically disordered system of localized states [2, 37]. Most of the models which consider hopping between localized states are based on Miller-Abrahams expression [38] for the probability of a tunnelling carrier jump over a distance $r$ separating an initial state at energy $E_i$ and a final state at energy $E_f$, written as

$$\mathrm{P}(r,E_i,E_f) = \nu_0 \exp(-\gamma r) \times \begin{cases} \exp\left(-\frac{E_f - E_i}{k_B T}\right), E_f > E_i \\ 1, E_f < E_i \end{cases} \tag{32}$$

where $\gamma$ is the inverse of the localization radius of the electronic wavefunction and $\nu_0$, is the attempt-to-jump frequency. The spatial dependence from Eq. (32) says that the carrier jump rate decreases exponentially with increasing distance between hopping sites. Moreover, the symmetry on position means that forward or backward jumps are equally probable and, therefore, within a hypothetic system where all hopping sites have the same energy, the trapping and releasing rates would be equal. On the other hand, the energetic dependence is not symmetrical. Upward jumps are thermally activated, whereas downward jumps dissipate the excess energy in the form of phonon emission. The consequence is that the upward jumps are much slower than downward jumps, resulting in totally different trapping and releasing rates for forward and backward jumps between two fixed states at different energy levels. The introduction of energetic disorder in a localized states hopping system, therefore, influences strongly the carrier transport behaviour.

The analytic treatment of carrier hopping in such a system described above can be quite complicated, since it is virtually impossible to consider the evaluation of all possible spatial and energetic configurations. However, approximate methods for a simplified picture of the problem are frequently applied. A particularly efficient method is based on the effective transport level [39], which effectively reduces the problem of hopping within a spatially and energetically disordered system into a trap-controlled transport mechanism.

Analytic models considering the random hopping of the carrier within an energetically and positionally disordered system were formulated to study the concentration and temperature dependence of the carrier mobility considering an arbitrary DOS distribution and in the high density of carriers regime [40, 41]. However, the direct application of the results to the set of equations of SCLC in Section 2.2.1 is still quite limited. Models based on Monte Carlo simulations were used to introduce the concept of spatially correlated energetic disorder in a system, assuming a Gaussian DOS distribution [42–44]. The results show that the obtained dependence of the carrier mobility on the electric field resembles the Poole-Frenkel-type behaviour in a quite wide electric field range.

## 3. Conclusion

The basic models of the operating mechanisms of PLEDs presented in this chapter can be employed in the interpretation of experimental results from dc electrical characterization of PLEDs, allowing the extraction of intrinsic polymer parameters like charge carrier mobility or structure-dependent parameters like the energy barrier height at a contact. To properly investigate the results from the electrical characterization of a PLED, however, a critical analysis procedure must be considered. First of all, knowledge of the undistorted energy bands of the polymer and the work function of the contacts is needed. From these information, it is possible to know if the assumption of ohmic contacts considered for a SCLC analysis is valid or not. If the contacts are not ohmic, a combined contribution from thermally activated injection and tunnelling must be considered and used to give the actual boundary conditions for the carrier density at the contacts. If thermionic emission and tunnelling cannot provide the necessary functional injection current, it can be estimated from the presented models considering interface recombination or spatial and energetic disorder in the first carrier jump in the semiconductor. For modelling a double-carrier device, where bimolecular recombination effects take place, the ideal approach is to first analyse, separately, the electrical transport in both single-carrier devices (electron-only and hole-only), with metallic electrodes as close as possible to the ohmic contact condition. Models considering spatial and energetic disorder in the polymer bulk, though are more accurate and complete from the theoretical viewpoint, are still difficult to be directly applied to the fitting of experimental dc current-voltage curves, being not applicable to all operating regimes.

## Acknowledgements

The authors are grateful to the research support from FAPESP-Brazil (grants 2013/24461-7 and 2008/57706-4) and to the contribution from Universidade Estadual Paulista–UNESP (PROPe/PROPG), which made possible the publication of this material.

## Author details

Lucas Fugikawa Santos[1*] and Giovani Gozzi[2]

*Address all correspondence to: lucas@sjrp.unesp.br

1 Physics Department, Campus of São José do Rio Preto, State University of São Paulo–UNESP, São Paulo, Brazil

2 Physics Department, Campus of Rio Claro, State University of São Paulo–UNESP, São Paulo, Brazil

## References

[1] Kao K.C.; Hwang W. Electrical Transport in Solids, With Particular Reference to Organic Semiconductors (International Series in the Science of the Solid States). 1st ed. Oxford: Pergamon Press; 1981. 660 pp.

[2] Pope M.; Swenberg C.E. Electronic Processes in Organic Crystals and Polymers. 2nd ed. Oxford: Oxford University Press; 1999. 1328 pp.

[3] Su W.P.; Schrieffer J.R.; Heeger A.J. Solitons in polyacetylene. Physical Review Letters. 1979;42(25):1698–1701. DOI: 10.1103/PhysRevLett.42.1698

[4] Su W.P.; Schrieffer J.R.; Heeger A.J. Solitons in polyacetylene. Physical Review B. 1980;22(4):2099–2111. DOI: 10.1103/PhysRevB.22.2099

[5] Heeger A.J.; Kivelson S.; Schrieffer J.R.; Su W.P. Solitons in conducting polymers. Reviews of Modern Physics. 1998;60(3):781–851. DOI: 10.1103/RevModPhys.60.781

[6] Burroughes J.H.; Bradley D.D.C.; Brown A.R.; Marks R.N.; Mackay K.; Friend R.H.; Burns P.L.; Holmes A.B. Light-emitting diodes based on conjugated polymers. Nature. 1990;347(6293):539–541. DOI: 10.1038/347539a0

[7] Tang C.W.; VanSlyke S.A. Organic electroluminescent diodes. Applied Physics Letters. 1987;51(12):913–915. DOI: 10.1063/1.98799

[8] Tsutsui T.; Takada N. Progress in emission efficiency of organic light-emitting diodes: basic understanding and its technical application. Japanese Journal of Applied Physics. 2013;52:110001. DOI: 10.7567/JJAP.52.110001

[9] Gather M.C.; Reineke S. Recent advances in light outcoupling from white organic light-emitting diodes. Journal of Photonics for Energy. 2015;5:057607. DOI: 10.1117/1.JPE.5.057607

[10] Lu C.; Pi S.; Meng H. Effect of defect-enhanced molecular oxygen adsorption on the imbalance of hole versus electron mobility in conjugated polymers. Physical Review B. 2007;75(19):195206. DOI: 10.1103/PhysRevB.75.195206

[11] Parker I.D. Carrier tunneling and device characteristics in polymer light-emitting diodes. Journal of Applied Physics. 1994;75(3):1656–1666. DOI: 10.1063/1.356350

[12] Santos L.F.; Bianchi R.F.; Faria R.M. Electrical properties of polymeric light-emitting diodes. Journal of Non-Crystalline Solids. 2004;338–340(1 Spec. Iss.):590–594. DOI: 10.1016/j.jnoncrysol.2004.03.048

[13] Campbell I.H.; Smith D.L. Physics of polymer light emitting diodes. In: Hadziioannou, G.; van Hutten, P.F., editors. Semiconducting Polymers. 1st ed. Weinheim: Willey-VCH; 2000. pp. 333–364.

[14] Davids P.S.; Campbell I.H.; Smith D.L. Device model for single carrier organic diodes. Journal of Applied Physics. 1997;82(12):6319–6325. DOI: 10.1063/1.366522

[15] Crone B.K.; Campbell I.H.; Davids P.S.; Smith D.L. Charge injection and transport in single-layer organic light-emitting diodes. Applied Physics Letters. 1998;73(21):3162–3164. DOI: 10.1063/1.122706

[16] Malliaras G.G.; Scott J.C. The roles of injection and mobility in organic light emitting diodes. Journal of Applied Physics. 1998;83(10):5399–5403. DOI: 10.1063/1.367369

[17] Scott J.C.; Malliaras G.G. Charge injection and recombination at the metal-organic interface. Chemical Physics Letters. 1999;299(2):115–119. DOI: 10.1016/S0009-2614(98)01277-9

[18] da Silva Filho D.A.; Olivier Y.; Coropceanu V.; Brédas J.L.; Cornil J. Theoretical aspects of charge transport in organic semiconductors. In: Bao, Z.; Locklin, J., editors. Organic Field Effect Transistors. 1st ed. Boca Raton: CRC Press Taylor and Francis; 2007. pp. 1–72.

[19] Abkowitz M.A.; Mizes H.A.; Facci J.S. Emission limited injection by thermally assisted tunneling into a trap-free transport polymer. Applied Physics Letters. 1995;66(10):1288–1290. DOI: 10.1063/1.113272

[20] Gartstein Y.N.; Conwell E.M. Field-dependent thermal injection into a disordered molecular insulator. Chemical Physics Letters. 1996;255(1–3):93–98. DOI: 10.1016/0009-2614(96)00359-4

[21] Arkhipov V.I.; Emelianova E.V.; Tak Y.H.; Bässler H. Charge injection into light-emitting diodes: theory and experiment. Journal of Applied Physics. 1998;84(2):848. DOI: 10.1063/1.368146

[22] Gozzi G.; Queiroz E.L.; Zucolotto V.; Faria R.M. Hopping–tunneling model to describe electric charge injection at metal/organic semiconductor heterojunctions. Physica Status Solidi (B). 2015;252(2):404–410. DOI: 10.1002/pssb.201451556

[23] Mott N.F.; Gurney R.W. Electronic Process in Ionic Crystals. 2nd ed. Oxford: Oxford University Press; 1948. 275 p.

[24] Lampert M.A.; Mark P. Current Injection in Solids. 1st ed. New York: New York Academic Press; 1970. 351 p.

[25] Parmenter R.H.; Ruppel W. Two-carrier space-charge-limited current in a trap-free insulator. Journal of Applied Physics. 1959;30(10):1548–1558. DOI: 10.1063/1.1734999

[26] Pai D.M. Transient photoconductivity in poly(N-vinylcarbazole). Journal of Chemical Physics. 1970;52(5):2285–2291. DOI: 10.1063/1.1673300

[27] Blom P.W.M.; de Jong M.J.M.; van Munster M.G. Electric-field and temperature dependence of the hole mobility in poly(p-phenylene vinylene). Physical Review B. 1997;55(2):R656–R659. DOI: 10.1103/PhysRevB.55.R656

[28] Gill W.D.; Kanazawa K.K. Transient photocurrent for field-dependent mobilities. Journal of Applied Physics. 1972;43(2):529–534. DOI: 10.1063/1.1661151

[29] Santos L.F.; Faria R.M.; de Andrade A.R.; Faria G.C.; Amorin C.A.; Mergulhao S. Transition from dispersive to non-dispersive transport of holes in poly(2-methoxy-5-(2′-ethyl-hexyloxy)-1,4-phenylene vinylene) light-emitting diodes investigated by time of flight measurements. Thin Solid Films. 2007;515(20–21):8034–8039. DOI: 10.1016/j.tsf.2007.03.052

[30] Rose A. Space-charge-limited currents in solids. Physical Review. 1955;97(6):1538–1544.

[31] Lampert M.A. Simplified theory of space-charge-limited currents in an insulator with traps. Physical Review. 1956;103(6):1648–1656. DOI: 10.1103/PhysRev.103.1648

[32] Mark P.; Helfrich W. Space-charge-limited currents in organic crystals. Journal of Applied Physics. 1962;33(1):205–215. DOI: 10.1063/1.1728487

[33] Blom P.W.M.; de Jong M.J.M.; Vleggaar J.J.M. Electron and hole transport in poly(p-phenylene vinylene) devices. Applied Physics Letters. 1996;68(23):3308–3310. DOI: 10.1063/1.116583

[34] Blom P.W.M.; de Jong M.J.M.; Breedijk S. Temperature dependent electron-hole recombination in polymer light-emitting diodes. Applied Physics Letters. 1997;71(7): 930–932. DOI: 10.1063/1.119692

[35] Blom P.W.M.; de Jong M.J.M. Electrical characterization of polymer light-emitting diodes. IEEE Journal of Selected Topics in Quantum Electronics. 1998;4(2):105–112. DOI: 10.1109/2944.669477

[36] Blom P.W.M.; Vissenberg M.C.J.M.; Huiberts J.N.; Martens H.C.F.; Schoo H.F.M. Optimum charge-carrier mobility for a polymer light-emitting diode. Applied Physics Letters. 2000;77(13):2057–2059. DOI: 10.1063/1.1313254

[37] Bässler H. Charge transport in disordered organic photoconductors – a monte-carlo simulation study. Physica Status Solidi B. 1993;175(1):15–56. DOI: 10.1002/pssb.2221750102

[38] Miller A.; Abrahams E. Impurity conduction at low concentrations. Physical Review. 1960;120(3):745–755. DOI: 10.1103/PhysRev.120.745

[39] Monroe D. Hopping in exponential band tails. Physical Review Letters. 1985;54(2):146–149. DOI: 10.1103/PhysRevLett.54.146

[40] Arkhipov V.I.; Heremans P.; Emelianova E.V.; Adriaenssens G.J.; Bassler H. Weak-field carrier hopping in disordered organic semiconductors: the effects of deep traps and partly filled density-of-states distribution. Journal of Physics-Condensed Matter. 2002;14(42):9899–9911. DOI: 10.1088/0953-8984/14/42/305

[41] Arkhipov V.I.; Emelianova E.V.; Heremans P.; Adriaenssens G.J. Equilibrium hopping conductivity in disordered materials. Journal of Optoelectronics and Advanced Materials. 2002;4(3):425–436.

[42] Gartstein Y.N.; Conwell E.M. High-field hopping mobility in molecular-systems with spatially correlated energetic disorder. Chemical Physics Letters. 1995;245(4–5):351–358. DOI: 10.1016/0009-2614(95)01031-4

[43] Gartstein Y.N.; Conwell E.M. Off-diagonal disorder and activation-energy of high-field hopping motion. Physical Review B. 1995;51(11):6947–6952. DOI: 10.1103/PhysRevB.51.6947

[44] Rakhmanova S.V.; Conwell E.M. Mobility variation with field in conducting polymer films. Synthetic Metals. 2001;116(1–3):389–391. DOI: 10.1016/S0379-6779(00)00443-4

Chapter 8

# Properties of Metal Oxide Pigments Surface Modified with Polyaniline Phosphate and Polypyrrole Phosphate in Corrosion Protective Organic Coatings

Andréa Kalendová, Tereza Hájková,
Miroslav Kohl and Jaroslav Stejskal

Additional information is available at the end of the chapter

http://dx.doi.org/10.5772/63290

**Abstract**

The objective of this work was to describe the properties of metal oxide–based pigments whose surface has been coated with a conductive polymer, if used as pigments in organic coating materials. The perovskite-type pigments were synthesized by high-temperature solid-phase reaction, and their surface was modified with a conductive polymer, specifically polyaniline phosphate or polypyrrole phosphate, by chemical oxidative polymerization. The pigments' structure and physicochemical properties were examined by XRD, XRF, and SEM. The composite pigments (pigment/conductive polymer) were dispersed in a solvent-type epoxy-ester resin binder to obtain a series of paints whose anticorrosion properties were assessed by means of corrosion tests in simulated corrosion atmospheres and by the linear polarization method. The anticorrosion and mechanical properties of the paints were compared with those of a paint containing the routinely used zinc phosphate hydrate pigment as a reference material. The pigments were found to possess a high anticorrosion efficiency, comparable to or higher than that of the commercially available zinc phosphate–based anticorrosion pigment. The highest anticorrosion efficiency was observed with the paints containing the Ca–Ti, Sr–Ti, and Sr–Mn perovskite pigments modified with polypyrrole phosphate.

**Keywords:** organic coating, anticorrosion pigment, polyaniline phosphate (PANI), polypyrrole phosphate (PPY), perovskite

INTECH
open science | open minds

## 1. Introduction

The most widespread method to protect the surfaces of metallic materials consists in coating them with paints possessing anticorrosion properties [1, 2]. The mechanism consists in an electrochemical reaction of the anticorrosion pigment with the metal itself or with the corrosive substances penetrating through the paint film from the outer atmospheric environment [3]. Corrosion-inhibiting pigments include materials that are actively engaged in the paint film lifetime extension [4]. Anticorrosion pigments acting on the chemical principle are soluble to some extent, contain water-soluble components that may maintain a constant pH in the paint layer, such as is more favorable for reducing the corrosion rate [5]. Electrochemically acting anticorrosion pigments passivate the substrate metal protected by the layer of the organic coating, acting either in the anodic region or in the cathodic region.

Many of the compounds are toxic or not entirely harmless/environment-friendly. So, new anticorrosion pigments are sought, equally efficient as the existing ones, but free from the harmful heavy metals (Pb, Cr (VI), Cd, Ni ...) [6–8]. Nontoxic pigment types started to be sought in the 1960s, but no equivalent substitutes for lead or chromium (VI)-based pigments have been found so far [9].

Among substances showing promise from the anticorrosion protection aspect are conductive polymers [10]. Added to paints on their own, conductive polymers cause certain troubles which must be addressed, for example, their effect on the liquid paint stability, paint adhesion if exposed to a high humidity/moisture, good dispersion in the binder and in the film, and the conductive polymer's efficient concentration in the organic binder to hinder corrosion of the substrate metal. So, it is frequently more convenient, rather than use the conductive polymer alone, to use composite pigment particles where a layer of the conductive polymer is deposited on an inorganic carrier particle. Metal oxides can be used as inorganic carrier particles.

Metal oxides of predominantly ionic nature of the MO (or $M_2O$) type include, for example, MgO, SrO, or NiO, which are not routinely used as ingredients in organic coating materials and paints. The $M_3O_4$ type is an exception, reminding of double oxides, such as the oldest anticorrosion pigment minium, $Pb_3O_4$, in which lead exists in two oxidation states. The other type includes oxides whose lattice, in contrast to the former type, is dissimilar to the structure of simple compounds. Most important in this group are three oxide types: ilmenite, $FeTiO_3$; spinel, $FeAl_2O_4$; and perovskite $CaTiO_3$ [11]. The basic formula of perovskites is $ABO_3$, where cation $A$ is relatively bulky and its valency is low (e.g., $Sr^{2+}$, $Ca^{2+}$, $La^{3+}$), whereas cation $B$ is relatively small (e.g., $Ti^{4+}$, $Zr^{4+}$, $W^{6+}$, $Fe^{3+}$, $Mn^{3+}$, $Zn^{2+}$). The perovskite-type oxides, whose properties can be modified through the selection of the structural lattice-forming elements, were selected owing to the stability of their physical and chemical properties [12, 13], insolubility, and thermal stability. It is also an asset of perovskites that they can be synthesized from a wide range of starting materials that are nontoxic and are reasonably environment-friendly. The $Ca^{2+}$and $Sr^{2+}$ cations were selected owing to their properties, which may be beneficial in suppressing corrosion of the metal surface beneath the paint film. The choice of those cations (Ca or Sr) in the perovskite structure can also be made use of to improve the inhibiting behavior of the pigments, viz., through their rate or ease with which the cation can be released from the

http://dx.doi.org/10.5772/63290

elementary lattice to act in the paint film. The perovskite carrier as well as the conductive polymer should modify the electric conductivity of the composite pigment particles and affect favorably the properties of the paint film when hindering electrochemical corrosion in environments with enhanced humidity/moisture, acid-nature corrosive substances, and corrosion initiators. If the perovskites prove to be good carriers for the conductive polymer layers (pigment/conductive polymer system), they may open up the door to additional interesting applications of the pigments as well as of the conductive polymers [10, 14].

## 2. Formulation of the anticorrosion organic coatings with polyaniline (PANI) phosphate

Thanks to its high electrical conductivity, PANI phosphate (hereinafter referred to as "polyaniline" or "PANI"), a conducting polymer, appears to be a possible inhibitor of corrosion reactions. Because all redox processes involving polyaniline are based on a transfer of electrons, the concentration of PANI is likely to be also important. If its inhibition properties and efficiency were proven in a paint film, it would suggest its prospective applications in organic binders where it would function as an organic corrosion inhibitor or in the form of an anticorrosion pigment.

Analyses of results of the laboratory tests of corrosion resistance concluded that the best anticorrosion efficiency was displayed by an epoxy-ester coating pigmented with polyaniline to pigment volume concentration, $PVC_{PANI}$ = 15 vol.%. This concentration facilitates the best anticorrosion effects of PANI in all the applied corrosion environments and approximately corresponds to the percolation threshold of conducting polyaniline dispersed in the nonconducting matrix (**Figure 1**) [15–17].

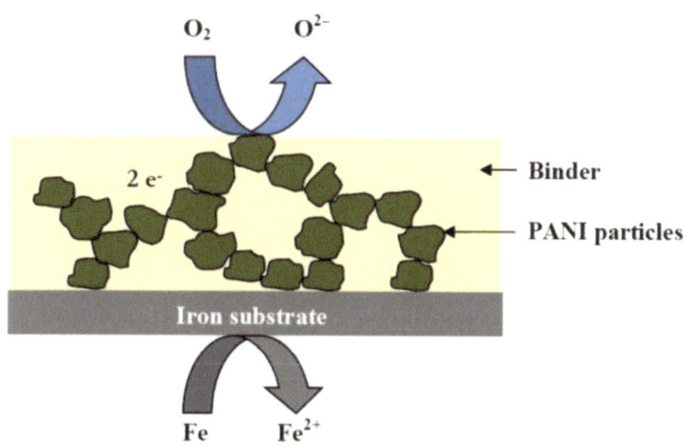


**Figure 1.** The possible role of polyaniline, at concentrations above the percolation threshold in the corrosion.

The pigments affect not only the anticorrosion properties of the coating materials but also the physicomechanical properties of the coating films. The role of percolation threshold of polymer-conductive materials can be connected with the role of the CPVC value of the powdery pigment and with the optimum performance properties of the pigmented paint film (**Figure 2**). When present in a volume concentration much less than the critical level, the pigment particles are isolated, mutually separated by the binder. Such a coating material is glossy and nearly impermeable by water vapor. As the PVC is increased, the pigment particles become closer to one another, although still separated by the binder. The film acquires hardness and tensile strength, while its elongability and gloss decrease. The properties of the film change depending on the type of pigment, till the PVC reaches its critical value (PVC = CPVC). At this point, the amount of binder is just sufficient to fill the space between the pigment particles. When the pigment concentration is further increased, the amount of binder is insufficient to fill the space between the pigment particles. The film contains air, and its properties change dramatically. Mechanical properties fail, and substrate metal corrosion and film blistering are appreciable in cases where the coating contains inorganic pigments. The optimum performance of pigmented coating is usually achieved below the CPVC value [18].

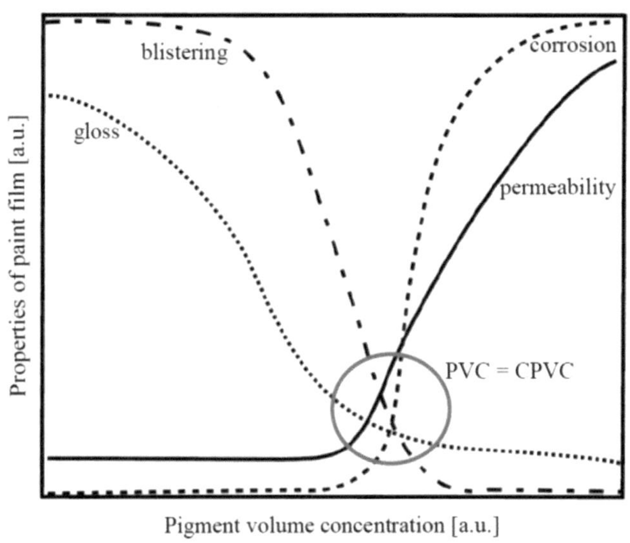


**Figure 2.** Coating properties depending on the degree of pigmentation (PVC). Notes: (a) The dependence of the properties of a pigmented film on PVC.

The goal of this study was to prepare a series of perovskites, coat their surfaces with layers of a conductive polymer–polyaniline phosphate (PANI) or polypyrrole phosphate (PPY) to obtain composite pigments (pigment/conductive polymer) with anticorrosion properties, and use them in paints for metallic substrate corrosion protection. The pigments with particles coated with a layer of a conductive polymer should exhibit properties of the conductive polymer, in particular, a higher conductivity for active protection of the metal against electrochemical corrosion, anticorrosion efficiency when exposing the system to corrosive substances with pH in the acid region, and passivation of the metal surface during corrosive damage of the organic coating.

## 2.1. Experimental

### *2.1.1. Laboratory preparation of the perovskite-type pigments*

Perovskites with a generally isometric particle [13, 19] shape were synthesized to serve as the pigment cores for coating with conductive polymers and to be added to a binder to form anticorrosion paints. Pigments possessing the simple perovskite structure, *Me*$TiO_3$, *Me*$MnO_3$ (*Me* = Ca, Sr), were synthesized by calcination (Eqs. (1)–(2)).

$$MeCO3 + TiO2 \rightarrow MeTiO3 + CO2 \qquad T = 1000\ °C,\ 1180\ °C \qquad (1)$$

$$2\ MeCO3 + Mn2O3 \rightarrow 2\ 0{,}MeMnO3 + CO2 \qquad T = 1000\ °C,\ 2x\ 1180\ °C \qquad (2)$$

The materials for the preparation of the pigments were as follows: titanium oxide (Precheza a.s. Prerov, CZ, composition: $TiO_2$ anatase); calcium carbonate (Omya a.s. Austria, composition: $CaCO_3$,natural calcite); strontium carbonate (Sigma-Aldrich Chemie, Germany, composition: $SrCO_3$); manganese (III) oxide (Sigma-Aldrich Chemie, Germany, composition: $Mn_2O_3$).

The pigments were synthesized by solid-phase reaction, viz., by high-temperature calcination of the homogenized mixtures of the starting materials [20] by following the general principles of preparation of high-temperature inorganic pigments [21]. The process of preparing the pigments consists of four operational steps: homogenization of starting compound mixtures, calcination procedure, leaching the calcination products by washing with water, and adapting the product to obtain the size of particles as necessary by a wet grinding process. The process was conducted as a two-stage procedure: the pigments were first calcined at 1000°C for 2 h and then at 1180°C. Since a suitable size of the pigment particles is a very important factor, the calcination step was followed by wet milling, performed in a planetary ball mill Pulverisette 6 (Netzsch, Germany). The pigment powder was placed in a milling container made from zircon-silicate ceramics and milled with rollers made from the corundum ceramics. The rotation speed was 400 rpm, and the process was conducted for 4–5 h. The milled pigments were followed rinsed with a multiply larger volume of distilled water followed by drying at 105°C in a hot-air dryer for 10 hours.

### *2.1.2. Perovskite structure examination*

The structure of the perovskites was examined by X-ray diffraction (XRD) analysis. The results gave evidence that the required structure had been attained. $ABO_3$ was found to be the majority phase in most of the pigments. Some of the pigments contained traces of the starting $TiO_2$or of reaction by-products. The pigment $CaTiO_3$ contained the main crystalline phase of $CaTiO_3$ and a small amount of rutile ($TiO_2$), $Ca(OH)_2$, and$CaCO_3$; $SrTiO_3$ contained the $SrTiO_3$ crystalline phase by $TiO_2$ (rutile) and $Sr_3Ti_2O_7$; $CaMnO_3$ contained the $CaMnO_3$ crystalline phase; and $SrMnO_3$ contained the $SrMnO_3$ crystalline phase and a small amount of $CaMn_2O_4$.

The results gave evidence that the pigments intended for surface modification with the conductive polymers had been obtained as intended, and are in agreement with previous studies [22].

*2.1.3. Laboratory preparation of pigments with conductive polymer surface layers*

Four perovskite-type pigments, $CaTiO_3$, $SrTiO_3$, $CaMnO_3$,and $SrMnO_3$, were subjected to surface treatment with the conductive polymers with a view to optimizing (enhancing) their anticorrosion properties.

*2.1.3.1. Preparation of the perovskites modified with a surface layer of polyaniline phosphate*

The pigment was suspended in 250 ml of 0.2 M aniline solution in 0.8 M *ortho*-phosphoric acid, and 250 ml of 0.25 M ammonium peroxodisulfate, also in 0.8 M *ortho*-phosphoric acid, was added to initiate the aniline polymerization process. The suspension was stirred for 1 h during which aniline polymerized on the surface of the pigment particles. Next day, the solids were filtered out and rinsed with 0.4 M phosphoric acid followed by acetone. The pigment particles coated with the polyaniline phosphate (PANI) overlayer were dried in air and then at 60°C in a laboratory drier [23].

*2.1.3.2. Preparation of the perovskites modified with a surface layer of polypyrrole phosphate*

The pigment was suspended in 250 ml of distilled water with 0.8 M *ortho*-phosphoric acid, and 0.2 M pyrrole was added. The system was stirred vigorously by using a glass stirrer, and an oxidant solution consisting of 0.25 M ammonium peroxodisulfate in 250 ml of distilled water was added. The whole solution was stirred for approximately 1 h. Next day, the solids were filtered out and rinsed with 0.4 M phosphoric acid followed by acetone. The pigment particles coated with the polypyrrole phosphate (PPY) overlayer were dried in air and then at 60°C in a laboratory drier [23].

*2.1.4. Characterization of the composite pigments containing a layer of a conductive polymer—PANI or PPY*

A total of eight perovskite-type pigments with their surfaces modified with PANI or PPY were prepared and subjected to X-ray fluorescence (XRF) analysis on a Philips PW 1404 X-ray spectrometer equipped with a Rh cathode, in conjunction with UniQuant software enabling 74 elements (from fluorine to uranium) semiquantitatively determined (10% relative error). XRD spectra of the synthesized perovskites were measured on an X'Pert PRO MPD 1880 X-ray diffractometer (PANanalytical, the Netherlands). The diffraction data were evaluated by means of the X'Pert programs (X'Pert HighScore Plus Software version 2.1b and X'Pert Industry Software version 1.1g); the phases were identified using data from the ICCD PDF2 diffraction database. The pigment surface and particle shape were examined on a JEOL-JSM 5600 LV scanning electron microscope (JEOL, Japan) in the secondary electron mode [24].

### *2.1.5. Determination of the physicochemical properties of the pigments with conductive polymer surface layers*

Determination of particle size and the distribution of pigment particle size were identified by means of Mastersizer 2000 (Malvern, Instruments Ltd., UK), which is able to measure the distribution of particle sizes from 0.01 to 2000 μm. Particle size is represented by the diameter of the equivalent sphere, that is, sphere whose laser radiation dispersion patterns are identical with those of the particle in question. The pigments' specific weights were determined by using a AccuPyc II 1340 gas pycnometer (Micromeritics, USA). Linseed oil absorption was measured by the pestle-mortar method. The outcome, called the oil number (in g/100 g), is a prerequisite for calculation of the CPVC and for the formulation of the paints [25, 26]. The determination of the pH level of aqueous extracts of the pigments was based on the ISO 789-9 standard. 10% pigment suspensions in redistilled water (pH = 7) were prepared and measured periodically during 28 days, after which they were filtered, and the ultimate (constant) pH value of the filtrate ($pH_p$), was recorded. Specific electrical conductivity of the 10% pigment suspensions ($\chi_P$) was based on the ISO 787-14 standard. The pH values of aqueous extracts ($pH_f$) prepared from 10% suspensions of loose paint films and the specific electric conductivity values of aqueous extracts ($\chi_f$) prepared from suspensions of loose paint films at PVC=1, 5, 10, and 15% were determined by the same method. The water-soluble fraction was measured gravimetrically by extraction of the powdered pigment, weighed with a precision of ± 0.01 g in distilled water at 20°C ($W_{20}$). This procedure was derived from the ČSN EN ISO 787-3 standard.

### *2.1.6. Assessment of the anticorrosion efficiency of the pigments with conductive polymer surface layers*

Model solvent-based epoxy-ester resin-based paints were formulated for investigation of the pigments' anticorrosion properties. Description of binder is as follows: a 60% solution of a medium high-molecular weight epoxy resin esterified with a mixture of fatty acids of dehydrated ricin oil and soy oil, trade name WorléeDur D 46, acid number 4, viscosity 2.5–5.0 Paṡs, flow time (DIN 53211-4200) 250 s. The PVCs in the paints were invariably 1, 5, 10, and 15%. The PVC/CPVC ratio was adjusted in all the model paints to 0.50 by means of the anticorrosion-neutral filler calcite $CaCO_3$. The total pigment plus filler concentration in the paint film was 50%, whereby a constant total concentration of the powder fractions in the dry paint film was assured, while varying only the proportion of the composite pigment. The paints were prepared by dispersing the powders in the liquid binder in a pearl mill Dispermat CV (VMA Getzmann GmbH Verfahrenstechnik, Germany). Co-octoate in a fraction of 0.3 wt.% was used as the siccative.

Test samples were prepared by applying the paint to steel panels (deep-drawn cold-rolled steel, manufactured by Q-panel, UK) 150 mm × 100 mm × 0.9 mm size, by using a box-type application ruler with a 250 μm slot, modified as per ISO 1514. The dry film thickness (DFT) was measured with a MiniTest 110 magnetic thickness gauge fitted with a F16-type probe (ElektroPhysik, Germany) in accordance with ISO 2808 [27]. A total of 10 test panels was prepared for each paint. A thin cut (groove) 7 cm long, which penetrated through the paint film and reached the substrate metal, was made by means of a sharp blade. The samples on the test panels were allowed to dry in standard conditions (temperature, 20°C; relative humidity, 50%) in a conditioned laboratory for 6 weeks. Paint films on polyethylene sheets were also prepared, peeled off when dry, and cut to pieces approximately 1 mm × 1 mm size. The unsupported films were used to prepare aqueous paint film suspensions in distilled water.

#### 2.1.7. *Laboratory corrosion tests*

##### 2.1.7.1. *Cyclic corrosion test in an atmosphere with NaCl mist and condensing moisture*

In this cyclic corrosion test, the test panels were exposed to the mist of a 5% NaCl solution at 35 ± 2°C for 10 h (first cycle stage) and to condensing distilled water at 40 ± 2°C for 1 h (second cycle stage), followed by drying at 23 ± 2°C (third cycle stage). The test encompassed 59 cycles, that is, its total time was 1416 h. The tests were conducted in a Liebisch S 400 salt chamber (Liebisch Labortechnik, Germany). The procedure was based on the ISO 7253 standard.

##### 2.1.7.2. *Cyclic corrosion test in an atmosphere of condensed moisture and $SO_2$*

This test consisted of 24-hour cycles comprising condensation of water with $SO_2$ at 36 ± 2°C for 8 h (first cycle stage), followed by drying at 23 ± 2°C for 16 h (second cycle stage). The test was conducted in a Liebisch V 400 corrosion chamber (Liebisch Labortechnik, Germany), and the samples were evaluated after completing 67 test cycles, that is, in 1608 h. The test procedure was based on the ISO 3231 standard.

#### 2.1.8. *Corrosion test evaluation methods*

After completing the corrosion tests, the paints were evaluated by methods derived from the ASTM D 714-87, ASTM D 610, and ASTM D 1654-92 standards. The corrosion phenomena evaluated included formation (size and frequency of occurrence) of blisters in the paint film surface and near a cut made in the film, percent fraction of substrate metal surface area affected by corrosion, and distance of propagation of substrate metal corrosion near the cut (in mm, both evaluated after removing the paint film).

By connecting all the three (four) methods for the evaluation of various manifestations of the corrosion substrate attacks and of protective film alone, we can obtain a single value of the protective efficiency. The results were converted to scores on a 100–0 scale, and a parameter called the overall anticorrosion efficiency of the paints was calculated by a mathematical relation [28]. The total anticorrosion efficiency from the cyclic corrosion tests ($E_{NaCl}$, $E_{SO2}$) was calculated as the arithmetic mean of the scores [29].

#### 2.1.9. *Linear polarization*

The linear polarization method is applied to corrosion monitoring. It is designed specifically for the determination of the polarization resistance $R_p$ and current density $I_{corr}$. Linear polarization was measured in a cell accommodating the reference electrode (saturated calomel electrode—SCE), counter-electrode (platinum electrode), and working electrode constituted by the sample. The method is based on the fact that a linear segment near the corrosion potential occurs on the polarization curve in linear coordinates.

http://dx.doi.org/10.5772/63290

A 1 $cm^2$ area of the working electrode in the measuring cell was exposed to a 3.5 wt.% NaCl solution. The cell was connected to a potentiostat/galvanostat (VSP-300/France). The paint films were exposed to the NaCl solution for 24 h, after which they were measured by the linear polarization method. The polarization region was from -10 mV/$E_{OC}$ to +10 mV/$E_{OC}$ at a rate of 0.166 mV/s. The following parameters were evaluated for each paint: spontaneous corrosion potential ($E_{corr}$), tafel region slopes ($\beta_a$ and $\beta_c$), current density ($I_{corr}$), polarization resistance ($R_p$), and corrosion rate ($v_{corr}$). Polarization resistance $R_p$ is defined as the inverse values of the current density $I$ against the curve of the spontaneous corrosion potential $E$ (at which $dE \rightarrow 0$) [29, 30].

*2.1.10. Comparison experiments*

The anticorrosion pigment based on the zinc phosphate hydrate $Zn_3(PO_4)_2xH_2O$ (PVC = 15%) was also tested as a reference material, allowing us to compare the results obtained with the pigments synthesized by us with those obtained with a commercially available product. Films of the coating materials free from any pigment were also used in some tests, in the linear polarization measurements.

## 2.2. Results evaluation

*2.2.1. Structure and morphology of the composite pigment particles*

Four perovskite-type pigments were subjected to surface treatment with the conductive polymers PANI and PPY. Like the bare perovskite pigments, the pigments coated with the conductive polymers were subjected to XRD and XRF analyses to elucidate their structure and composition. The results of the XRF data of the initial untreated pigments are listed in **Table 1**.

| Parameter [wt.%] | $CaMnO_3$ | $SrMnO_3$ | $CaTiO_3$ | $SrTiO_3$ |
|---|---|---|---|---|
| $Al_2O_3$ | 3.00 | 0.70 | 3.20 | 3.40 |
| $SiO_2$ | 0.25 | 0.16 | 0.41 | 0.11 |
| CaO | 43.61 | – | 39.44 | – |
| MnO | 53.14 | 57.72 | – | – |
| SrO | – | 41.42 | – | 52.98 |
| $TiO_2$ | – | – | 56.95 | 43.51 |

**Table 1.** Results of XRF analysis of the untreated pigments (the data are in wt.%; elements present at concentrations lower than 0.01% are omitted).

The pigments contained $Al_2O_3$ in the order of tenths to units percent (0.7–3.4%) due to wear of the corundum milling bodies in the milling equipment. For the same reason, the samples also contained trace amounts (0.11–0.41%) of $SiO_2$. The two substances are neutral with respect to the chemical properties of the anticorrosion pigments. Analysis results (XRF) of surface-treated pigments are given in **Tables 2** and **3**. The surface-modified (composite) pigments contained the respective oxides ($TiO_2$, $Mn_2O_3$, SrO, CaO), and also some amounts of compounds from the surface treatment procedure: $P_2O_5$and $SO_3$(associated with the polypyrrole phosphate/polyaniline phosphate layer). The surface treatment procedure was associated with weight loss of the oxides (of Ti, Mn, Ca, Sr) due to the nature of the medium (strongly acidic, with acids). What is important, however, is the fact that the pigment matrices for the conductive polymers and the mixed oxide contents were ensured (**Tables 2** and **3**). Once enveloped in a PANI or PPY layer, the oxides were protected from additional dissolution in an aqueous solution of phosphoric acid.

| Parameter [wt.%] | $CaTiO_3$/PANI | $SrTiO_3$/PANI | $CaMnO_3$/PANI | $SrMnO_3$/PANI |
|---|---|---|---|---|
| $Al_2O_3$ | 2.40 | 2.40 | 2.21 | 1.10 |
| $SiO_2$ | 0.25 | – | 0.11 | 0.13 |
| $P_2O_5$ | 9.50 | 10.70 | 8.20 | 17.80 |
| $SO_3$ | 10.10 | 17.50 | 41.80 | 39.10 |
| CaO | 27.70 | 0.10 | 33.80 | 0.07 |
| $TiO_2$ | 48.60 | 35.00 | – | – |
| MnO | – | – | 13.20 | 7.53 |
| SrO | – | 30.00 | – | 33.40 |

**Table 2.** Results of XRF analysis of the treated pigments/PANI (the data are in wt.%; elements present at concentrations lower than 0.01 % are omitted).

| Parameter [wt.%] | $CaMnO_3$/PPY | $SrMnO_3$/PPY | $CaTiO_3$/PPY | $SrTiO_3$/PPY |
|---|---|---|---|---|
| $Al_2O_3$ | 2.20 | 1.30 | 2.35 | 2.10 |
| $SiO_2$ | 0.12 | 0.18 | 0.13 | – |
| $P_2O_5$ | 4.90 | 5.40 | 2.65 | 5.50 |
| $SO_3$ | 37.20 | 42.70 | 16.90 | 11.50 |
| CaO | 34.10 | 0.08 | 29.70 | 0.12 |
| $TiO_2$ | – | – | 46.80 | 42.80 |
| ZnO | – | – | – | – |
| SrO | – | 42.00 | – | 36.30 |
| MnO | 20.80 | 7.32 | – | – |

**Table 3.** Results of XRF analysis of the treated pigments/PPY (the data are in wt.%; elements present at concentrations lower than 0.01% are omitted).

The composite pigments contained amorphous fractions of the conductive polymer and crystalline fractions of the carrier pigment. In addition to the amorphous fractions, which were invariably present, the composite pigments contained the following phases: $CaTiO_3$/PANI: a crystalline phase of $CaTiO_3$, a small amount of a by-phase of rutile ($TiO_2$), and a small amount of corundum ($Al_2O_3$); $CaTiO_3$/PPY: a crystalline phase of $CaTiO_3$, minor phases of rutile ($TiO_2$) and $CaSO_4.2H_2O$, and a small amount of anatase ($TiO_2$); $SrTiO_3$/PANI: a crystalline phase of $SrTiO_3$ and small amounts of corundum ($Al_2O_3$) and rutile ($TiO_2$); $SrTiO_3$/PPY: a crystalline phase of $SrTiO_3$ and a small amount of $SrSO_4$; $CaMnO_3$/PANI: crystalline phases of $CaMnO_3$ and $CaSO_4.2H_2O$; $CaMnO_3$/PPY: a crystalline phase of $CaMnO_3$ and a smaller fraction of $CaSO_4.2H_2O$ and $CaSO_4$; $SrMnO_3$/PANI: a crystalline phase of $SrMnO_3$and$SrSO_4$; and $SrMnO_3$/PPY: a crystalline phase of $SrMnO_3$ and a minor phase of $SrSO_4$.

**Figure 3.** Morphology of perovskite particles with PANI and PPY as observed by SEM (magnification 10,000×). Notes: (a) $CaTiO_3$/PANI; (b) $SrTiO_3$/PANI; (c) $SrMnO_3$/PANI; (d) $CaMnO_3$/PANI.

**Figure 4.** Morphology of perovskite particles with PANI and PPY as observed by SEM (magnification 10,000×). Notes: (a) $CaTiO_3$/PPY; (b) $SrTiO_3$/PPY; (c) $SrMnO_3$/PPY; (d) $CaMnO_3$/PPY.

The pigment particle morphology is illustrated by scanning electron micrographs for the perovskites and for the surface-modified perovskites in **Figures 3** and **4**. The micrographs were taken in the secondary electron imaging (SEI) mode. The composite pigments had a tendency to form clusters [30]. The particles of the initial perovskite pigments had a regular nodular shape, which remained unaffected by surface treatment with conducting polymers.

*2.2.2. Physicochemical properties of the powdered pigments*

The physicochemical properties of the powdery pigments are given in **Tables 4** and **5**, including density, linseed oil consumption, CPVC, pH, specific electric conductivity, particle size distribution, and water-soluble fractions $W_{20}$. **Table 5** lists the specific electric conductivities of the pigments in dependence on their concentrations in the paint films. The particle size distribution values *D* (0.5), *D* (0.9), and *D* (0.1) show that the size of 50, 90, or 10% particles, respectively, in the volume is smaller than the specified value. *D* (4.3) is the mean particle size.

| Pigment | [a]Density (g/cm³) | [a]Oil consumption (g/100 g pigment) | CPVC (%) | Particle size Percent fraction below the specified size *D* (0.1) (μm) | *D* (0.5) (μm) | *D* (0.9) (μm) | *D* (4.3) (μm) |
|---|---|---|---|---|---|---|---|
| $CaTiO_3$/PANI | 2.97 | 37 | 46 | 1.11 | 17.29 | 23.63 | 19.37 |
| $SrTiO_3$/PANI | 2.76 | 49 | 41 | 4.52 | 12.23 | 24.89 | 20.43 |
| $CaMnO_3$/PANI | 2.17 | 57 | 43 | 2.45 | 8.70 | 16.06 | 13.87 |
| $SrMnO_3$/PANI | 1.42 | 55 | 80 | 2.65 | 13.61 | 39.16 | 38.86 |
| $CaTiO_3$/PPY | 2.83 | 22 | 60 | 1.93 | 21.79 | 40.29 | 34.44 |
| $SrTiO_3$/PPY | 1.44 | 49 | 57 | 1.86 | 15.93 | 17.67 | 20.45 |
| $CaMnO_3$/PPY | 2.54 | 31 | 54 | 1.28 | 9.17 | 28.25 | 13.24 |
| $SrMnO_3$/PPY | 2.87 | 33 | 50 | 2.21 | 15.56 | 33.22 | 29.34 |
| $Zn_3(PO_4)_3xH_2O$ | 3.28 | 34 | 45 | 0.13 | 4.78 | 9.96 | 5.34 |
| $CaCO_3$ | 2.72 | 14 | 71 | 0.20 | 5.60 | 9.99 | 6.14 |

[a]Parameters are given as arithmetic averages within 10 measured values.

**Table 4.** Physicochemical properties of the powdered pigments.

The mean particle size ($D_{4.3}$) of the perovskites coated with layers of PANI and of PPY lay within the similar ranges of 13.87–38.86 μm and 13.24–34.44 μm, respectively (**Table 4**). The densities of the perovskites coated with layers of PANI and of PPY lay within the ranges of 1.42–2.97 gċcm$^{-3}$ and 1.44–2.87 gċcm$^{-3}$, respectively (**Table 4**), that is, less than the densities of the initial inorganic perovskites. The densities of the PANI and PPY powders themselves are 1.58 gċcm$^{-3}$ and 1.76 gċcm$^{-3}$, respectively [18].

| Pigment/coating | Aqueous extracts of the pigments | | | Aqueous extracts of the paint films |
|---|---|---|---|---|
| | $W_{20}$ (%) | [a]$pH_p$ | [b]Specific electric conductivity (μS/cm) | [a]$pH_f$ |
| $CaTiO_3$/PANI | 9.46 | 4.34 | 2220 | 6.54 |
| $SrTiO_3$/PANI | 5.09 | 3.00 | 1450 | 6.16 |
| $CaMnO_3$/PANI | 12.78 | 5.62 | 2260 | 7.44 |
| $SrMnO_3$/PANI | 9.78 | 5.03 | 994 | 7.27 |
| $CaTiO_3$/PPY | 10.69 | 5.29 | 2390 | 7.28 |
| $SrTiO_3$/PPY | 10.07 | 6.02 | 1008 | 7.65 |
| $CaMnO_3$/PPY | 18.47 | 6.26 | 2100 | 7.73 |
| $SrMnO_3$/PPY | 3.36 | 6.17 | 949 | 7.46 |
| $Zn_3(PO_4)_3xH_2O$ | 0.26 | 6.65 | 50 | 7.28 |
| $CaCO_3$ | 0.51 | 8.47 | 10 | 8.47 |
| Nonpigmented film | – | – | – | 3.71 |

[a]pH was measured with an accuracy ±0.01.

[b]Conductivity was measured with an accuracy ±0.5%.

**Table 5.** Physicochemical properties of the pigments.

Oil consumption of the composite pigments with PANI and with PPY lies within the ranges of 37–57 g and 22–49 g per 100 g of the pigment, respectively (**Table 4**). The values were higher than those measured for the initial perovskite pigments due to the presence of the porous layers of the conductive polymers. The oil consumption data for the composite pigments depended on the pigment particle heterodispersity; in fact, the oil consumption is generally dependent on the particle size and particle shape (i.e., on the specific surface area of the particle) [31].

Knowledge of the critical pigment volume concentration (CPVC) value was a prerequisite for a correct formulation of the pigmented organic coating material [26]. The CPVC value depends on the density and on the pigment's oil number. The CPVC levels were calculated to be about from 41 to 80% for the surface-modified perovskites with PANI and from 50 to 60% for the perovskite pigments surface modified with PPY. From the above data, it follows that the CPVCs of the composite pigments are lower for PANI than for PPY as the surface-modifying conductive polymer (**Table 4**).

### 2.2.3. *Water-soluble contents*

The water-soluble contents ($W_{20}$) of the pigments are listed in **Table 5**. The water-soluble content was higher in the pigments modified with the conductive polymers than in the initial inorganic perovskites due to the presence of minor by-phases, such as $CaSO_4$ or $SrSO_4$. The lowest content of substances soluble in cold water, $W_{20}$=3.36%, was found in the $SrMnO_3$/PPY system, and the highest content, $W_{20}$= 18.47%, in the $CaMnO_3$/PPY system due to the presence of the by-phase of $CaSO_4$. The differences in the water-soluble contents between the pigments coated with PPY and the pigments coated with PANI were units percent only: for example, the maximum value observed with PPY was $W_{20}$= 18.47%, and the maximum value observed with PANI was $W_{20}$= 12.78%. The increased water-soluble content of the modified pigments compared to the nonmodified pigments also supports the concept of deprotonation of the conductive polymer layers in the former [32]. A high water-soluble content is indicative of a potentially increased occurrence of osmotic blisters on the surfaces of the paint films [18]. It is concluded that the pigments are not "hazardous" because of this parameter [23].

### 2.2.4. *pH values of aqueous extracts of the pigments and of loose paint films*

The observed pH values of the pigment powders ($pH_p$) and of loose paint films containing the pigments ($pH_f$) are listed in **Table 5**. The values of extracts of the nonmodified perovskite pigments ($pH_p$) lay within the region of Ph range of 9–12, and the values of extracts of the pigments coated with the conductive polymer layers ($pH_p$) lay within the range of 3.0–6.3. Hence, the pigment surface treatment brought about a pH shift toward more acid values: modification with PPY shifted the pH to a slightly acidic region, and modification with PANI, to the acid region. This shift can be explained in terms of deprotonation of the phosphate salts of PANI/PPY in aqueous solutions. The presence of the PANI or PPY salt in the composite pigment reduced the individual differences in the pH values between the initial perovskites. The pH values of extracts of the loose pigmented paint films lay within the region of $pH_f$ (6.2–7.7); within this region, the paint films containing PPY occupied the basic side $pH_f$ (73–7.7); within this region, the paint films containing PANI occupied the basic side $pH_f$ (6.2–7.4). The pH of the extract of the nonpigmented film lay in the acid region, at $pH_f$ = 3.7. The pH values for the paints containing the composite pigments were also affected by the presence of $CaCO_3$ in the paint, or by release of the basically reacting calcium and strontium cations ($Ca^{2+}$, $Sr^{2+}$). The pH of the extract of the paint film containing calcite at PVC = 50% lay in the slightly basic region, at pH = 8.47. This is beneficial with respect to suppression of corrosion on the metal surface beneath the paint film.

### 2.2.5. *Specific conductivities of aqueous extracts of the pigments and of the loose paint films*

Some electric conductivity is necessary for the anticorrosion pigments to passivate the metal surface beneath the paint film. The specific conductivities of extracts of the pigment powders modified with the conductive polymers lay within the regions of 949–2390 $\mu S\dot{c}cm^{-1}$ for PPY and 994–2260 $\mu S\dot{c}cm^{-1}$ for PANI and increased in time due to the release of the soluble components into the aqueous environment. The specific electric conductivities did not differ appreciably between the individual surface-modified pigments: the differences lay within one order of magnitude—from 2390 to 949$\mu S\dot{c}cm^{-1}$. Hence, the specific conductivity was one to two orders of magnitude higher for the composite pigments than for the initial perovskites.

The specific conductivity of the aqueous extracts of the loose films ($\chi_f$) increased with increasing conductive polymer content of the paint (**Table 6**).

http://dx.doi.org/10.5772/63290

| Pigment modified with PANI | PVC (%) | [a]Specific electric conductivity (μS/cm$^{-1}$) | Pigment modified with PPY | PVC (%) | [a]Specific electric conductivity (μS/cm$^{-1}$) |
|---|---|---|---|---|---|
| **$CaTiO_3$/PANI** | 1 | 313 | **$CaTiO_3$/PPY** | 1 | 542 |
| | 5 | 318 | | 5 | 584 |
| | 10 | 348 | | 10 | 626 |
| | 15 | 465 | | 15 | 992 |
| **$SrTiO_3$/PANI** | 1 | 332 | **$SrTiO_3$/PPY** | 1 | 446 |
| | 5 | | 431 | 5 | 454 |
| | 10 | | 589 | 10 | 461 |
| | 15 | | 778 | 15 | 521 |
| **$CaMnO_3$/PANI** | 1 | 381 | **$CaMnO_3$/PPY** | 1 | 530 |
| | 5 | 443 | | 5 | 645 |
| | 10 | 1353 | | 10 | 887 |
| | 15 | 1894 | | 15 | 1204 |
| **$SrMnO_3$/PANI** | 1 | 364 | **$SrMnO_3$/PPY** | 1 | 398 |
| | 5 | 421 | | 5 | 503 |
| | 10 | 513 | | 10 | 558 |
| | 15 | 851 | | 15 | 639 |
| **$Zn_3(PO_4)_3xH_2O$** | 15 | 176 | **Nonpigmented film** | - | 50 |

[a]Specific electric conductivity was measured with an accuracy ±0.5%.

**Table 6.** Specific electric conductivity of aqueous extracts of loose paint films containing the composite pigments.

The specific conductivities were lowest for the paints with PVC = 1%, from 313 to 381 μSċcm$^{-1}$ for the pigments modified with PANI and from 398 to 542 μSċcm$^{-1}$ for the pigments modified with PPY. The specific conductivities were higher at PVC = 5% and lay within the ranges of 318 to 443 μSċcm$^{-1}$ for PANI and from 503 to 645 μSċcm$^{-1}$ for PPY. When the pigment concentrations were further increased to PVC = 10% and to PVC = 15%, the specific conductivities also increased, viz. to 348–1353 and 465–1894 μSċcm$^{-1}$, respectively, for PANI, and to 461–887 and 521–1204 μSċcm$^{-1}$, respectively, for PPY. The higher specific conductivities of the aqueous extracts of the paint films with the pigments modified with the conductive polymers as compared to the untreated pigments were due to the presence of free charge carriers on the polymeric chain, providing charge transfer across the chain. The positive charge at the chain is counterbalanced by the anion of the acid used for the protonation, that is, phosphoric acid (and the phosphate anion derived from it) in this case. The layers of the conductive polymers undergo partial deprotonation in aqueous solutions [31]. The extent of release and deprotonation of increased amounts of the conductive polymers in the paint films in aqueous systems was higher for paints with higher pigment concentrations (PVC).

| Pigment | PVC (%) | Paint assessment | | Substrate metal assessment | | Calculated anticorrosion efficiency $E_{NaCl}$ |
|---|---|---|---|---|---|---|
| | | Degree of blistering ASTM D 714-87 | | Corrosion in a cut ASTM D 1654-92 | Surface corrosion ASTM D 610-85 | |
| | | In a cut | Metal base | (mm) | (%) | |
| **$CaTiO_3$/PANI** | 1 | 2F | 4F | 0.5–1.0 | 0.3 | 79 |
| | 5 | – | – | 0.0–0.5 | 0.03 | 98 |
| | 10 | – | | 1.5–2.0 | 1 | 90 |
| | 15 | – | 2F | 0.5–1.0 | 0.5 | 85 |
| **$SrTiO_3$/PANI** | 1 | 2M | – | 1.5–2.0 | 0.1 | 79 |
| | 5 | 2M | – | 0.5–1.0 | 0.03 | 82 |
| | 10 | 4M | 4F | 0.5–1.0 | 0.3 | 74 |
| | 15 | 4F | 2MD | 1.5–2.0 | >50 | 44 |
| **$CaMnO_3$/PANI** | 1 | 2F | 4F | 1.5–2.0 | 0.3 | 75 |
| | 5 | – | 6F | 0.5–1.0 | 3 | 84 |
| | 10 | – | 4F | 2.0–2.5 | 10 | 76 |
| | 15 | – | 4F | 1.5–2.0 | 50 | 65 |
| **$SrMnO_3$/ PANI** | 1 | – | 2M | 1.0–1.5 | 1 | 77 |
| | 5 | – | 4F | 2.0–2.5 | 3 | 79 |
| | 10 | 2F | 2F | 2.5–3.0 | 10 | 65 |
| | 15 | – | 2D | 3.0–4.0 | 50 | 42 |
| **$CaTiO_3$/PPY** | 1 | – | – | 0 | 0.3 | 98 |
| | 5 | – | – | 0 | 0.1 | 99 |
| | 10 | – | 4F | 0.0–0.5 | 0.3 | 90 |
| | 15 | – | – | 1.5–2.0 | 33 | 75 |
| **$SrTiO_3$/PPY** | 1 | 4F | – | 0 | 0.01 | 93 |
| | 5 | – | 6F | 0–0.5 | 0.01 | 92 |
| | 10 | 2F | 4F | 1.0–1.5 | 0.03 | 79 |
| | 15 | – | 2F | 1.0–1.5 | 0.1 | 86 |
| **$CaMnO_3$/PPY** | 1 | – | – | 0 | 0.01 | 100 |
| | 5 | – | – | 0–0.5 | 0.01 | 99 |
| | 10 | – | 8F | 0–0.5 | 1 | 90 |
| | 15 | 4F | 6F | 1.0–1.5 | 33 | 63 |
| **$SrMnO_3$/PPY** | 1 | – | 6F | 0 | 0.1 | 93 |
| | 5 | – | 8M | 0-0.5 | 0.3 | 85 |
| | 10 | 2F | 8F | 0–0.5 | 3 | 79 |
| | 15 | 2F | 8F | 0–0.5 | 3 | 77 |
| **$Zn_3(PO_4)_3xH_2O$** | 15 | 2M | – | 3.0–4.0 | 3 | 68 |
| **Nonpigmented film** | – | 6M | – | 1.5–2.0 | 50 | 60 |

**Table 7.** Results of accelerated corrosion tests of the paints containing the composite pigments in the NaCl mist atmosphere (exposure 1416 h, DFT = 95 ± 10 μm).

### 2.3. Exposure of the paint films in the atmosphere with the mist of a neutral NaCl solution

The results of the accelerated corrosion test in which the steel panels coated with the paints were exposed to a salt mist atmosphere for 1416 h are included in **Table 7**.

Osmotic blisters were observed on the surface of the paint films containing the composite pigments $CaMnO_3$/PANI and $SrMnO_3$/PANI at higher pigment concentrations, due to the presence of water-soluble substances (e.g., 12.78% in $CaMnO_3$) and higher concentration of the composite pigments playing a role in the film exposed to the mist of the neutral aqueous solution of chloride ions and to increased humidity/moisture. Therefore, the protective effect of the paint films was lower for the two pigment systems at PVC 15%, also due to the higher degree of corrosion (affected fraction > 50%) of the metal surface beneath the paint film. So, a compromise pigment concentration must be sought for the above paints in order to achieve the optimum anticorrosive effect [31].

PPY was found superior to PANI at PVC = 1%. The highest possible efficiency $E_{NaCl}$= 100% at PVC = 1 (and also at PVC = 5%) was achieved by the paints containing the $CaTiO_3$/PPY and $CaMnO_3$/PPYsystems, followed by the $SrMnO_3$/PPY and $SrTiO_3$/PPY systems ($E_{NaCl}$= 93 in both cases). In the paints with PVC = 5%, PPY was again superior to PANI in this respect. The overall anticorrosion efficiency values ($E_{NaCl}$) were only slightly poorer at PVC = 5% than at PVC = 1%. The highest corrosion-inhibiting efficiency at PVC = 5% was observed with the paints containing the $CaTiO_3$/PPY system and the $CaMnO_3$/ PPY system ($E_{NaCl}$ = 99 in both cases). Slightly lower corrosion resistance was observed with the paint with $SrTiO_3$/PPY ($E_{NaCl}$= 92). The above paints exhibited a high ability to protect the substrate metal against surface corrosion and against corrosion near the cut, and virtually no blisters were observed on the paint films.

#### 2.3.1. *Findings from the corrosion-inhibiting efficiency of pigmented paint films in NaCl atmosphere*

1. Paints providing high substrate metal surface protection against corrosion (affected area fraction ≤ 0.0–0.03%) contained the following composite pigments at the following concentrations: $CaTiO_3$/PANI at PVC = 5% (corroded surface fraction, 0.03%); $SrTiO_3$/PANI at PVC = 5% (corroded surface fraction, 0.03%); $SrTiO_3$/PPY at PVC = 1 and 5% (corroded surface fraction, 0.01%), and at PVC = 10% (corroded surface fraction, 0.03%); $CaMnO_3$/PPY at PVC = 1 and 5% (corroded surface fraction, 0.01%).

2. Paints providing high substrate metal protection against corrosion in the cut (i.e., corrosion propagation within 0.5 mm from the cut) contained the following composite pigments at the following concentrations: $CaTiO_3$/PPY at PVC = 1 and 5% (corrosion propagation 0 mm), and at PVC = 10% (corrosion propagation ≤ 0.5 mm); $SrTiO_3$/PPY at PVC = 1 and 5% (corrosion propagation ≤ 0.5 mm, respectively); $CaMnO_3$/PPY at PVC = 1, 5, and 10% (corrosion propagation ≤ 0.5 mm, respectively); $SrMnO_3$/PPY at PVC = 1, 5, 10, and 15% (corrosion propagation ≤ 0.5 mm, respectively).

3. Paint films exhibiting high resistance to surface blistering (score 0–8 F) contained the following composite pigments at the following concentrations: $CaTiO_3$/PPY at PVC = 1, 5, and 15% (no blisters); $SrTiO_3$/PPY at PVC = 1% (no blisters); $CaMnO_3$/PPY at PVC = 1–10% (no blisters, no blisters, 8 F); and $SrMnO_3$/PPY at PVC = 10 and 15% (8 F and 8 F).

4. Paint films exhibiting an anticorrosion effect at a pigment concentration as low as PVC = 1% included the following composite pigments: $CaTiO_3$/PPY ($E_{NaCl}$ = 98); $SrTiO_3$/PPY ($E_{NaCl}$ = 93); $CaMnO_3$/PPY ($E_{NaCl}$ = 100); and $SrMnO_3$/PPY ($E_{NaCl}$ = 93). The pigments with PANI exhibited lower efficiencies: $CaTiO_3$/PANI ($E_{NaCl}$ = 79); $SrTiO_3$/PANI ($E_{NaCl}$ = 79); $CaMnO_3$/PANI ($E_{NaCl}$ = 75); and $SrMnO_3$/PANI ($E_{NaCl}$ = 77).

5. Paint films that attained overall anticorrosion efficiency levels higher than the reference paint, that is, $E_{NaCl}$> 68, contained the following composite pigments at the following concentrations: $CaTiO_3$/PANI at PVC = 1, 5, 10, and 15%; $SrTiO_3$/PANI at PVC = 1, 5, and 10%; $CaMnO_3$/PANI at PVC = 1, 5, and 10%; $SrMnO_3$/PANI at PVC = 1, 5%; $CaTiO_3$/PPY at PVC = 1, 5, 10, and 15%; $SrTiO_3$/PPY at PVC = 1, 5, 10, and 15%; $CaMnO_3$/PPY at PVC = 1, 5, and 10%; and $SrMnO_3$/PPY at PVC = 1, 5, 10, and 15%.

6. The following paints containing pigments at PVC = 15% provided anticorrosion efficiency better than or roughly identical with that provided by the reference paint (zinc phosphate at PVC = 15%, $E_{NaCl}$ = 68): $CaTiO_3$/PANI ($E_{NaCl}$ = 85); $CaTiO_3$/PPY ($E_{NaCl}$ = 75); $SrTiO_3$/PPY ($E_{NaCl}$ = 86); and $SrMnO_3$/PPY ($E_{NaCl}$ = 77).

### 2.4. Corrosion-inhibiting efficiency of pigmented paint films with $SO_2$ and water condensation

The results of the accelerated corrosion test of the paint films in a condensation chamber with an $SO_2$ mist are listed in **Table 8**. The following PVCs were found optimal with respect to the anticorrosion efficiency ($E_{SO2}$) of the paints containing the pigments modified with PANI: $CaTiO_3$/PANI: PVC = 10% ($E_{SO2}$ = 99); $SrTiO_3$/PANI: PVC = 5 and 10% ($E_{SO2}$= 100); $CaMnO_3$/PANI: PVC = 1% ($E_{SO2}$= 95); and $SrMnO_3$/PANI: PVC = 1% ($E_{SO2}$= 92). The analogous values for the pigments modified with PPY were as follows: $CaTiO_3$/PPY: PVC = 1% ($E_{SO2}$ = 100); $SrTiO_3$/PPY: PVC = 10 and 15% ($E_{SO2}$ = 97); $SrMnO_3$/PPY: PVC = 1% ($E_{SO2}$ = 99); and $CaMnO_3$/PPY: PVC = 5% ($E_{SO2}$ = 88).

| Pigment | PVC (%) | Paint assessment | | Substrate metal assessment | | Calculated anticorrosion efficiency $E_{SO2}$ |
|---|---|---|---|---|---|---|
| | | Degree of blistering ASTM D 714-87 | | Corrosion in a cut ASTM D 1654-92 | Surface corrosion ASTM D 610-85 | |
| | | In a cut | Metal base | (mm) | (%) | |
| $CaTiO_3$/PANI | 1 | – | – | 3.0–4.0 | 0.1 | 89 |
| | 5 | – | – | 1.0–1.5 | 0.1 | 94 |
| | 10 | – | – | 0–0.5 | 0.03 | 99 |
| | 15 | – | – | 0.5–1.0 | 0.1 | 96 |
| $SrTiO_3$/PANI | 1 | – | – | 1.5–2.0 | 0.03 | 94 |
| | 5 | – | – | 0 | 0.03 | 100 |

| Pigment | PVC (%) | Paint assessment | | Substrate metal assessment | | Calculated anticorrosion efficiency $E_{SO2}$ |
|---|---|---|---|---|---|---|
| | | Degree of blistering ASTM D 714-87 | | Corrosion in a cut ASTM D 1654-92 | Surface corrosion ASTM D 610-85 | |
| | | In a cut | Metal base | (mm) | (%) | |
| | 10 | – | – | 0 | 0.03 | 100 |
| | 15 | – | – | 0–0.5 | 3 | 90 |
| $CaMnO_3$/PANI | 1 | – | – | 1.0–1.5 | 0.01 | 95 |
| | 5 | 6F | – | 2.0–2.5 | 0.03 | 86 |
| | 10 | 6F | – | 1.5–2.0 | 0.3 | 86 |
| | 15 | 4F | – | 3.0–4.0 | 0.3 | 81 |
| $SrMnO_3$/PANI | 1 | – | – | 2.0–2.5 | 0.03 | 92 |
| | 5 | – | – | 1.5–2.0 | 1 | 90 |
| | 10 | – | – | 1.5–2.0 | 3 | 88 |
| | 15 | – | – | 1.0–1.5 | 10 | 86 |
| $CaTiO_3$/PPY | 1 | – | – | 0 | 0.03 | 100 |
| | 5 | – | – | 0–0.5 | 0.03 | 99 |
| | 10 | – | – | 0.5–1.0 | 0.03 | 96 |
| | 15 | – | – | 1.0–1.5 | 0.03 | 95 |
| $SrTiO_3$/PPY | 1 | – | – | 2.0–2.5 | 0.01 | 93 |
| | 5 | – | – | 1.5–2.0 | 0.1 | 94 |
| | 10 | – | – | 0.5–1.0 | 0.03 | 97 |
| | 15 | – | – | 0.5–1.0 | 0.03 | 97 |
| $CaMnO_3$/PPY | 1 | 4F | – | 2.0–2.5 | 0.03 | 85 |
| | 5 | 4F | – | 1.0–1.5 | 0.03 | 88 |
| | 10 | 4F | – | 2.0–2.5 | 0.1 | 85 |
| | 15 | 2F | – | 3.0–4.0 | 0.1 | 81 |
| $SrMnO_3$/PPY | 1 | – | – | 0–0.5 | 0.03 | 99 |
| | 5 | – | – | 0–0.5 | 0.3 | 98 |
| | 10 | 4F | – | 1.0–1.5 | 0.3 | 86 |
| | 15 | 4F | – | 2.5–3.0 | 3 | 77 |
| $Zn_3(PO_4)_3.xH_2O$ | 15 | 8MD | 0 | 1.5–2.0 | 0.03 | 79 |
| Nonpigmented film | – | 6MD | – | 2.5–3.0 | 0.01 | 75 |

**Table 8.** Results of accelerated corrosion tests of the paints containing composite pigments in the $SO_2$ atmosphere (exposure 1608 h, DFT = 95 ± 10 μm).

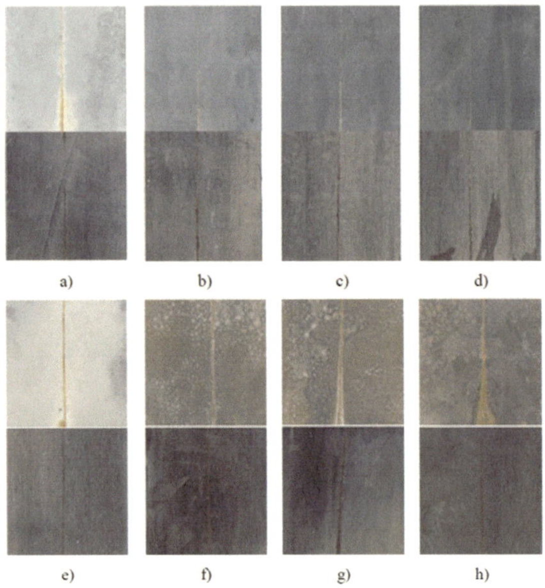


**Figure 5.** Results of exposure of a steel panel coated with the paint containing $CaTiO_3$/PANI and $CaTiO_3$/PPY in the $SO_2$atmosphere.

### *2.4.1. Findings from the corrosion-inhibiting efficiency of pigmented paint films in $SO_2$atmosphere*

1. Paint films providing high protection against substrate metal surface corrosion (corroded area fraction ≤ 0.03%) contained the following composite pigments at the following concentrations: $CaTiO_3$/PANI at PVC = 10% (corroded surface fraction, 0.03%); $SrTiO_3$/PANI at PVC = 1–10% (corroded surface fraction, 0.03%); $CaMnO_3$/PANI at PVC = 1% (corroded surface fraction, 0.01%), PVC = 5% (corroded surface fraction, 0.03%); $SrMnO_3$/PANI at PVC = 1% (corroded surface fraction, 0.03%); $CaTiO_3$/PPY at PVC = 1–15% (corroded surface fraction, 0.03%); $SrTiO_3$/PPY at PVC = 1% (corroded surface fraction, 0.01%), PVC = 10 and 15% (corroded surface fraction, 0.03%); $CaMnO_3$/PPY at PVC = 1–5% (corroded surface fraction, 0.03%); and $SrMnO_3$/PPY at PVC = 1% (corroded surface fraction, 0.03%).

2. Paints providing high protection against corrosion propagation from the cut (corrosion propagation to ≤ 0.5 mm) contained the following composite pigments at the following concentrations: $CaTiO_3$/PANI at PVC = 10% (corrosion propagation to 0.5 mm); $SrTiO_3$/PANI at PVC = 5 and 10% (corrosion propagation 0 mm), PVC = 15% (corrosion propagation to 0.5 mm); $CaTiO_3$/PPY at PVC = 1% (corrosion propagation 0 mm), PVC = 5% (corrosion propagation to 0.5 mm), $SrMnO_3$/PPY at PVC = 1–5% (corrosion propagation to 0.5 mm). No blisters on the paint surface or near the cut were observed with the following composite pigments at any of the pigment concentrations applied: $CaTiO_3$/PANI, $SrTiO_3$/PANI, $SrMnO_3$/PANI, $CaTiO_3$/PPY, $SrTiO_3$/PPY.

3. The following pigments were efficient at a concentration as low as PVC = 1%: $CaTiO_3$/PANI ($E_{SO2}$= 89), $SrTiO_3$/PANI ($E_{SO2}$= 94), $CaMnO_3$/PANI ($E_{SO2}$= 95), $SrMnO_3$/PANI ($E_{SO2}$= 92), $CaTiO_3$/PPY ($E_{SO2}$= 100), $SrTiO_3$/PPY ($E_{SO2}$= 93), $CaMnO_3$/PPY ($E_{SO2}$= 85), and $SrMnO_3$/PPY ($E_{SO2}$= 99). All of the above overall anticorrosion efficiency levels are higher than that of the reference paint (i.e., $E_{SO2}$> 79). Paints with PVC = 15% that were more efficient than the reference paint (zinc phosphate as the pigment at PVC = 15%, $E_{SO2}$ = 79) contained the following composite pigments: $CaTiO_3$/PANI ($E_{SO2}$ = 96); $SrTiO_3$/PANI ($E_{SO2}$ = 90); CaM-n$O_3$/PANI ($E_{SO2}$ = 81); $SrMnO_3$/PANI ($E_{SO2}$ = 86); $CaTiO_3$/PPY ($E_{SO2}$ = 95); $SrTiO_3$/PPY ($E_{SO2}$ = 97); $CaMnO_3$/PPY ($E_{SO2}$ = 81).The majority of the paint films containing the composite pigments at any of the PVC levels applied attained an overall anticorrosion efficiency better than or comparable to that of the reference paint. The anticorrosion efficiencies of the paints containing $CaTiO_3$/PANI and $CaTiO_3$/PPY are graphically documented in **Figure 5**.

### 2.5. Assessment of the corrosion protection of coatings with perovskites surface-modified with PANI and PPY

Pigment surface modification with PPY was beneficial in the inhibition of blistering both on the paint film surface and in the area of the cut in atmosphere with NaCl, observed also at higher pigment concentrations. Paint resistance to blistering was highest at PVC = 5% or up to 10%. Also beneficial was the pigment surface treatment with PPY in metal surface protection against corrosion, observed also at PVC > 10%.

The ability of the paint films containing pigments modified with PPY to inhibit corrosion of the substrate metal near the cut, also at higher PVC levels, can be explained in terms of a higher specific electric conductivity of the paint films containing PPY as compared to the paint films containing PANI, where this ability is lower. At the same time, this higher specific conductivity was not associated with a local loss of adhesion or with blistering. PPY was superior to PANI, also at PVC = 10%, and largely also at PVC = 15%. The highest anticorrosion efficiency at PVC = 10% was observed with the paints with the $CaTiO_3$/PPY and $CaMnO_3$/PPY systems ($E_{NaCl}$= 90 in either case).

Perovskites providing the best results in atmosphere with NaClif used as the cores for coating with the conductive polymers, particularly with PPY, were those with the Ca–Ti, Sr–Ti, and Ca–Mn combinations. This can be explained by the optimum combination of the specific electric conductivity of the pigments, their solubility in the wet aqueous environment, and the potential of the pigmented paint film adequate to hinder corrosion of the substrate metal. The results of the accelerated corrosion test in the atmosphere of the NaCl mist were supported by the linear polarization measurements.

The occurrence of blister was very low in the atmosphere with $SO_2$and water condensation (in comparison with the atmosphere with the NaCl mist) when using the paints containing the composite pigments even at the higher pigment concentration levels. Conductive layer of PPY was superior to the layer of PANI also in this test; in fact, the PPY layer–perovskite carrier system was ideal as a corrosion inhibitor. The pigment particle treatment with PPY was beneficial particularly to corrosion inhibition on the substrate metal surface, up to the highest

PVC levels applied (PVC = 10–15%). Also beneficial was this treatment with respect to suppression of paint film blistering, and also across a wide range of PVC levels.

Corrosion of the substrate metal surface was also low across the whole PVC series in the atmosphere with $SO_2$. This effect can be explained in terms of active electrochemical protection of the steel panel by the paint film containing the active form of the conductive polymer (**Figure 6**). The reason why the anticorrosion effect was higher in the atmosphere with $SO_2$ than in the atmosphere with NaCl lies in the fact that the former atmosphere is appreciably acid and the conductive polymers are more efficient in it, or the nonconductive forms of the composite pigments are maintained in/converted to the conductive forms. So, the electrochemical and chemical mechanisms of action of the pigments and conductive polymers can play their role [32, 33]. Therefore, the pigment surface modification with either of the conductive polymers resulted in more pronounced anticorrosion effects compared to the NaCl mist; also, the differences in the efficiencies of the paints between the different PVC levels of the composite pigments were less marked. The surface modification of the pigments with either of the two conductive polymers (PANI and PPY) was beneficial with respect to resistance against paint surface blistering (at any PVC) and also with respect to anticorrosion protection of the substrate metal.

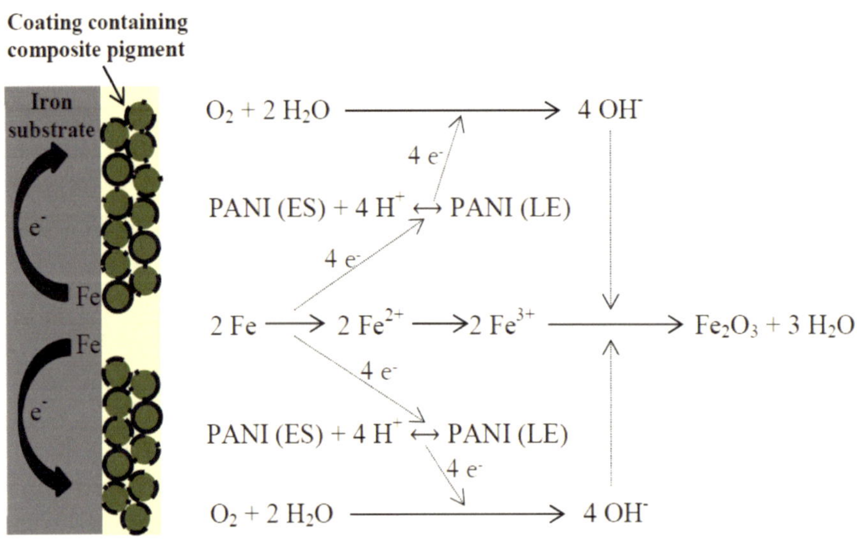


**Figure 6.** Mechanism of corrosion protection by polyaniline coating (PANI-LE = leucoemeraldine, PANI-ES = emeraldine salt).

## 2.6. Linear polarization

The parameters measured, that is, the spontaneous corrosion potential, polarization resistance, and corrosion rate, provide information about the paint films' corrosion resistance (**Table 9**). A paint containing zinc phosphate at PVC = 15% served as the reference material in the linear polarization measurements.

| Pigment in the paint | PVC (%) | $E_{corr}$ (mV) | $I_{corr}$ (μA) | $\beta_c$ (mV) | $\beta_a$ (mV) | $R_p$ (Ω) | $v_{corr}$ (mm/year) |
|---|---|---|---|---|---|---|---|
| **$CaTiO_3$/PANI** | 1 | -431 | $7.10^{-4}$ | 43.6 | 40.1 | $1.10^{7}$ | $8.10^{-7}$ |
| | 5 | -382 | $2.10^{-4}$ | 18.8 | 19.1 | $2.10^{7}$ | $1.10^{-5}$ |
| | 10 | -463 | $4.10^{-3}$ | 27.4 | 28.3 | $2.10^{6}$ | $5.10^{-4}$ |
| | 15 | -468 | $8.10^{-3}$ | 25.9 | 27.2 | $7.10^{5}$ | $8.10^{-4}$ |
| **$SrTiO_3$/PANI** | 1 | -286 | $7.10^{-4}$ | 16.4 | 19.1 | $5.10^{7}$ | $8.10^{-7}$ |
| | 5 | -593 | $1.10^{-3}$ | 22.6 | 19.3 | $4.10^{6}$ | $1.10^{-5}$ |
| | 10 | -587 | $5.10^{-2}$ | 31.6 | 31.7 | $1.10^{5}$ | $5.10^{-4}$ |
| | 15 | -635 | $7.10^{-2}$ | 32.7 | 29.8 | $1.10^{5}$ | $8.10^{-4}$ |
| **$CaMnO_3$/PANI** | 1 | -687 | $9.10^{-5}$ | 23.3 | 37.7 | $7.10^{7}$ | $1.10^{-6}$ |
| | 5 | -558 | $7.10^{-3}$ | 33.2 | 35.0 | $1.10^{6}$ | $8.10^{-5}$ |
| | 10 | 554 | $1.10^{-3}$ | 12.4 | 11.6 | $2.10^{6}$ | $1.10^{-5}$ |
| | 15 | -563 | $3.10^{-2}$ | 36.5 | 35.0 | $2.10^{5}$ | $4.10^{-4}$ |
| **$SrMnO_3$/PANI** | 1 | -585 | $3.10^{-3}$ | 18.1 | 21.2 | $2.10^{6}$ | $3.10^{-5}$ |
| | 5 | -551 | $4.10^{-3}$ | 36.0 | 37.6 | $2.10^{6}$ | $5.10^{-5}$ |
| | 10 | -590 | $2.10^{-2}$ | 24.8 | 21.6 | $3.10^{5}$ | $2.10^{-4}$ |
| | 15 | -651 | $6.10^{-3}$ | 21.1 | 14.8 | $6.10^{5}$ | $7.10^{-5}$ |
| **$CaTiO_3$/PPY** | 1 | -175 | $1.10^{-7}$ | 27.2 | 26.6 | $5.10^{10}$ | $1.10^{-11}$ |
| | 5 | -192 | $4.10^{-7}$ | 7.8 | 6.2 | $3.10^{9}$ | $5.10^{-9}$ |
| | 10 | -354 | $2.10^{-6}$ | 35.9 | 34.9 | $4.10^{9}$ | $2.10^{-8}$ |
| | 15 | -600 | $8.10^{-4}$ | 35.3 | 38.6 | $1.10^{7}$ | $9.10^{-6}$ |
| **$SrTiO_3$/PPY** | 1 | -556 | $7.10^{-8}$ | 23.9 | 22.2 | $7.10^{10}$ | $8.10^{-10}$ |
| | 5 | -553 | $5.10^{-8}$ | 9.0 | 8.8 | $4.10^{10}$ | $6.10^{-10}$ |
| | 10 | -544 | $4.10^{-4}$ | 34.5 | 35.0 | $2.10^{7}$ | $4.10^{-6}$ |
| | 15 | -521 | $3.10^{-4}$ | 32.1 | 33.1 | $2.10^{7}$ | $5.10^{-6}$ |
| **$CaMnO_3$/PPY** | 1 | -120 | $6.10^{-8}$ | 17.8 | 15.3 | $4.10^{10}$ | $7.10^{-10}$ |
| | 5 | -372 | $1.10^{-3}$ | 36.4 | 24.3 | $6.10^{6}$ | $1.10^{-5}$ |
| | 10 | -483 | $8.10^{-3}$ | 33.0 | 31.0 | $7.10^{5}$ | $9.10^{-5}$ |
| | 15 | -576 | $3.10^{-2}$ | 28.7 | 29.8 | $2.10^{5}$ | $4.10^{-4}$ |
| **$SrMnO_3$/PPY** | 1 | -574 | $6.10^{-6}$ | 25.8 | 13.6 | $7.10^{8}$ | $7.10^{-7}$ |
| | 5 | -500 | $2.10^{-6}$ | 22.0 | 27.3 | $3.10^{9}$ | $2.10^{-8}$ |
| | 10 | -492 | $2.10^{-3}$ | 23.7 | 21.4 | $2.10^{6}$ | $2.10^{-5}$ |
| | 15 | -509 | $4.10^{-4}$ | 16.6 | 16.9 | $8.10^{6}$ | $5.10^{-6}$ |
| **$Zn_3(PO_4)_3xH_2O$** | 15 | -473 | $1.10^{-4}$ | 17.0 | 16.1 | $3.10^{7}$ | $1.10^{-6}$ |
| **Nonpigmented film** | 0 | 42 | $2.10^{-3}$ | 21.6 | 18.9 | $3.10^{6}$ | $1.10^{-5}$ |

**Table 9.** Results of linear polarization measurements of the paints containing composite pigments, DFT = 60 ± 10 μm.

The nonpigmented coating, with a spontaneous corrosion potential of 42 mV, exhibited polarization resistance of $3 \times 10^6\ \Omega$ and corrosion rate $1 \times 10^{-5}$mm/year. The reference paint with zinc phosphate at PVC = 15% exhibited a lower spontaneous corrosion potential, -473 mV, higher polarization resistance, $3 \times 10^7\ \Omega$, and corrosion rate one order of magnitude lower, $1 \times 10^{-6}$mm/year. The two materials served as reference materials for the paint films containing the pigments that modified the conductive polymers: $CaTiO_3$, $SrTiO_3$, $CaMnO_3$, $SrMnO_3$, $Ca_2ZnWO_6$, and $Ca_2ZnMoO_6$ at PVC = 1, 5, 10, and 15%.

The paint films with the $CaTiO_3$/PANI system exhibited spontaneous corrosion potential increase (-382 to -468 mV) against the reference paint with zinc phosphate. In comparison with the nonpigmented coating, the corrosion rate of this paint was lower only at PVC = 1 5 %, viz., $v_{corr} = 8 \times 10^{-7}$ and $1 \times 10^{-5}$mm/year, respectively. The paint films with the $CaTiO_3$/PPY system at PVC = 1, 5, and 10% exhibited spontaneous corrosion potentials higher than the reference paint film with zinc phosphate, viz., -175 to -354 mV. The spontaneous corrosion potential was lower, -600 mV, only at PVC = 15%. The corrosion rate of the paint film at PVC = 1% was nearly one-half that of the above reference paint, viz., $1 \times 10^{-11}$ mm/year. The paint films containing this pigment at PVC = 5 and 10% also exhibited appreciably lower corrosion rates, $5 \times 10^{-9}$ and $2 \times 10^{-8}$ mm/year, than the reference materials (nonpigmented coating and paint with zinc phosphate).

The paint films with $SrTiO_3$/PANI exhibited increase in the spontaneous corrosion potential against that of the reference zinc phosphate paint at PVC = 1%, viz., -286 mV, whereas the reverse was true at PVC = 5, 10, and 15% (-635 to -587 mV). Also, the paint film with the pigment at PVC = 1% was the only one in the increasing PVC series to exhibit a corrosion rate lower ($8 \times 10^{-7}$ mm/year) than the corrosion rate of the two reference materials. The paint films with $SrTiO_3$/PPY exhibited spontaneous corrosion potential decrease (-556 to -521 mV) compared to the reference paint film with zinc phosphate at any of the pigment concentrations used, and the corrosion rates at PVC = 1 and 5%, viz., $8 \times 10^{-10}$ and $6 \times 10^{-10}$ mm/year, respectively, were lower than the corrosion rate of the reference paint.

All of the paint films with the $CaMnO_3$/PANIand $SrMnO_3$/PANI systems exhibited spontaneous corrosion potential decrease against the reference paint with zinc phosphate; only the paint film with $CaMnO_3$/PANI at PVC = 1% exhibited a corrosion rate at the same level as the reference paint film, viz., $1 \times 10^{-6}$ mm/year. All of the paint films with the $CaMnO_3$/PPY and $SrMnO_3$/PPY systems also exhibited spontaneous corrosion potential decrease against the reference paint with zinc phosphate (-576 to -483 mV), except for the paint films with $CaMnO_3$/PPY at PVC = 1 and 5%, where the spontaneous corrosion potential values were higher (-120 and -372 mV, respectively). The paint films with $CaMnO_3$/PPY at PVC = 1% exhibited a lower corrosion rate than the reference paint with zinc phosphate, viz., $7 \times 10^{-10}$ mm/year, and the same was true of the paints with $SrMnO_3$/PPY at PVC = 1 and 5% ($7 \times 10^{-7}$ and $2 \times 10^{-8}$mm/year, respectively).

### *2.6.1. Conclusion from measurements of linear polarization*

Perovskites with PPY layer appeared to be superior to perovskites with PANI layer, also in the linear polarization measurements, and lower concentrations of the pigments coated with PPY were more efficient than high concentrations. Paint films with the following composite pigments exhibited better resistance and lower corrosion rates than the reference paint with the zinc phosphate pigment: $CaTiO_3$/PPY at PVC = 1 and 5% (1 × $10^{-11}$ and 5 × $10^{9}$ mm/year, respectively); $SrTiO_3$/PPY at PVC = 1 and 5% (8 × $10^{-10}$ and 6 × $10^{-10}$ mm/year, respectively); $CaMnO_3$/PPY at PVC = 1% (7 × $10^{-10}$ mm/year); $SrMnO_3$/PPY at PVC = 1 and 5% (7 × $10^{-7}$ and 2 × $10^{-8}$ mm/year, respectively); $CaTiO_3$/PANI at PVC = 1 (8 × $10^{-7}$ mm/year); and $SrTiO_3$/PANI at PVC = 1% (8 × $10^{-7}$ mm/year).

## 3. Conclusion

This study was devoted to the anticorrosion and adhesive-barrier properties of paint films containing perovskite pigments whose surface had been modified with polyaniline phosphate or polypyrrole phosphate. The findings from the tests can be summarized as follows:

1. PPY as the conductive polymer for pigment particle surface modification was found superior to PANI with respect to the corrosion-inhibiting efficiency of the composite pigments. It is an advantage that a low PVC in the epoxy-ester resin based paint is adequate to attain a high anticorrosion efficiency. The pigments with PPY exhibited more favorable physicochemical properties (water-soluble content, pH of the extracts, etc.), which did not detract from the binder's barrier efficiency. For some of the pigments, their optimum concentration in the paints was as low as PVC = 1%, which is beneficial also from the financial aspect.

2. The highest anticorrosion efficiency in the environment with a neutral NaCl mist was exhibited by paint films containing the Ca–Ti or Ca–Mn perovskites, specifically $CaTiO_3$/PPY and $CaMnO_3$/PPY, both at PVC = 1 and 5%. The same pigments, plus the Sr–Ti and Ca–Mn perovskites, were also most efficient in the environment with $SO_2$. All of the paint films attained high anticorrosion efficiencies in the $SO_2$ environment. Outstanding results were obtained with the paint films containing pigments coated with a layer of PANI, in particular, $SrTiO_3$/PANI at PVC = 5 and 10%. PPY was found superior to PANI in the environment with the NaCl mist, while the differences between the two conductive polymers were not that marked in the environment with $SO_2$, and both polymers can be used with advantage. A 1% PVC will be sufficient to attain a high anticorrosion efficiency with the pigments described above.

3. Corrosion in the cut diminished with increasing pigment concentration in the paint (particularly in the atmosphere with $SO_2$), irrespective of the conductive polymer coating the perovskite core. This result is in line with the catalytic passivation mechanism in the paint film damage area. Steel surface corrosion beneath the paint film was low up to high pigment concentrations, which was facilitated by the anodic protection mechanism [31, 32], in which the composite pigments containing metal oxides are assumed to be able to form a protective layer on the metal surface, thereby protecting the metal against corrosion.

4. The most efficient paints, based on the linear polarization measurements and the accelerated corrosion test in the neutral salt mist atmosphere, were those containing the $CaTiO_3$/PPY composite pigment and the $SrTiO_3$/PPY composite pigment.

5. All of the paints with the pigments modified with the conductive polymers exhibited good physicomechanical resistance levels—to the extent that the paints are applicable to surfaces exposed to mechanical stresses.

6. The surface-modified pigments are promising from several aspects, particularly owing to their high anticorrosion efficiency compared to that of the reference paint, low pigment concentration in the paints adequate to attain a high anticorrosion effect, and their environmental harmlessness. The results obtained suggest that the pigments deserve further research.

## Author details

Andréa Kalendová[1*], Tereza Hájková[1], Miroslav Kohl[1] and Jaroslav Stejskal[2]

*Address all correspondence to: andrea.kalendova@upce.cz

1 Faculty of Chemical Technology, University of Pardubice, Pardubice, Czech Republic

2 Institute of Macromolecular Chemistry, Academy of Sciences of the Czech Republic, Prague, Czech Republic

## References

[1] Criado M, Sobrados I, Bastidas JMX. Steel corrosion in simulated carbonated concrete pore solution its protection using sol–gel coatings. Progress in Organic Coatings 2015;88:228–236. DOI:10.1016/j.porgcoat.2015.06.002.

[2] Yang W, Li Q, Xiao QX. Improvement of corrosion protective performance of organic coating on low carbon steel by PEO pretreatment. Progress in Organic Coatings 2015;89:260–266. DOI:10.1016/j.porgcoat.2015.09.003.

[3] Vakili H, Ramezanzadeh B, Amini R. The corrosion performance and adhesion properties of the epoxy coating applied on the steel substrates treated by cerium-based conversion coatings. Corrosion Science 2015;94:466–475. DOI:10.1016/j.corsci.2015.02.028.

[4] Naderi R, Attar MM. The role of zinc aluminum phosphate anticorrosive pigment in protective performance and cathodic disbondment of epoxy coating. Corrosion Science 2010;52:1291–1296. DOI:10.1016/j.corsci.2015.02.028.

[5] Havlík J, Kalendová A, Veselý D. Electrochemical, chemical and barrier action of zinc dust/anticorrosive pigments containing coatings. Journal of Physics and Chemistry of Solids 2007;68:1101–1105. DOI: 10.1016/j.jpcs.2006.11.016.

[6] Zhang X, Sloof WG, Hovestad AX. Characterization of chromate conversion coatings on zinc using XPS and SKPFM. Surface and Coatings Technology 2005;197:168–176. DOI:10.1016/j.surfcoat.2004.08.196.

[7] Ahmed NM, Mohamed MG, Mabrouk MRX. Novel anticorrosive pigments based on waste material for corrosion protection of reinforced concrete steel. Construction and Building Materials 2015;98:388–396. DOI:10.1016/j.conbuildmat.2015.08.111.

[8] Svoboda M, Mleziva J. Properties of coatings determined by anticorrosive pigments. Progress in Organic Coatings 1984;12:251–297. DOI:10.1016/0033-0655(84)80011-4.

[9] Deya MC, Blustein G, Romagnoli RX. The influence of the anion type on the anticorrosive behaviour of inorganic phosphates. Surface and Coatings Technology 2002;150:133–142. DOI:10.1016/S0257-8972(01)01522-5.

[10] Armelin E, Alemán C, Iribarren JI. Anticorrosion performances of epoxy coatings modified with polyaniline: a comparison between the emeraldine base and salt forms. Progress in Organic Coatings 2009;65:88–93. DOI:10.1016/j.porgcoat.2008.10.001.

[11] Lang DJ, Wiek A, Bergmann MX. Transdisciplinary research in sustainability science: practice, principles, and challenges. Sustainability Science 2012;7:25–43. DOI:10.1007/s11625-011-0149-x.

[12] Saad Y, Álvarez-Serrano I, Lóprez MLX. Structural and dielectric characterization of new lead-free perovskites in the ($SrTiO_3$)–($BiFeO_3$) system. Ceramics International 2016.;42:8962–8973 DOI:10.1016/j.ceramint.2016.02.028.

[13] Feng LM, Jiang LQ, Zhu MX. Formability of $ABO_3$ cubic perovskites. Physics and Chemistry of Solids 2008;69:967–974. DOI:10.1016/j.jpcs.2007.11.007.

[14] Ramadass N. $ABO_3$-type oxides, their structure and properties: a bird's eye view. Materials Science and Engineering 1978;36:231–239. DOI:10.1016/0025-5416(78)90076-9.

[15] Kalendová A, Veselý D, Sapiurina IX. Anticorrosion efficiency of organic coatings depending on the pigment volume concentration of polyaniline phosphate. Progress in Organic Coatings 2008;63:228–237. DOI:10.1016/j.porgcoat.2008.06.005.

[16] Scher H, Zallen R. Critical density in percolation processes. Chemical Physics 1970;53:3759–3761. DOI:10.1063/1.1674565.

[17] Křivka I, Prokeš J, Tobolková EX. Application of percolation concepts to electrical conductivity of polyaniline-inorganic salt composites. Materials Chemistry 1999;9:2425–2428. DOI:10.1039/A904687I.

[18] Kalendová A, Veselý D, Kohl MX. Anticorrosion efficiency of zinc-filled epoxy coatings containing conducting polymers and pigments. Progress in Organic Coatings 2015;78:1–20. DOI:10.1016/j.porgcoat.2014.10.009.

[19] Choudhary VR, Banerjee S, Uphade BS. General: activation by hydrothermal treatment of low surface area $ABO_3$-type perovskite oxide catalysts. Applied Catalysis A 2000;197:183–186. DOI:10.1016/S0926-860X(99)00485-8.

[20] Trojan M, Brandova D, Solc Z. Study of the thermal preparation and stability of tetrametaphosphates of bivalent metals. Thermochimica Acta 1987;110:343–358. DOI: 10.1016/0040-6031(87)88244-8.

[21] Alizahed R, Beaudoin JJ, Ramachandran VSX. Applicability of the Hedvall effect to study the reactivity of calcium silicate hydrates. Advances in Cement Research 2009;21:59–66. DOI:10.1680/adcr.2008.00008.

[22] Kalendová A, Hejdová M, Veselý D. Investigation of the anticorrosion properties of perovskites in epoxy-ester resin based organic coating materials. Anti-Corrosion Methods and Materials 2015;62:197–211. DOI:10.1108/ACMM-01-2014-1344.

[23] Kalendová A, Veselý D, Kohl MX. Effect of surface treatment of pigment particles with polypyrrole and polyaniline phosphate on their corrosion inhibiting properties in organic coatings. Progress in Organic Coatings 2014;77:1465–1483. DOI:10.1016/j.porgcoat.2014.04.012.

[24] Vesely, D. Properties of calcined kaolins in anticorrosion paints depending on PVC, chemical composition and shape of particles. Progress in Organic Coatings 2012;74:82–91. DOI:10.1016/j.porgcoat.2011.11.017.

[25] Kalendová A, Veselý D, Kalenda P. Properties of paints with hematite coated muscovite and talc particles. Applied Clay Science 2010;48:581–588. DOI:10.1016/j.clay.2010.03.007.

[26] Kalendová A, Veselý D. Study of the anticorrosive efficiency of zincite and periclase-based core-shell pigments in organic coatings. Progress in Organic Coatings 2009;64:5–19. DOI:10.1016/j.porgcoat.2008.07.003.

[27] Veselý D, Kalendová A, Kalenda P. A study of diatomite and calcined kaoline properties in anticorrosion protective coatings. Progress in Organic Coatings 2010;68:173–179. DOI:10.1016/j.porgcoat.2010.02.007.

[28] Veselý D, Kalendová A, Manso MV. Properties of calcined kaolins in anticorrosion paints depending on PVC, chemical composition and shape of particles. Progress in Organic Coatings 2012;74:82–91. DOI:10.1016/j.porgcoat.2011.11.017.

[29] Kouřil M, Novák P, Bojko M. Limitations of the linear polarization method to determine stainless steel corrosion rate in concrete environment. Cement & Concrete Composites 2006;28:220–225. DOI:10.1016/j.cemconcomp.2006.01.007.

[30] Millard SG, Law D, Bungey JHX. Environmental influences on linear polarisation corrosion rate measurement in reinforced concrete. NDT&E International 2001;34:409–417. DOI:10.1016/S0963-8695(01)00008-1.

[31] Kohl M, Kalendová A. Assessment of the impact of polyaniline salts on corrosion properties of organic coatings. Koroze a ochrana materiálu 2014;58:113–119. DOI: 10.1515/kom-2015-0004.

[32] Rout TK, Jha G, Singh AKX. Development of conducting polyaniline coating: a novel approach to superior corrosion resistance. Surface and Coatings Technology 2003;167:16–24. DOI:10.1016/S0257-8972(02)00862-9.

[33] Wessling B. Scientific and commercial breakthrough for organic metals. Synthetic Metals 1997;85:1313–1318. DOI:10.1016/S0379-6779(97)80254-8.

Chapter 9

# Exfoliated Nanocomposites Based on Polyaniline and Tungsten Disulfide

Barrit C.S. Lane , Rabin Bissessur ,
Alaa S. Abd-El-Aziz , Wael H. Alsaedi ,
Douglas C. Dahn , Emma McDermott and
Andrew Martin

Additional information is available at the end of the chapter

http://dx.doi.org/10.5772/63457

**Abstract**

Nanocomposite materials consisting of polyaniline (PANI) and exfoliated $WS_2$ were synthesized. The $WS_2$ was prepared by reacting tungstic acid with thiourea at 500°C under nitrogen flow. Samples were prepared with a $WS_2$ content of 1, 5, 7.5, 10, 12.5, 15, 20, 37, and 64% by mass. An improvement in the electronic conductivity value of the PANI was observed through the incorporation of exfoliated $WS_2$. The electronic conductivity of PANI-15%$WS_2$ was 24.5 S/cm, an eightfold increase when compared to pure PANI. Powder X-ray diffraction (XRD), transmission electron microscopy (TEM) and electron paramagnetic resonance (EPR) provided evidence that the nanocomposites are in an exfoliated state. XRD and TEM showed that the nanocomposites were completely amorphous, suggesting lack of structural order in these materials, while their EPR signals were considerably narrower compared to pure PANI, indicating the formation of genuine exfoliated systems. Furthermore, our research showed that $WS_2$ can be used as a filler to improve activation energy of decomposition of the polymer. By using the Ozawa method, we studied the decomposition kinetics for the nanocomposites, as well as for the pure polymer. The activation energy for the decomposition of pure PANI was found to be 131.2 kJ/mol. Increasing the amount of $WS_2$ to 12.5% in the PANI increases the activation energy of decomposition to 165.4 kJ/mol, an enhancement of 34.2 kJ/mol over the pure polymer.

**Keywords:** nanocomposite, polyaniline, transition metal dichalcogenide, exfoliated systems, graphene analogous material

# 1. Introduction

There has recently been a significant amount of interest concerning the development of a wide variety of inorganic-polymer-based nanocomposite materials. The preparation of these nanocomposites involves the incorporation of various inorganic nanoparticles into the matrix of a certain polymer using methods such as sol-gel processing, blending or mixing the polymer and filler material, as well as *in situ* polymerization [1]. A variety of different nanoparticles including clays, metal oxides, transition metal dichalcogenides (TMDs) and semiconductor metallic crystals have been incorporated into polymeric materials. The resulting nanocomposites typically exhibit improved properties derived from the presence of both the polymer and filler material. Depending on the polymer and the filler material, the improved properties synergistically derived from the two components may include enhancement in thermal stability, electrical conductivity and mechanical strength.

Graphene is a material that has attracted a tremendous amount of interest in the scientific community. It is an allotrope of carbon prepared by exfoliation of graphite into individual two-dimensional layers. Interest in graphene arises from the fact that it possesses a number of unique electrical, thermal, and mechanical properties due to its two-dimensional hexagonal structure [2]. Since the discovery of graphene, much effort has been placed into other layered materials that can be exfoliated into individual sheets in a similar fashion to graphite. One of the most widely studied class of graphite analogues are the transition metal dichalcogenides (TMDs) such as $MoS_2$, $WS_2$, $MoSe_2$ and $WSe_2$. These layered materials possess hexagonal lattice structures that can be exfoliated into single layers, similar to graphite. Due to its ease of exfoliation, $MoS_2$ has become the most researched TMD for its use in a variety of devices. However, it is interesting to note that $WS_2$ is much harder to exfoliate, although it is structurally similar to $MoS_2$. The challenge with the exfoliation of $WS_2$ has inspired many researchers to examine other methods of exfoliation. Recently, Matte et al. prepared $WS_2$ nanosheets in an exfoliated state by grinding tungstic acid with an excess of thiourea and heating the reaction mixture to 500°C in an inert atmosphere.

Since the discovery of its conductive nature, polyaniline (PANI) has been one of the most extensively studied electronically conductive polymers for a number of reasons, such as its relative ease of synthesis, low cost of production, high conductivity and impressive environmental stability [3, 4]. Due to these reasons, PANI is considered to be the most promising conductive polymer when compared to others such as polypyrrole, polythiophene and polyacetylene [4]. Over the past 30 years, PANI has become a material of great importance due to its many applications such as in rechargeable batteries, microwave and radar devices, nonlinear optical and light-emitting devices, sensors, catalysts and solar cells [5–12]. In this chapter, we discuss the synthesis of exfoliated $WS_2$ and the preparation of inorganic-polymer nanocomposites through the incorporation of $WS_2$ into the polymer matrix of PANI via an *in situ* polymerization technique. Characterization of the exfoliated $WS_2$ and PANI-$WS_2$ nanocomposites is also discussed.

### 1.1. Polyaniline

Polyaniline (PANI) is among the most widely studied polymeric materials due to its electronically conductive nature. The discovery of PANI can be traced back almost 200 years to the experiments of Runge, who reported a colour change from dark green to black upon heating a mixture of copper(II) chloride and aniline nitrate [13]. Although the existence of this polymer has been known for quite some time, it was not until over 100 years after its discovery that PANI was shown to demonstrate significant electrical conductivity [14].

Generally speaking, the synthesis of PANI can be achieved using two different methods: chemical polymerization and electrochemical polymerization. The chemical synthesis of PANI involves the polymerization of aniline using strong oxidizing agents in acidic media. The electrochemical polymerization of PANI can be carried out in a one compartment cell or a two compartment cell in an aqueous medium or a suitable solvent [15]. It has been reported that PANI synthesized using the electropolymerization method results in a sample with lower conductivity, lower crystallinity, higher solubility, and higher bandgap energy when compared to the chemically synthesized polymer [4]. Although it is apparent that there are some advantages associated with the electrochemical synthesis of PANI, the chemical polymerization method is favoured for the large scale production of the polymer [15].

Most frequently, the oxidants used in the chemical synthesis of PANI include ammonium peroxydisulfate (APS) and Fe(III) compounds; however, a wide variety of oxidizing agents have recently been employed. These oxidizing agents include a number of different transition metal compounds such as Mn(III), Mn(IV), Cr(VI), V(V) and Cu(II). Other reported oxidants include $KIO_3$, $H_2O_2$ and benzoyl peroxide [16]. There have been many reports on the chemical synthesis of PANI; however, the use of APS in its preparation typically results in the highest electrical conductivity [17].

Depending on the type of oxidizing agent and the reaction conditions, there are two fundamentally different ways that aniline can be oxidized. If the oxidizing agent has a high oxidation potential and does not possess a reactive oxygen atom, it is able to remove electrons or hydrogen from the aniline monomer which results in the formation of dianilines in reduced and/or oxidized form, oxidatively cyclized dianiline products, as well as linear/branched aniline oligomers and PANI. Oxidizing agents with a reactive oxygen atom are able to either donate the oxygen to the aniline monomer resulting in the formation of oxygen-containing products, and/or remove electrons/hydrogen from the aniline monomer [16]. There are a number of oxidants which can act as either oxygen donors or electron acceptors depending on the reaction conditions. In the case of oxygen donor species, high temperatures and alkaline conditions result in the formation of favourable products. However, the opposite is true for the electron acceptor mechanism, where low temperatures and acidic conditions produce favourable results. Peroxydisulfate salts such as APS are able to either donate oxygen or accept electrons; however, the latter is the preferred mechanism for the production of PANI. The reaction of APS with aniline in alkaline conditions at an elevated temperature prompts the donation of oxygen by the APS which results in the production of insoluble oxygen-/sulphur-containing oligoanilines, whereas the same reaction in acidic conditions at a low temperature leads to the production of PANI as the major product through acceptance of electrons by APS [14].

PANI exists in four different forms depending on the level of oxidation of the polymer. The fully reduced form of the polymer is known as leucoemeraldine, the half oxidized form is referred to as the emeraldine base, and the fully oxidized form is pernigraniline. Although PANI is known to be a conductive polymer, all of these three forms are insulating. The conductive nature of PANI arises only when it is in the emeraldine salt form, where repeating units of the completely oxidized form are followed by units of the completely reduced form [14]. When doped with HCl, PANI demonstrates a significant increase in conductivity by approximately 10 orders of magnitude. Although this mechanism is not entirely understood, it has been proposed that protonation of the emeraldine base of the polymer leads to a spinless bipolaron structure, which rearranges and splits into two polaron units resulting in the emeraldine salt form [18].

### 1.2. Transition metal dichalcogenides

Graphene has gained significant attention over the past number of years due to its incredible properties. Research into 2D graphene has led to the award of the Nobel Prize in Physics in 2010. Graphene possesses unique electrical, thermal, and mechanical properties and is widely researched for its potential applications in different devices [2]. The unique properties of graphene are due to its two-dimensional hexagonal structure. The discovery of graphene has triggered a considerable amount of interest in other materials that possess layered structures that can be exfoliated into single layers.

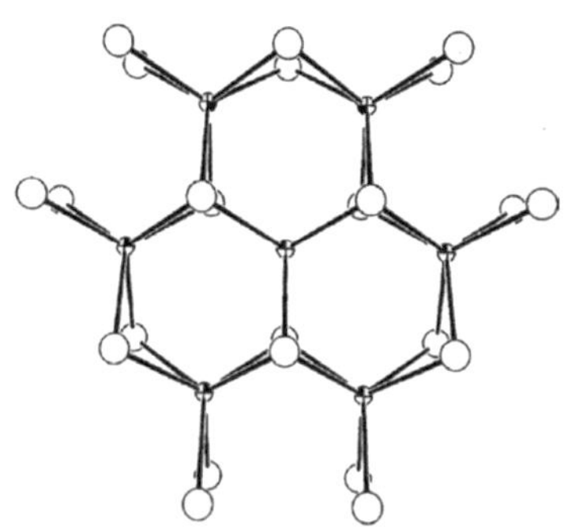

**Figure 1.** Structure of graphene analogue $WS_2$.

Layered materials that have recently been studied include TMDs, $BC_3$, silicene, transition metal carbides and nitrides, and a number of coordination polymers, each of which exists in a structure composed of stacked layers analogous to graphite [19]. These graphite analogues can be cleaved into single layers that are more or less structurally similar to that of graphene. Among these graphene analogous materials, the TMDs are among the most widely studied. Similar to graphene, they possess a two-dimensional hexagonal 'honey-comb' like lattice structure as shown in **Figure 1**. These materials are composed of a transition metal that is covalently sandwiched between two chalcogen atoms which make up the two-dimensional layer. Weak van der Waals forces exist between the two-dimensional sheets of the TMDs and are responsible for the layered structure of the materials.

There are approximately 60 different TMDs, 40 of which exist in the layered form [20]. Variations in the compositions of TMDs result in a variety of different properties possessed by the compound. TMDs can exist in three different forms which are a result of altering the transition metal or the chalcogen (S, Se, Te) causing changes in the coordination or oxidation state of the material. These forms include metallic, semimetallic and semiconductor. For example, $NbSe_2$ is a metallic material, whereas $WS_2$ is semiconducting [21].

Out of the many TMDs, those containing a group six transition metal are among the most highly researched. Recently, $MoS_2$ as well as the chemically similar $WS_2$ have been materials of interest due to the fact that they are layered semiconductors with tunable bandgaps depending on size [22, 23]. Much like graphene, these materials possess unique properties when in an exfoliated state. Due to its ease of exfoliation, $MoS_2$ has been the more highly researched TMD for its use in a variety of devices.

There are a number of methods for preparing exfoliated $MoS_2$, the most common being exfoliation by the n-butyllithium technique [24]. First, the bulk $MoS_2$ is mixed with an n-butyllithium/hexane solution in order to achieve lithium-intercalation. The $Li_xMoS_2$ is then submerged in water, resulting in the production of LiOH and the evolution of $H_2$ gas causes separation of the $MoS_2$ layers.

Although $WS_2$ is similar in chemical structure to $MoS_2$, exfoliation of $WS_2$ using the n-butyllithium technique is much more difficult [25, 26]. This difficulty arises from the fact that when $WS_2$ is mixed with the n-butyllithium hexane solution, very little lithium intercalation occurs. When performing lithiation of $MoS_2$, the resulting $Li_xMoS_2$ typically has a lithium content of $x \approx 1$. $WS_2$ under identical conditions yields $Li_xWS_2$, with an x value of under 0.4. As a result, there is negligible $H_2$ formation when $Li_xWS_2$ is submerged in water causing the degree of $WS_2$ exfoliation to be very low [26]. Consequently, this has inspired researchers to examine other methods of $WS_2$ exfoliation. A solvothermal method has been used to increase the amount of lithium intercalation in $WS_2$ [27]. Other reported methods include using a strong acid treatment to separate the $WS_2$ layers, as well as using different synthetic techniques to prepare the $WS_2$ in an already exfoliated state [28–30].

Recently, Wu et al. reported a method of preparing $WS_2$ nanosheets in an exfoliated state. This method involved ball-milling $WO_3$ in the presence of sulphur and heating to 500–800°C under argon atmosphere [29]. Shortly after, Matte et al. reported a method for the preparation of exfoliated $WS_2$ which involved the use of thiourea as the source of sulphur and tungstic acid as the source of tungsten. This process simply involves grinding of tungstic acid with an excess of thiourea (tungstic acid–thiourea 1:48 mole ratio), and heating the mixture to 500°C under nitrogen atmosphere [30].

Once in an exfoliated state, $WS_2$ has a wide variety of potential applications including photoconductors, catalysts, lubricants and lithium batteries [29]. Although there are many potential applications for this material, the amount of work done on $WS_2$ has been quite limited compared to other TMDs.

### 1.3. Inorganic-polymer nanocomposites

There has been significant research into the development of a wide variety of inorganic-polymer-based nanocomposites. Inorganic-polymer nanocomposites are materials that consist of inorganic nanoparticles mixed with a polymer where the size of the inorganic nanoparticle is less than 100 nm, for example layered materials, incorporated into a polymer host [31]. Research into these types of materials is primarily concerned with the enhancement of the polymer properties through the incorporation of different particles with varying characteristics. Examples of polymer properties that are considered include thermal stability, electrical conductivity and mechanical strength. A variety of different nanoparticles such as clays, metal oxides, transition metal dichalcogenides and semiconductor metallic crystals have been previously incorporated into polymeric matrices. There are a number of different methods used to prepare inorganic-polymer nanocomposites such as blending or mixing the polymer and filler material, sol–gel processing, and *in situ* polymerization [1].

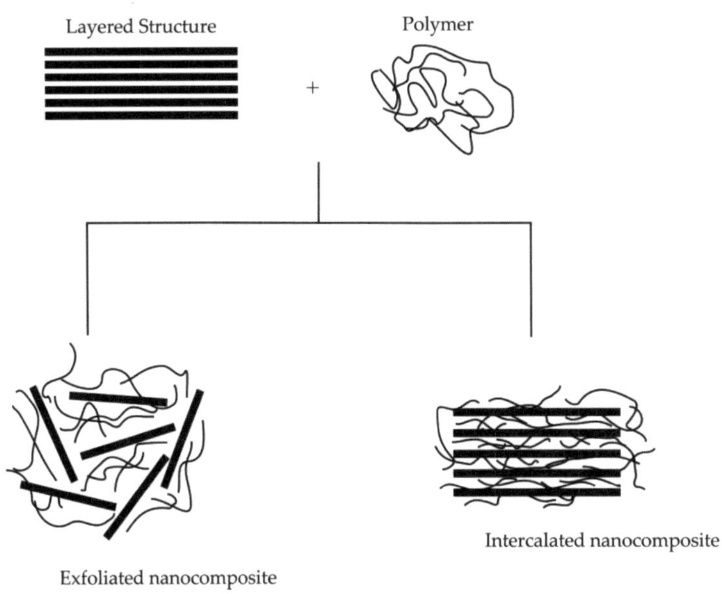


**Figure 2.** Intercalated and exfoliated nanocomposites.

Polymer-based nanocomposites containing layered materials can exist in two major forms: exfoliated nanocomposites and intercalated nanocomposites (**Figure 2**). Exfoliated nanocomposites consist of single or few layered nanosheets incorporated within a polymer matrix. Conversely, intercalated nanocomposites are materials formed when the layered species are stacked with polymer chains, resulting in alternating inorganic and polymeric layers.

XRD characterization of the nanocomposite is very useful for determining its structural characteristics. In particular, XRD can be used to detect three types of nanocomposites. For an immiscible nanocomposite, the powder pattern is similar to that of the starting layered structure, demonstrating no change in d-spacing, and therefore suggesting that no separation of the layers took place. For an intercalated nanocomposite, its XRD pattern will be different from that of the pristine lamellar structure. Typically, an increase in d-spacing is observed due to the incorporation of the polymer in the gallery spaces of the inorganic host. Finally, in the case of an exfoliated nanocomposite, no peaks are expected in the XRD scan. This is due to the fact that there is no regular spacing between the layers of the material, indicating that the layers are exfoliated and distributed throughout the polymer matrix [31]. It is also worth noting that the XRD scan of the exfoliated nanocomposite assumes that the incorporated polymer is completely amorphous. However, a polymer possessing a crystalline structure could show diffraction peaks in the XRD pattern of its corresponding exfoliated nanocomposite.

## 2. Synthetic methodology

### 2.1. Synthesis of exfoliated $WS_2$

Tungstic acid was purchased from Sigma-Aldrich and used as received. Typical synthesis of exfoliated $WS_2$ involved grinding and mixing tungstic acid in excess of thiourea with the use of a mortar and pestle. The mole ratio of the tungstic acid to thiourea was 1:48. This mixture was placed in a ceramic reaction vessel and inserted into a ceramic tube installed in a split furnace. The mixture was then heated to approximately 500°C for 3.5 h under nitrogen atmosphere and then allowed to cool overnight under nitrogen purge. This yielded a black solid product which was ground to a fine powder for use in future reactions.

### 2.2. Synthesis of polyaniline-$WS_2$ nanocomposites

A sample of distilled aniline along with 1 M HCl was placed in a large Erlenmeyer flask. The solution was placed in an ice bath and cooled to approximately 0°C with mechanical stirring. A sample of the previously synthesized $WS_2$ was suspended in 10–20 mL of deionized water and probe-sonicated for approximately 20 min at 30% amplitude. The $WS_2$ suspension was then added to the aniline solution and stirred mechanically while keeping the temperature between 0°C and 5°C.

A solution of ammonium peroxydisulfate (APS) was prepared by dissolving a sample of APS in 1 M HCl. This solution was cooled in an ice bath and then slowly added to the aniline-$WS_2$ mixture. The mole ratio of aniline to APS was 1:1. The reaction mixture was left to stir for 1.5 h before the product was collected by vacuum filtration, washed thoroughly with 1 M HCl, and left to dry overnight under suction.

For the synthesis of bulk polyaniline, the same procedure as outlined above was used, but without addition of any $WS_2$.

## 3. Instrumentation

Powder X-ray diffraction (XRD) was performed using a Bruker AXS D8 Advance instrument equipped with a graphite monochromator, variable divergence slit, variable antiscatter slit and scintillation detector. Cu (K$\alpha$) radiation ($\lambda$ = 1.524 Å) was used for sample measurements carried out at room temperature. Samples were run in air from 2 to 60° (2θ).

Thermogravimetric analysis (TGA) was performed on a TA Instruments TGA Q500 in dry air or nitrogen purge. For measurements performed under nitrogen, the furnace was allowed to purge for 15 min using a 60.00 mL/min purge flow rate before the measurement was started. The heating rates used were varied between 5 and 40°C/min. The range of heating used was from 20 to 800°C.

Scanning electron micrographs were obtained on an LVEM5 benchtop instrument, operating at 5 kV. Powdered samples were placed on carbon-taped stubs prior to analysis.

High-resolution transmission electron microscopy (HRTEM) was performed on a Hitachi 7500 Bio-TEM, using an accelerating voltage of 80 kV. The powdered samples were dispersed in deionized water with the help of ultrasonication, and the dispersed samples were cast on carbon-coated copper grids.

Electrical conductivity measurements were performed using the four-probe van der Pauw technique on a home built system. Samples were prepared as thin circular pressed pellets, 12.7 mm in diameter, with a thickness between 0.5 and 1.0 mm. Pellets were attached to a sample holder with either vacuum grease or double-sided tape, and wires were secured to the pellets using silver or carbon paste. Room-temperature conductivity measurements were performed in air. For some samples, variable-temperature conductivity measurements were also made, with the pellets in vacuum.

EPR spectra were recorded using a Bruker Elexsys E580 pulse spectrometer operating in CW mode. The solid samples were placed in Suprasil EPR sample tubes (4 mm o.d.) and were sealed.

## 4. Results and discussion

### 4.1. Characterization of exfoliated $WS_2$ nanoparticles

In order to determine whether or not the synthesized product was in fact $WS_2$ in an exfoliated state, powder XRD diffraction data were collected. The diffractogram is shown below in **Figure 3(b)**.

XRD data show that the synthesized $WS_2$ is highly amorphous in nature, with a low crystallinity percentage of 17.2%, as determined by the XRD software. The diffractogram shows the presence diffraction peaks at approximately 33° (2θ) and 58° (2θ) corresponding to the (100) and the (110) planes, respectively. In contrast, the XRD diffractogram of pristine layered $WS_2$

(**Figure 3a**), shows the characteristic (002) line just above 14° (2θ). The lack of the (002) line and the presence of the (100) and (110) lines in the diffractogram of the synthesized $WS_2$ indicate that it is, in fact, in an exfoliated state. The XRD data correlate well with the literature for exfoliated $WS_2$ synthesized via the n-butyllithium method [25].

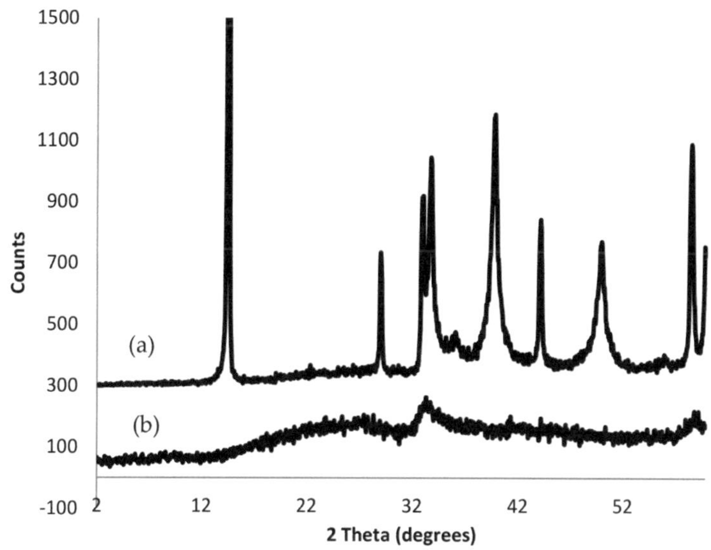


**Figure 3.** (a) XRD diffractogram of (a) pristine layered $WS_2$ (Aldrich), (b) synthesized $WS_2$.

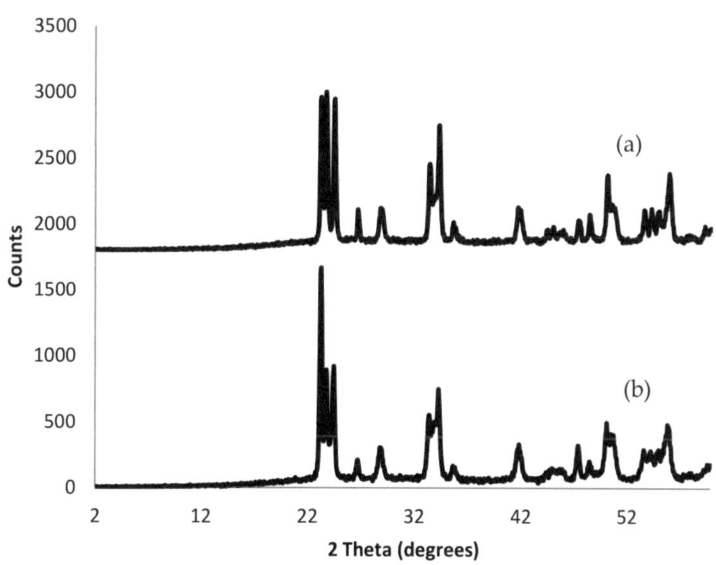


**Figure 4.** XRD of products obtained after TGA in air: (a) product from synthesized exfoliated sample, (b) product from pristine layered sample.

TGA was performed in air on the synthesized exfoliated $WS_2$ and on pristine layered $WS_2$ (Aldrich). Both TGA runs resulted in the formation of a light green powder, demonstrating oxidation of both samples. It was assumed that in both cases, $WO_3$ was formed as a result of oxidation of the initial materials. To confirm this, XRD data were collected on both light green samples and the results were compared to the literature data for pure $WO_3$. It was noted that the diffraction peaks shown in both diffractograms (**Figure 4**) correlated well with one another, and closely resembled the literature powder pattern of $WO_3$ [32]. It was also found that the light green products from exfoliated $WS_2$ and pristine layered $WS_2$ demonstrated a similar high percentage in crystallinity of 79.4% and 76.6%, respectively. This confirms that both $WS_2$ samples were oxidized to $WO_3$, and thus possess a similar chemical composition.

Electrical conductivity measurements were performed on pressed pellets of the synthesized exfoliated $WS_2$ using the van der Pauw technique. The synthesized exfoliated $WS_2$ was found to be non-conductive, while pristine layered $WS_2$ exhibits a conductivity value of $3.6 \times 10^{-3}$ S/cm.

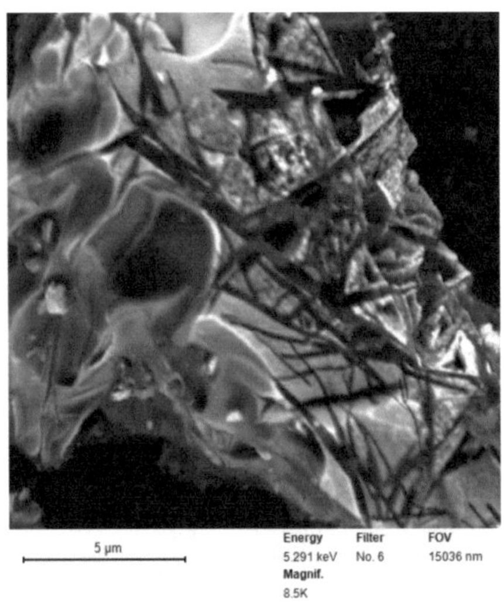


**Figure 5.** SEM micrograph of exfoliated $WS_2$.

SEM was performed on the exfoliated $WS_2$ in order to gain insight into its surface morphology. As shown in the SEM micrograph (**Figure 5**), the material appears to be amorphous, which is consistent with the XRD data.

TEM was used to further characterize the material in order to gain a more in-depth understanding of the structure of the material. As viewed under TEM, exfoliated $WS_2$ appears to be very thin and membrane like. The membranes do not stack, clearly indicating the disordered state of the material. There is also a certain degree of folding/wrinkling of the membranes. It could also be observed that the exfoliated $WS_2$ is quite porous, and might potentially have a very large surface area (**Figure 6**).

http://dx.doi.org/10.5772/63457

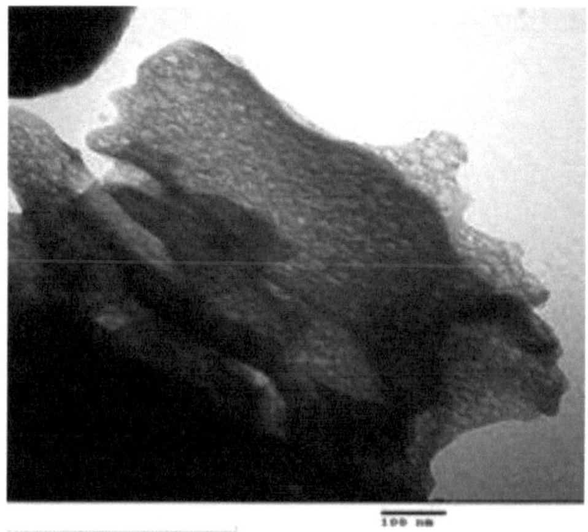


Figure 6. TEM of exfoliated $WS_2$.

## 4.2. Characterization of polyaniline

XRD was used to characterize the synthesized PANI. The diffractogram features a very broad peak, indicating a low degree of crystallinity of the material. The XRD analysis software calculated a low percent crystallinity of 11.6%, clearly indicating a highly disordered structure (**Figure 7**). This is consistent with the fact that PANI is an amorphous polymer, and is in agreement with the literature [33].

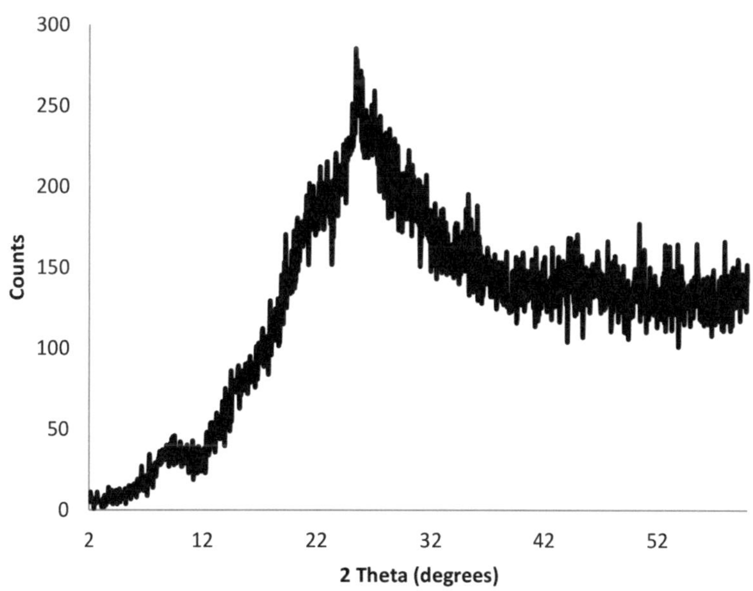


**Figure 7.** XRD diffractogram of pure PANI.

FTIR was also used to characterize the PANI. The FTIR spectrum correlates well with the literature, demonstrating wavenumbers within the same range [33].

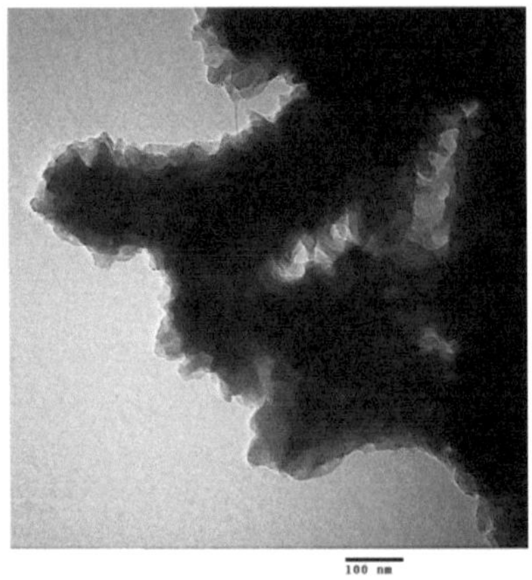


**Figure 8.** TEM micrograph of bulk PANI.

Electron microscopy was also used to further characterize the bulk PANI. Bulk PANI as observed under SEM and TEM shows that it is completely featureless, and highly disordered. The TEM micrograph of bulk PANI is shown in **Figure 8**.

### 4.3. XRD characterization of PANI-$WS_2$ nanocomposites

Powder X-ray diffraction data were collected on the prepared PANI-$WS_2$ nanocomposites (**Figure 9**).

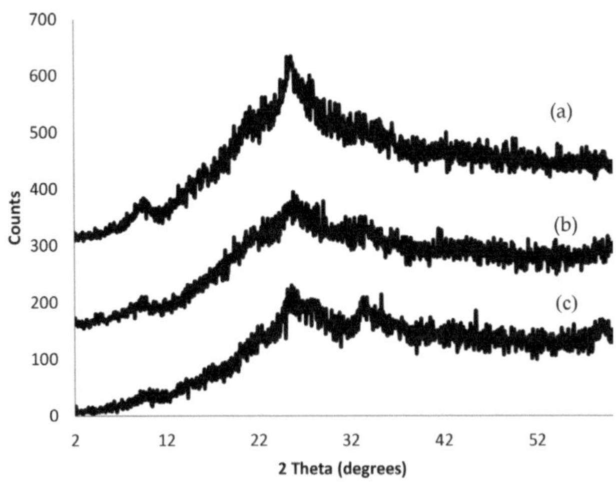


**Figure 9.** Diffractograms of nanocomposites (a) PANI-10%$WS_2$, (b) PANI-20%$WS_2$, (c) PANI-37%$WS_2$.

XRD data provided evidence for the formation of exfoliated nanocomposites. The XRD scans for all prepared compositions appeared to be highly amorphous and lacked any sharp crystalline peaks, indicating that the materials possessed a low degree of structural order. It was also observed that as the $WS_2$ content is increased, the appearance of the characteristic exfoliated $WS_2$ diffraction peaks just above 33° and 58° became more apparent. These peaks were not detected in the XRD scans of samples with lower $WS_2$ content, which closely resemble the scan of the bulk polymer.

### 4.4. Transmission electron microscopy on PANI-$WS_2$ nanocomposites

Transmission electron microscopy was further used to characterize the nanocomposites. All nanocomposites as viewed under TEM showed that they are in a completely amorphous, disordered state, and these findings are in very good agreement with the powder X-ray diffraction data. Thus, TEM further confirms the formation of exfoliated systems. As an illustration, the TEM micrographs of PANI-10%$WS_2$ and PANI-64%$WS_2$ are shown in **Figure 10(a)** and **(b)**.

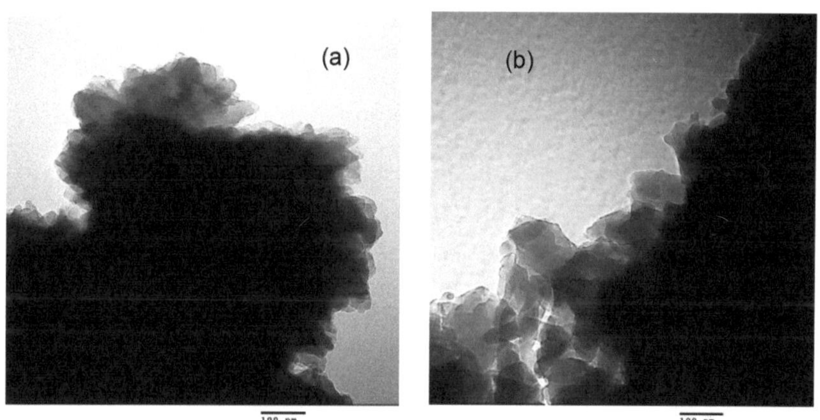


**Figure 10.** TEM micrographs of (a) PANI-10%$WS_2$, (b) PANI-64%$WS_2$.

### 4.5. EPR analysis of PANI-$WS_2$ nanocomposites

Electron paramagnetic resonance spectroscopy was used to characterize the PANI-$WS_2$ nanocomposites, as well as the pure PANI and exfoliated $WS_2$. As an illustration, the overlaid EPR spectra of pure PANI, exfoliated $WS_2$ and PANI-20%$WS_2$ nanocomposite are shown in **Figure 11**. Exfoliated $WS_2$ was found to be EPR silent, and pure PANI showed an EPR signal with a g-value of 2.0023. However, the PANI-20%$WS_2$ nanocomposite demonstrated an intense peak that was significantly narrower than that of the pure PANI. The dramatic difference in EPR signal between the pure PANI and that of the PANI-20%$WS_2$ material implies that the $WS_2$ and PANI in the nanocomposite are not simply a physical mixture, but are actually mixed at the molecular level. Other compositions showed similar behaviour.

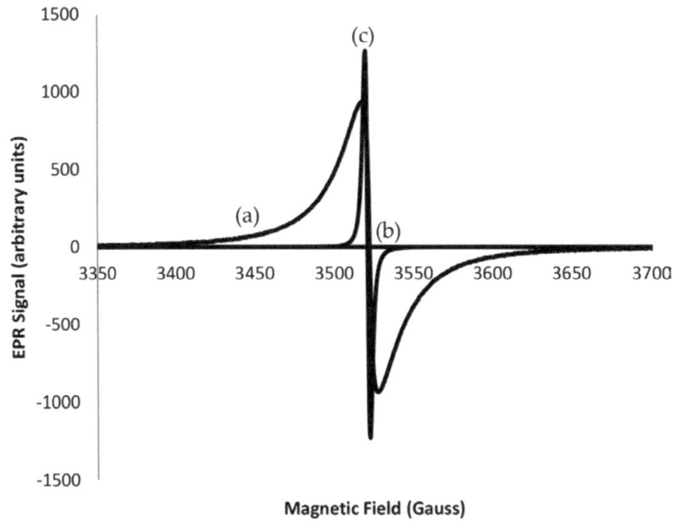


**Figure 11.** EPR spectra of (a) pure PANI, (b) exfoliated $WS_2$, and (c) PANI-20%$WS_2$.

### 4.6. Decomposition kinetics of PANI-$WS_2$ nanocomposites

The decomposition kinetics of the PANI-$WS_2$ nanocomposites (1, 5, 7.5, 10, 12.5, 15, 20, 37 and 64%) and pure PANI were determined using the Ozawa method [34, 35]. The Ozawa method involved performing a series of TGAs at different heating rates (5, 10, 20 and 40°C/min), obtaining the conversion curves from the thermograms, and then a series of Ozawa plots at different conversion values. The activation energy of decompositions ($E_a$) were then determined from the Ozawa plots. The results are summarized in **Table 1** and displayed in **Figure 12**. The data demonstrate that the presence of $WS_2$ in PANI enhances its activation energy of decomposition. The maximum enhancement (about 34 kJ/mol) was observed at 12.5% by mass of $WS_2$.

| % by mass of $WS_2$ | $E_a$ (kJ/mol) | Standard deviations (kJ/mol) |
|---|---|---|
| 0 | 131.2 | ±7.7 |
| 1 | 146.6 | ±5.1 |
| 5 | 133.1 | ±9.3 |
| 7.5 | 133.2 | ±15.9 |
| 10 | 152.3 | ±4.8 |
| 12.5 | 165.4 | ±5.5 |
| 15 | 151.2 | ±5.2 |
| 20 | 129.5 | ±8.4 |
| 37 | 144.8 | ±12.7 |
| 64 | 152.5 | ±11.5 |

**Table 1.** Activation energies of PANI-$WS_2$ nanocomposite samples.

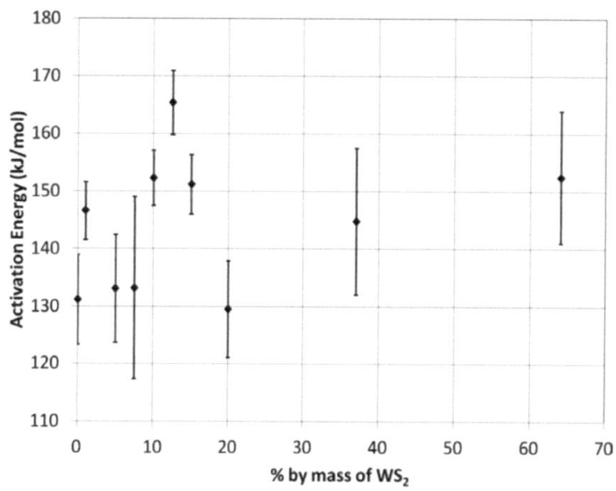


**Figure 12.** Plot of activation energy values versus mass% of $WS_2$.

## 4.7. Electrical conductivity of PANI-$WS_2$ nanocomposites

Room temperature conductivity values for the nanocomposite samples are summarized below in **Table 2**, and plotted in **Figure 13**. The accuracy of the conductivity of a given sample is about ±10%. For some of the nanocomposite compositions, several pellets were measured and each conductivity value is listed. Considerable variation is observed in some cases, but this is not surprising considering that the samples were pellets made from pressed powders. Conductivity can be affected many factors such as the pressure used in forming the pellets, the quality of inter-grain contact, sample aging, and the humidity of the air. In addition, different batches of PANI may have slightly different oxidation or doping levels.

| $WS_2$ content in nanocomposite | Room-temperature conductivity (S/cm) |
|---|---|
| 0% (Pure PANI) | 3.1, 3.7, 5.0, 5.3 |
| 1% | 0.60, 0.68 |
| 5% | 0.022, 0.024, 0.051 |
| 7.5% | 0.30 |
| 10% | 13 |
| 12.5% | 8.8, 16 |
| 15% | 24, 25 |
| 20% | 9.1 |
| 37% | 8.5, 10 |
| 64% | 8.0 |

**Table 2.** Nanocomposite conductivity at room temperature. Multiple values indicate readings on different samples of the same nominal composition.

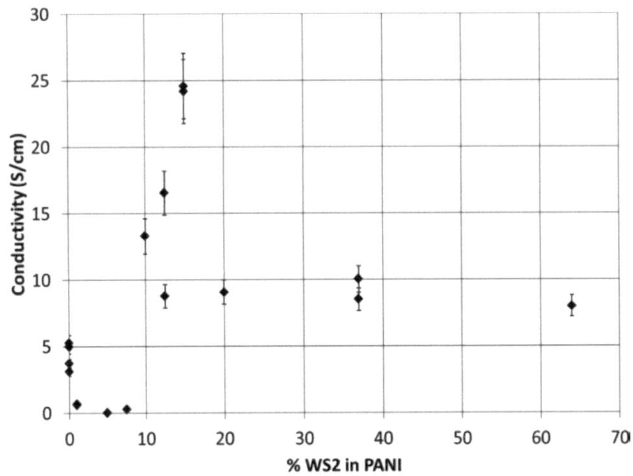


**Figure 13.** Nanocomposite conductivity at room temperature.

Two pellets of exfoliated $WS_2$ were tested but showed no detectable conductivity. $WS_2$ is known to be a semiconductor, and when a pellet pressed from bulk $WS_2$ powder (Aldrich) was measured a conductivity of $3.6 \times 10^{-3}$ S/cm was found. We conclude that either our exfoliated $WS_2$ is much less conducting than bulk $WS_2$, or that the inter-grain contact in our pellets of exfoliated $WS_2$ is so poor as effectively prevent conduction through the pellet.

As seen in **Table 2** and **Figure 13**, as the percentage of $WS_2$ in the nanocomposites is increased, the conductivity drops initially, reaching a minimum at about 5% by mass. This drop is not surprising, since we are adding poorly-conducting $WS_2$ to conducting PANI. However, as the percentage of $WS_2$ is increased further, the conductivity rises again, and samples containing at least 10% $WS_2$ demonstrate a significant enhancement in electronic conductivity over the pure polymer. The 15% samples exhibited the highest electronic conductivity at about 24 S/cm. When a two-component composite material is more conducting than either of its components, this suggests that interaction between the components has altered the electrical properties of one or both of the components. For example, a layered nanocomposite consisting of polypropylaniline intercalated between layers of FeOCl was a much better conductor than either FeOCl or polypropylaniline alone [36]. This is believed to be due to electron transfer from the polymer to the FeOCl layers. The nature of the interaction causing conductivity enhancement in the PANI/$WS_2$ system above 10% $WS_2$ is not yet understood, and will be a topic for future research.

Variable-temperature conductivity measurements were attempted on a number of the samples. In the experimental system used, the sample is in vacuum and can be cooled to 50 K or below. The pressed pellet nanocomposite samples were quite fragile and often cracked on exposure to vacuum or due to thermal stresses on cooling. For this reason, it was only possible to obtain variable-temperature data on the pure PANI, as well as 12.5% and 20% $WS_2$–PANI nanocomposite samples. The results are shown in **Figure 14**.

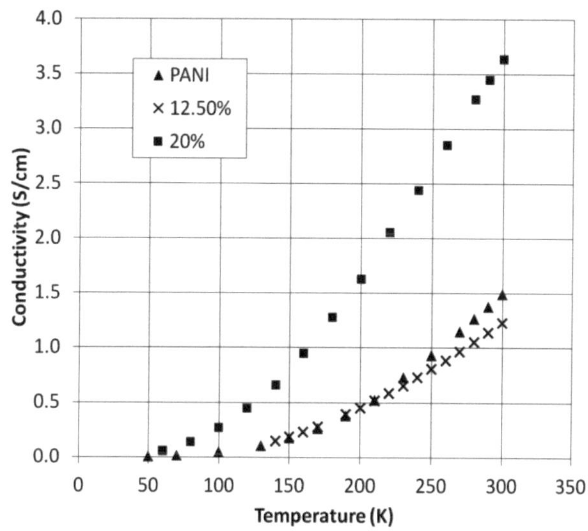


**Figure 14.** Variable temperature conductivity of PANI, PANI-12.5%$WS_2$ nanocomposite, and PANI-20%$WS_2$ nanocomposite.

It was also observed that conductivity dropped when the pellets were placed in vacuum, possibly due to the loss of small amounts of residual water. For example, the PANI sample shown in **Figure 14** had a conductivity of 3.1 S/cm in air. This dropped to 2.0 S/cm after 4 h in vacuum at room temperature, and to 1.5 S/cm after 24 h. After this, the variable-temperature measurements were made, and then the sample was returned to air for four days, after which the conductivity was 2.9 S/cm, close to the original value. Similar conductivity decreases in vacuum were observed in the nanocomposite samples, especially the 12.5% sample, which was in vacuum for a longer time than the others.

The variable-temperature data are consistent with the variable-range hopping (VRH or Mott law) model for conduction in disordered materials. In the VRH model, the resistivity $\rho$ is given as a function of absolute temperature $T$ by

$$\rho(T) = \rho_0\, exp\left[ (\frac{T_0}{T})^{\frac{1}{d+1}} \right] \quad (1)$$

where $d$ is the dimensionality of the material, and $T_0$ and $\rho_0$ are parameters that are nearly independent of temperature. Both three-dimensional and two-dimensional Mott law behaviours have been reported in a number of studies of conducting polymer and nanocomposite systems [36–38]. If Eq. (1) with $d = 3$ is valid, a plot of ln $\rho$ as a function of $T^{-1/4}$will be linear. Our data are plotted in this format in **Figure 15**. The straight-line behaviour seen in the figure shows that the three-dimensional VRH model is a reasonable description of electrical conduction in these materials.

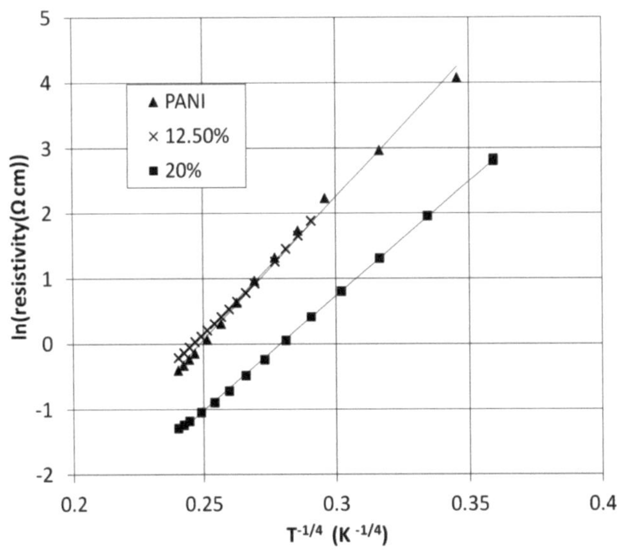


**Figure 15.** Three-dimensional VRH model of PANI, PANI-12.5%$WS_2$ nanocomposite, and PANI-20%$WS_2$ nanocomposite.

If $d = 2$ in Eq. (1) a plot of ln $\rho$ as a function of $T^{-1/3}$will be linear. Our data also appears nearly linear in such a plot, and the 12.5% data with its more limited temperature range even appears nearly linear in a plot of ln $\rho$ as a function of $T^{-1/2}(d = 1)$. We conclude that our variable-temperature conductivity data are consistent with VRH conduction, although the dimensionality $d$ cannot be determined precisely.

## 5. Conclusion

A significant amount of information was generated on the synthesized PANI-$WS_2$ nanocomposite materials. Evidence has been provided to show that the $WS_2$ synthesized from thiourea and tungstic acid is in an exfoliated state. XRD, TEM and EPR provide evidence that genuine exfoliated nanocomposites have been prepared through the incorporation of the exfoliated $WS_2$ into PANI. Furthermore, the incorporation of exfoliated $WS_2$ into PANI resulted in significant changes in the thermal and electrical properties of the polymer.

## Acknowledgements

Financial support was provided by the Natural Sciences and Engineering Research Council of Canada (NSERC) and the University of Prince Edward Island is gratefully acknowledged. We thank Professor Art van der Est (Chemistry Department, Brock University) for collecting the EPR data.

## Author details

Barrit C.S. Lane[1], Rabin Bissessur[1*], Alaa S. Abd-El-Aziz[1], Wael H. Alsaedi[1], Douglas C. Dahn[2], Emma McDermott[1] and Andrew Martin[2]

*Address all correspondence to: rabissessur@upei.ca

1 Department of Chemistry, University of Prince Edward Island, Charlottetown, Prince Edward Island, Canada

2 Department of Physics, University of Prince Edward Island, Charlottetown, Prince Edward Island, Canada

## References

[1] Haldorai Y, Shim J-J, Lim KT. Synthesis of Polymer-Inorganic Filler Nanocomposites in Supercritical $CO_2$. Journal of Supercritical Fluids. 2012; 71: 45–63. doi:10.1016/j.supflu.2012.07.007.

[2] Singh V, Joung D, Zhai L, Das S, Khondaker SI, Seal S. Graphene Based Materials: Past, Present and Future. Progress in Materials Science. 2011; 56: 1178–1271. doi:10.1016/j.pmatsci.2011.03.003.

[3] Stejskal J, Gilbert RG. Polyaniline. Preparation of a Conducting Polymer (IUPAC Technical Report). Pure and Applied Chemistry. 2002; 74: 857–867. doi:10.1351/pac200274050857.

[4] Bhadra S, Singha NK, Khastgir D. Electrochemical Synthesis of Polyaniline and its Comparison with Chemically Synthesized Polyaniline. Journal of Applied Polymer Science. 2007; 104: 1900–1904. doi:10.1002/app.

[5] MacDiarmid AG, Yang LS, Huang WS, Humphrey BD. Polyaniline: Electrochemistry and Application to Rechargeable Batteries. Synthetic Metals. 1989; 18: 393–398. doi: 10.1016/0379-6779(87)90911-8.

[6] de Chanterac H, Roduit P, Belhadj-Tahar N, Fourrier-Lamer A, Djigo Y, Aeiyach S, Lacaze PC. Electromagnetic Absorption of Polyanilines at Microwave Frequencies. Synthetic Metals. 1992; 52: 183–192. doi: 10.1016/0379-6779(92)90306-4.

[7] Mäkelä T, Sten J, Hujanen A, Isotalo H. High Frequency Polyaniline Shields. Synthetic Metals. 1999; 101: 707. doi:10.1016/S0379-6779(98)01095-9.

[8] Halvorson C, Cao Y, Moses D, Heeger AJ. Third Order Nonlinear Optical Susceptibility of Polyaniline. Synthetic Metals.1993; 57: 3941-3944. doi: 10.1016/0379-6779(93)90538-8.

[9] Wang HL, MacDiarmid AG, Wang YZ, Gebier DD, Epstein AJ. Application of Polyaniline (Emeraldine Base, EB) in Polymer Light-Emitting Devices. Synthetic Metals. 1996; 78: 33–37. doi:10.1016/0379-6779(95)03569-6.

[10] Dutta D, Sarma TK, Chowdhury D, Chattopadhyay A. A Polyaniline-Containing Filter Paper that Acts as a Sensor, Acid, Base, and Endpoint Indicator and Also Filters Acids and Bases. Journal of Colloid and Interface Science. 2005; 283: 153–159. doi:10.1016/j.jcis. 2004.08.051.

[11] Drelinkiewicz A, Waksmundzka-Gòra A, Sobczak, JW, Stejskal J. Hydrogenation of 2-ethyl-9, 10-anthraquinone on Pd-polyaniline($SiO_2$) Composite Catalyst: The Effect of Humidity. Applied Catalysis A: General. 2007; 333: 219. doi:10.1016/j.apcata. 2007.09.011.

[12] Chang M-Y, Wu C-S, Chen Y-F, Hsieh B-Z, Huang W-Y, Ho K-S, Hsieh T-H, Han Y-K. Polymer Solar Cells Incorporating One-Dimensional Polyaniline Nanotubes. Organic Electronics. 2008; 9: 1136–1139. doi:10.1016/j.orgel.2008.08.001.

[13] Runge von FF. Ueber einige Produkte der Steinkholendestillation. Annalen der Physik Und Chemie. 1834; 107: 65–78. doi:10.1002/andp.18341070502.

[14] Macdiarmid AG, Chiang JC, Richter AF, Epstein AJ. Polyaniline: A New Concept in Conducting Polymers. Synthetic Metals. 1987; 18: 285–290. doi: 10.1016/0379-6779(87)90893-9.

[15] Marie E, Rothe R, Antonietti M, Landfester K. Synthesis of Polyaniline Particles via Inverse and Direct Miniemulsion. Macromolecules. 2003; 36: 3967–3973. doi:10.1021/ma0257550.

[16] Ćirić-Marjanović G. Recent Advances in Polyaniline Research: Polymerization Mechanisms, Structural Aspects, Properties and Applications. Synthetic Metals. 2013; 177: 1–47. doi:10.1016/j.synthmet.2013.06.004.

[17] Gök A, Sarı B, Talu M. Synthesis and Characterization of Conducting Substituted Polyanilines. Synthetic Metals. 2004; 142: 41–48. doi:10.1016/j.synthmet.2003.07.002

[18] Varela-Álvarez A, Sordo JA. A Suitable Model for Emeraldine Salt. Journal of Chemical Physics. 2008; 128: 174706-1–174706-6. doi:10.1063/1.2913246.

[19] Tang Q, Zhou Z. Graphene-Analogous Low-Dimensional Materials. Progress in Materials Science. 2013; 58: 1244–1315. doi:10.1016/j.pmatsci.2013.04.003.

[20] Wilson JA, Yoffe AD. The Transition Metal Dichalcogenides Discussion and Interpretation of the Observed Optical, Electrical and Structural Properties. Advances in Physics. 1969; 18: 193–335. doi:10.1080/00018736900101307.

[21] Ibrahem MA, Lan T-w, Huang JK, Chen Y-Y, Wei K-H, Li L-J, Chu CW. High Quantity and Quality Few-Layers Transition Metal Disulfide Nanosheets from Wet-Milling Exfoliation. RSC Advances. 2013; 3: 13193–13202. doi:10.1039/c3ra41744a.

[22] Mak KF, Lee C, Hone J, Shan J, Heinz TF. Atomically Thin $MoS_2$: A New Direct-Gap Semiconductor. Physical Review Letters. 2010; 105: 136805-1–136805-4. doi:10.1103/PhysRevLett.105.136805.

[23] Albe K, Klein A. Density-Functional-Theory Calculations of Electronic Band Structure of Single-Crystal and Single-Layer $WS_2$. Physical Review B. 2002; 66: 073413-1–073413-3. doi:10.1103/PhysRevB.66.073413.

[24] Xu B-H, Lin B-Z, Sun D-Y, Ding C. Preparation and Electrical Conductivity of Polyethers/$WS_2$ Layered Nanocomposites. Electrochimica Acta. 2007; 52: 3028–3034. doi:10.1016/j.electacta.2006.09.046.

[25] Miremadi BK, Morrison SR. The Intercalation and Exfoliation of Tungsten Disulfude. Journal of Applied Physics. 1988; 63: 4970–4974. doi:10.1063/1.340441.

[26] Yang D, Frindt RF. Li-Intercalation and Exfoliation of $WS_2$. Journal of Physics and Chemistry of Solids. 1996; 57: 1113–1116. doi:10.1016/0022-3697(95)00406-8.

[27] Xu B-H, Lin B-Z, Chen Z-J, Li X-L, Wang Q-Q. Preparation and Electrical Conductivity of Polypyrrole/$WS_2$ Layered Nanocomposites. Journal of Colloid and Interface Science. 2009; 330: 220–226. doi:10.1016/j.jcis.2008.10.033.

[28] Bhandavat R, David L, Singh G. Synthesis of Surface-Functionalized $WS_2$ Nanosheets and Performance as Li-Ion Battery Anodes. Journal of Physical Chemistry Letters. 2012; 3 (11): 1523–1530. doi:10.1021/jz300480w.

[29] Wu Z, Wang D, Zan X, Sun A. Synthesis of $WS_2$ Nanosheets By a Novel Mechanical Activation Method. Materials Letters. 2010; 64: 856–858. doi:10.1016/j.matlet.2010.01.040.

[30] Ramakrishna Matte HSS, Gomathi A, Manna AK, Late DJ, Datta R, Pati SK, Rao CNR. $MoS_2$ and $WS_2$ Analogues of Graphene. Angewandte Chemie International Edition. 2010; 122: 4153–4156. doi:10.1002/anie.201000009.

[31] Paul DR, Robeson LM. Polymer Nanotechnology: Nanocomposites. Polymer. 2008; 49: 3187–3204. doi:10.1016/j.polymer.2008.04.017.

[32] Leghari SAK, Sajjad S, Zhang J. A Time Saving and Cost Effective Route for Metal Oxides Activation. RSC Advances. 2014; 4: 5248–5253. doi:10.1039/c3ra46518g.

[33] Xu H, Bissessur R, Dahn DC. Nanomaterials Based on Polyanilines and $MoSe_2$. Journal of Inorganic and Organometallic Polymers and Materials. 2014; 24: 219–225. doi:10.1007/s10904-013-9981-z.

[34] Ozawa, T. Kinetic Analysis of Derivatives Curves in Thermal Analysis. Journal of Thermal Analysis 1970; 2: 301–324. doi:10.1007/BF01911411.

[35] Scully K, Bissessur R. Decomposition Kinetics of Nylon-6/Graphite and Nylon-6/Graphite Oxide Composites. Thermochimica Acta 2009; 490:32–36. doi:10.1016/j.tca.2009.01.029.

[36] Scully SF, Bissessur R, Dahn DC, Xie G. In Situ Polymerization/Intercalation of Substituted Anilines into Iron (III) Oxychloride. Solid State Ionics. 2010; 181: 933–938. doi: 10.1016/j.ssi2010.05.015.

[37] Kanatzidis MG, Marcy HO, McCarthy WJ, Kannewurf CR, Marks TJ. In Situ Intercalative Polymerization Chemistry of FeOCl. Generation and Properties of Novel, Highly Conductive Inorganic/Organic Polymer Microlaminates. Solid State Ionics. 1989; 32–33: 594–608. doi:10.1016/0167-2738(89)90272-5.

[38] Reghu M, Cao Y, Moses D, Heeger AJ. Counterion-Induced Processibility of Polyaniline: Transport at the Metal-Insulator Boundary. Physical Review B. 1993; 47: 1758–1764. doi:10.1103/PhysRevB.47.1758.

Chapter 10

# Phenylenevinylene Systems: The Oligomer Approach

Juan C. Cárdenas, Cristian Ochoa-Puentes and

Cesar A. Sierra

Additional information is available at the end of the chapter

http://dx.doi.org/10.5772/63394

**Abstract**

Among conducting polymers, poly-*p*-phenylenevinylenes (PPVs) have attained a special place in polymer electronics. The optoelectronic properties initially exposed by PPVs in organic light-emitting diodes (OLEDs) turned these organic electronic conjugated systems from the solo academic interest into a technologically very promising area. The easiness of the tuning of their optoelectronic properties through synthetic modifications make PPVs an outstanding and suitable compound for technological applications and fundamental science development. Unfortunately, the synthesis and structural optoelectronic characterization of novel PPVs is a long and difficult task that sometimes yields unclear results. However, phenylenevinylene oligomers (oPV) can be synthesized and characterized in a very straightforward manner, and their performance in novel applications can be directly related to their structural analogue polymer, methodology designated as the oligomer approach. Herein, we describe the oligomer approach using the Mizoroki-Heck reaction as a synthetic route for oPVs and PPVs, and the importance of an extensive characterization for novel applications, such as photocatalysis and matrix-assisted laser desorption/ionization (MALDI) matrices, where these electronic conjugated systems have very promising applications.

**Keywords:** phenylenevinylene, oligomer approach, optoelectronic polymers, Mizoroki-Heck reaction, conjugated systems applications

## 1. Summary

This chapter describes the physicochemical characteristics of conjugated polymers and the growing importance of the polymer electronics in our actual and future day life, followed by advantages and disadvantages of the most common synthetic routes reported to get any imaginable electronic conjugated chemical structure.

As a result of the explanation and development of the oligomer approach in this chapter, applied to the synthesis of phenylenevinylene oligomers (oPV) and polymers by the Mizoroki-Heck reaction and the usefulness of this methodology to define precise chemical structure as target compound in a specific application, the reader will have a perspective of the convenience of working with oligomers instead of polymers as a previous and relevant condition, in terms of time, economy and simplicity, before a target poly-*p*-phenylenevinylene (PPV) system is designed for a specific application. Furthermore, since the most important methods of oPV synthesis are described and discussed, the reader can also select the methodology of his or her convenience according to the oPV structure and application of interest. Finally, taking into account that the optoelectronic properties of several oPVs will be exposed, the reader may develop a preview of the influence of some substituent groups over the main chain of the oPV.

## 2. Introduction to optoelectronic active polymers

Polymers are organic or inorganic macromolecules of natural, industrial and technological importance because of the wide range of physicochemical and mechanical attributes that they possess. Since the early twentieth century, due to their insulating properties, polymers were widely used as packaging and protecting materials, but the discovery in recent decades of their electrical conduction capability led them to become "active" materials within highly attractive applications such as light-emitting diodes, photovoltaic cells, chemical sensors, among others, to which it is known as polymer electronics. Alan J. Heeger, Alan G. MacDiarmid and Hideki Shirakawa were awarded with the Nobel Prize in chemistry in 2000 for the discovery and development of polymer electronics [1]. There are many advantages in using these materials over conventional electronic devices that are based on inorganic semiconductors. These polymer devices can be designed with large areas, they are mechanically flexible and very light, their operation requires less energy and their production is economically profitable.

Active polymers have a common structural feature—presence of an extensive electronic conjugation. In this conjugation, the alternation of single and double bonds creates the overlapping of electrons in *p* orbitals (unhybridized) over the entire polymer backbone, generating an electronic delocalization along the polymer structure. This delocalization provides the route to the mobility of charges along the polymer (**Figure 1**).

For interpreting the physical chemistry of conducting polymers, the use of the band theory is widespread. In this, induced or not induced "defects" lead to the formation of an energy difference between the HOMO (valence band) and the LUMO (conduction band) orbitals of the polymer, what is referred to as "band gap". Because of this, conducting polymers are considered as semiconductors and since the band gap depends on the molecular structure of the electronically conjugated repeating unit, there is a great challenge, and also a possibility, of controlling this energy difference by designing at the molecular level through the implementation of various synthetic methodologies. Therefore, optoelectronic properties of the polymer can be finely adjusted according to the needs of any technological application.

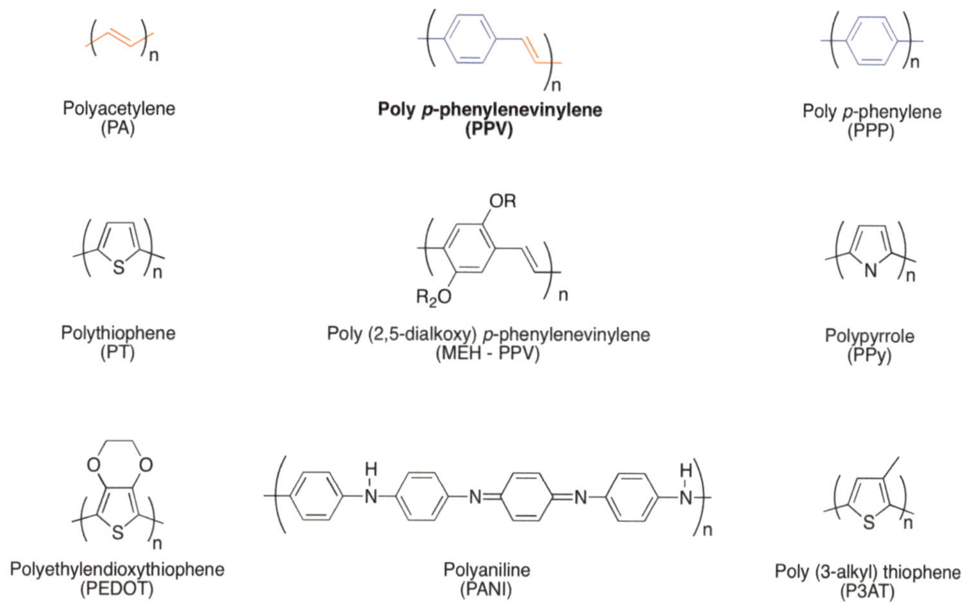


**Figure 1.** Chemical structures of some conducting polymers.

Among conducting polymers, PPVs (**Figure 1**) have a distinctive place in polymer electronics, which is confirmed by the large amount of scientific literature dedicated to this polymer over the last years. The impressive electroluminescence exhibited by this polymer in a diode with a very simple architecture [Al/PPV/ITO, where aluminium (Al) and indium tin oxide (ITO) act as electrodes] turned polymer electronics from an area of pure academic interest to a very promising technological area [2]. A few years later, the first solar cell based on MEH-PPV and C61-phenyl butyric acid methyl ester (PCBM) was reported [3–5]. Also in the 1990s, the first conjugated polyelectrolyte based on a sulfonated PPV was developed and it proved to be a highly sensitive fluorescent sensor [6].

Chemically, PPVs can be considered as a copolymer that combines the repeating units of polyacetylene (PA) and poly *p*-phenylene (PPP). In this way, their properties are located in the middle of these two polymers; thus, they are more chemically stable than the PA, while are not as robust as the PPP. Furthermore, while the PA is black and has a smaller band gap ($E_g$ = 1.4 eV), PPV films without substituents are yellow due to absorption around 420 nm and with a band gap significantly higher ($E_g \geq 1.9$ eV) [3–5]. Additionally, the optical properties of PPVs can be influenced by conformational factors. This can be observed by the solvatochromism, wherein the length of the effective electronic conjugation can be modified according to the chemical nature of the solvent. This situation is exploited in the process of spin-casting, leading to the modulation of the device properties [7]. Another way of modulating the optoelectronic properties in conducting polymers is by the chemical nature of the substituents present on the electronically conjugated structure, which will be explained in detail in the following sections.

All these physicochemical attributes belonging to the polymer electronics have been successfully exploited by the industry through the generation of novel technological applications that generally seek to make concrete contributions in order to improve the living standards of humanity without forgetting the growing global preoccupation for the conservation and improvement of our environment. Just to mention a couple of examples, recently LG chem, the electronic giant company, developed a house lamp based on organic light-emitting diodes (OLEDs) with a brightness comparable to conventional LED lights and with a lifetime of 40,000 hours [8]; also, the Swager group at MIT developed a chemical sensor based on conjugated polymers that efficiently determines the ripeness of fruits at very low cost, which can be used as regular plastic bags in supermarkets [9].

## 3. Synthetic routes for obtaining PPVs

Although there have been several synthetic methods reported for the preparation of PPVs, we discuss only those that are notable for their easy implementation and good results in terms of molecular weight and stereochemistry control of the products, since these features will govern the solubility, crystallinity, processability and optoelectronic properties of the PPVs obtained. It is noteworthy that when designing a PPV for a given application, the need to add substituents to the PPV backbone should be taken into account, since this is necessary to produce a soluble and processable material.

### 3.1. Gilch polymerization

The Gilch reaction (**Scheme 1**) is an economic synthetic methodology that facilitates obtaining PPVs with very high molecular weight. This method, first developed in 1965, employs an $\alpha,\alpha'$-dichloro-*p*-xylene **1** as precursor. This, by treatment with a strong base such as potassium tert-butoxide, undergoes elimination of HCl to form $\alpha$-chloro-*p*-quinodimethane **2**, which polymerizes via radicals to produce the intermediate **3**. This poly ($\alpha$-chloro-*p*-xylene), in the presence of excess base, yields a high molecular weight PPV through E2 elimination [10–13].

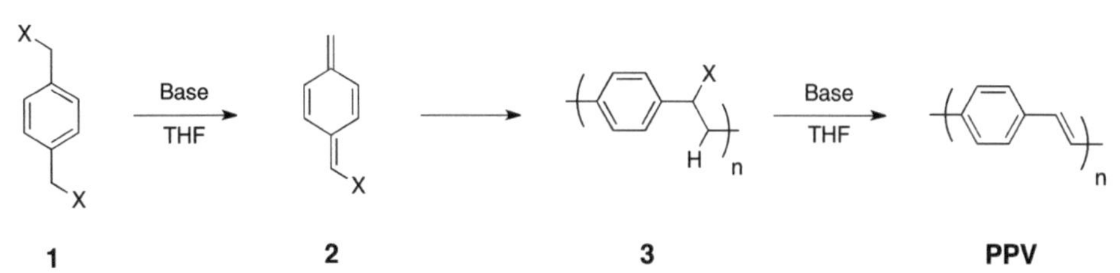


**Scheme 1.** Gilch synthetic route.

The polymerization process occurs rapidly with few defects and low dispersity, but it must be carried out at temperatures as low as -78°C, which is considered an experimental and economical difficulty. The rapid polymerization also leads to a *cis/trans* ratio large enough to produce a loss in the photoluminescent properties of the PPV yielding blue-shifted spectra, simultaneously making the obtained polymer not attractive for some technological applications [14].

### 3.2. Wessling polymerization

This method, implemented in 1966, utilizes a pre-monomer in which the two chlorine atoms over the *α,α*'-dichloro-*p*-xylene **1** are replaced by sulfonium groups to get compound **4**. Base-catalyzed polymerization of compound **4** (**Scheme 2**) leads to a poly-*p*-xylene precursor functionalized with sulfonium groups **5**, which is soluble in water and alcohols. Heating of compound **5** leads to the elimination of HCl to yield the target PPV [15].

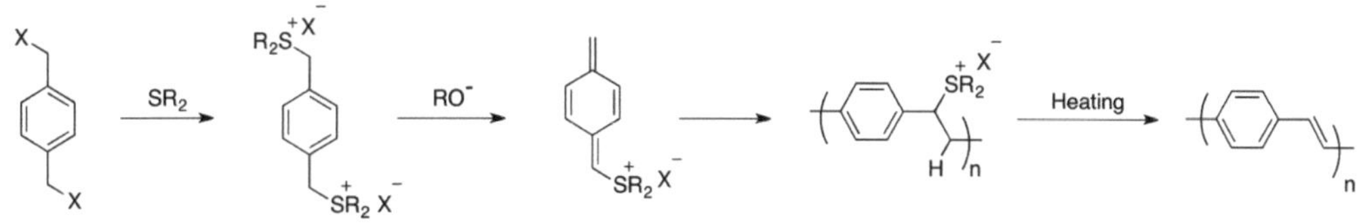


**Scheme 2.** Wessling polymerization route.

However, it is important to highlight that despite the good yields of this reaction, the production of HCl in the process might damage the substrates [(e.g., (ITO)] during the in situ polymerization for the construction of a device like OLEDs, thus demanding a more complex design for device production.

### 3.3. Wittig polycondensation

Perhaps the most simple, direct and widespread methodology to produce completely conjugated PPVs and derivatives is the Wittig polycondensation reaction (**Scheme 3**). However, although this represents a very favorable approach to get PPVs with or without substituents, the Wittig reaction generally produces only low molecular weight materials, with a mixture of *cis* and *trans* vinyl bonds. This mixture of *cis* and *trans* segments is very inconvenient to achieve homogeneous optoelectronic properties that are required in some polymer electronic applications [16, 17].

Consequently, the use of phosphorous ylides (**7**, **Scheme 4**) instead of phosphonium ylides (**6**, **Scheme 3**), modification of the Witting polycondensation mechanism and known as the Wittig-Horner reaction (**Scheme 4**), not only increased the amount of *trans* bonds in PPVs, but also increased the molecular weights to exceed 10 KDa. These features as well as the versatility in the selection of monomer have led to the wide use of Wittig-Horner reaction for the preparation of PPVs. However, despite the higher molecular weights obtained with this modification, the stereoselectivity of the reaction remains insufficient to obtain configurationally pure *trans* PPV systems [18].

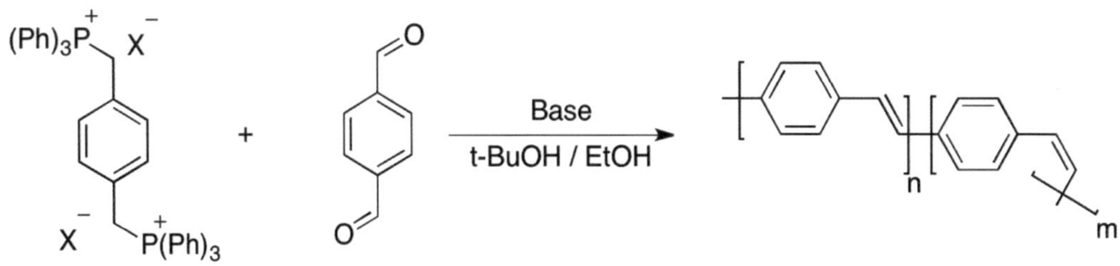


**6**

**Scheme 3.** Classical Wittig polymerization.

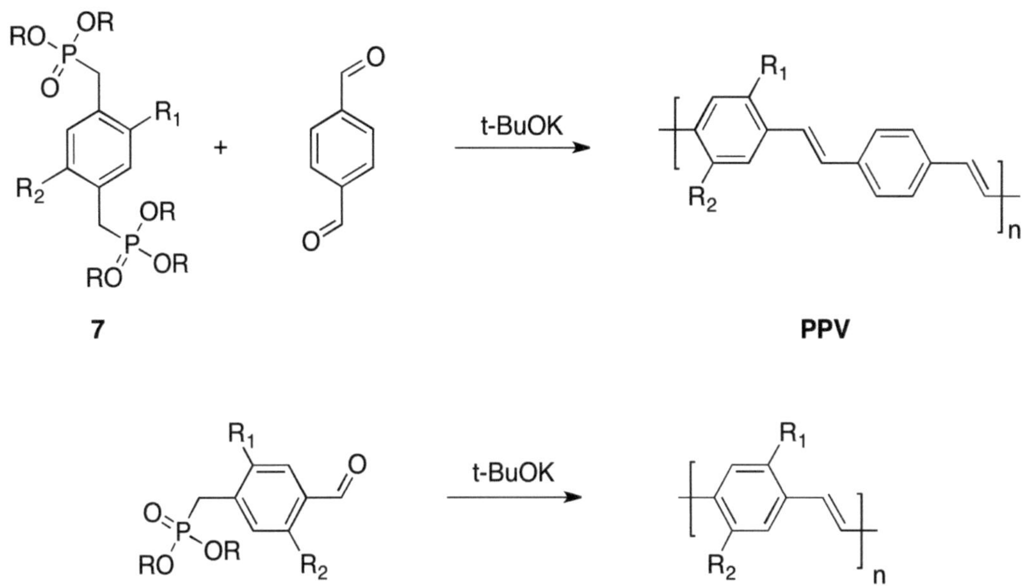


**7** **PPV** **PPV**

**Scheme 4.** Wittig–Horner polymerization.

### 3.4. Knoevenagel polycondensation

This reaction is especially useful for producing electronically deficient PPVs by exploiting the characteristic acidity of the benzylic hydrogens derived from *p*-xylene with strong electron-withdrawing groups in the $\alpha$ position. The formation of the respective carbanion over compound **8** in the presence of a strong base allows the condensation with terephthaldehyde derivatives **9** to produce PPVs of low to high molecular weights depending on the easiness of the activation of the carbonyl group due to the substituents on the aromatic ring **9** (**Scheme 5**). To perform this reaction in a straightforward manner, it is important to handle the synthesis conditions carefully, avoiding the competitive Michael addition of the propagator nucleophile to a cyanovinyl unit of another polymeric chain [19–21].

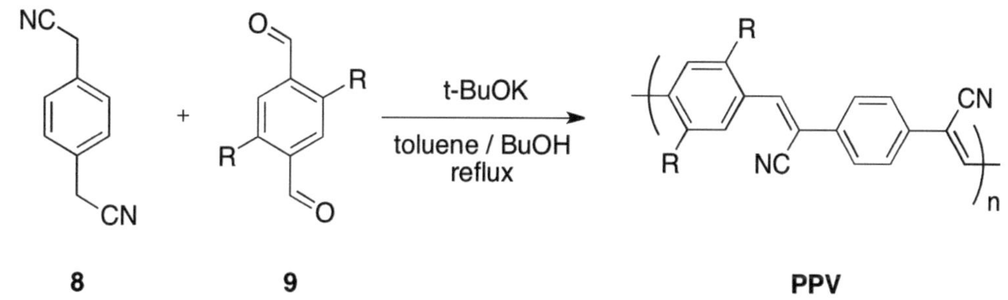


**Scheme 5.** Knoevenagel polymerization.

### 3.5. The Mizoroki-Heck reaction

This cross-coupling reaction, catalyzed by palladium, uses olefin derivatives **10** and unsaturated halides **11** (**Scheme 6**) as precursors for the formation of C-C bonds. Unlike the methods described earlier, the Mizoroki-Heck reaction employs mild bases, and although the traditional reaction conditions use refluxing dimethylformamide (DMF), novel catalytic systems allow the reaction to take place at room temperature in a variety of solvents [22]. Unfortunately, though the reaction conditions are usually simple and there are no specific structural limitations for the precursors, most of the literature reported for the synthesis of PPVs with this methodology produced polymers with very low molecular weights [23, 24]. However, many advances in the development of catalytic systems have been employed to improve the degree of polymerization under this protocol [25–28]. Notably, the most important feature of this reaction is that it allows obtaining configurationally pure *trans* PPVs.

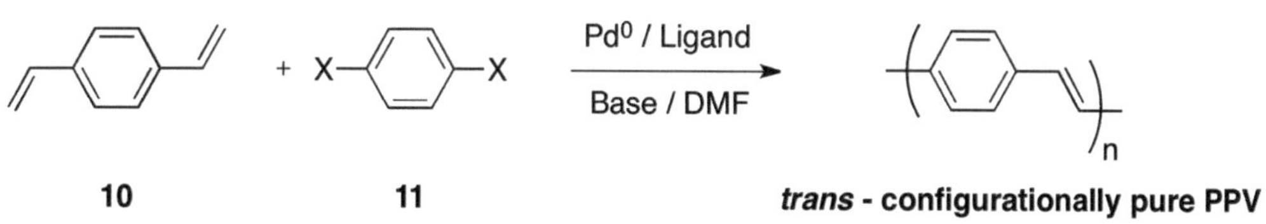


**Scheme 6.** The Mizoroki–Heck cross-coupling reaction polymerization.

Among the improvements to the reaction conditions, the use of palladium (0) sources as well as the utilization of phosphite ligands instead of phosphines, are the most important changes for the reaction optimization. Additionally, solvothermal conditions have been explored with excellent increase in the product yield and enhancing the easiness through the purification process.

However, it can be inferred that the synthesis and characterization of small molecules is much simpler than that for polymers; therefore, to make significant and rapid progresses for obtaining high molecular weight PPVs, it will be easier if this is done initially with oligomers, and subsequently with these optimized synthetic conditions, which will lead to the formation of structurally analogous polymer. In a similar way, to explore the optoelectronic properties in PPVs for a targeted application, it is easier to start with the study of the properties of oligomers, and then synthesize only one or few polymers with the chemical structure that resembles the structure for the oligomers with the best performance for the application in question. These and other favorable aspects of the oligomer approach will be discussed in detail in the next section using only the Heck methodology as synthetic route for PPVs or oPVs, since this methodology has proven to be the right choice to get conjugated electronic systems with high stereochemical control.

## 4. The oligomer approach

Whichever is the chosen methodology for the synthesis of conjugated polymers, synthetic and characterization processes are challenging and this sometimes causes the scaling of production not to occur at the speed demanded by industry or at the pace required for impacting basic research on time. Therefore, different strategies have been designed to solve these issues, the most reported being "the oligomer approach".

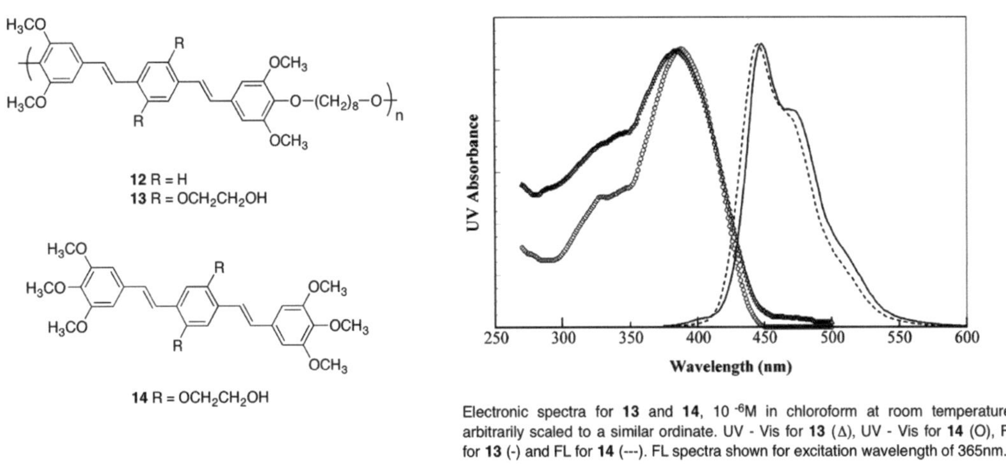


**Figure 2.** Segmented PPV and analogue oPV structures (taken from reference 29).

Of course, the synthesis and characterization of oligomers is usually much easier and more efficient than the synthesis of their respective analogue polymers, which leads to faster and cheaper results in terms of the design of a polymer with particular applications. Analysis of the optoelectronic properties of synthesized oligomers helps us determine which electronically conjugated structure is the most appropriate for the intended application, and thus, only few polymers structurally analogous to the oligomer will be synthetized. These analogous polymers, according to several reports, demonstrated to maintain the same optoelectronic properties of the oligomer previously synthetized. This homogeneity of the properties between oligomer and its polymer analogue is more evident when the target polymer is a segmented structure (**Figure 2**) instead of a fully conjugated polymer (e.g., MEH-PPV in **Figure 1**) [4, 15, 29], since the optoelectronically active segment in the segmented polymer can be truly reproduced in the oligomer (**Figure 2**). In segmented conjugated polymers, the chain tensions that cause torsions on the polymer backbone are suffered and assumed by the flexible aliphatic segments. Therefore, the conjugated segment responsible for the optoelectronic properties in the polymer conserves an unaltered chemical structure, closely similar or equal to the oligomer analogue.

Another advantage of working with segmented polymers is that it is possible to make theoretical studies regarding the relationship between the structure and the optoelectronic properties. The computational cost of working with oligomers is much more reachable than that for working with polymers. Furthermore, in the electronically conjugated polymers, the segmented part and its properties can be easily modelled in computer and these data can be validated by the experimental results obtained with the previously synthesized structurally analogous oligomers, which can often include even the crystal structure (**Figure 3**) [30]. In applications like OLEDs, the optical and electrical behavior comes from the polymer in solid state; unfortunately, getting the solid-state structure of a polymer is not an easy task. Then, it is more accurate to get the conformation and packing properties from the crystal structure of oligomers, as can be seen in **Figure 3** for an oPV, and use this information to make more precise assumptions, observations, conclusions and structure-property relationships to the structurally analogous polymer.

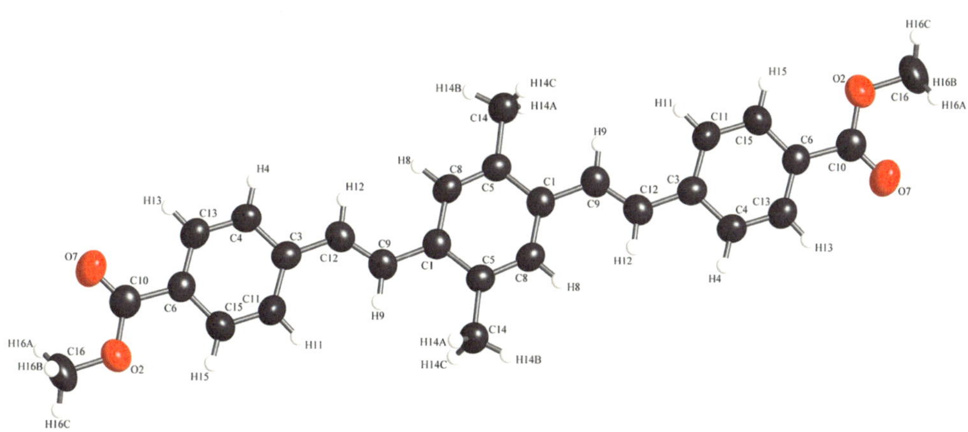


**Figure 3.** Crystal (top) and supramolecular structures (bottom) of compound **22** in **Table 1**.

As an example of the development of the oligomer approach, in order to solve the problem of low molecular weights obtained in the polymerization of PPVs through the Heck reaction reported by many research groups around the world [22], recent investigations showed that the use of a catalytic system composed of triphenylphosphite and Pd(dba)$_2$ in the presence of ionic liquids, significantly increases the reaction yields during the synthesis of oPVs and shows the catalyst reusability throughout several cycles [25]. Thus, these reaction conditions were applied to synthesize several oPVs [26–28] and according to the results obtained regarding the reaction yields as well as the optoelectronic properties, a segmented PPV with a degree of polymerization close to 20 (twice superior the size obtained previously with conventional Heck conditions) was obtained [27]. The molecular weight obtained for this PPV following the oligomer approach methodology yielded polymer films of sufficient quality to fabricate OLEDs. This improved synthetic route has also been used for the synthesis of several series of oPVs with a clear enhancement on the reaction yields and incorporating substituents that are very difficult to use as part of the precursors by other synthetic methodologies [31].

| Structure | UV and FL spectra |
|---|---|
| 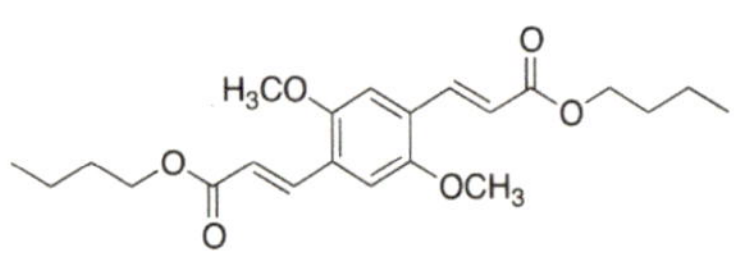  **15** | 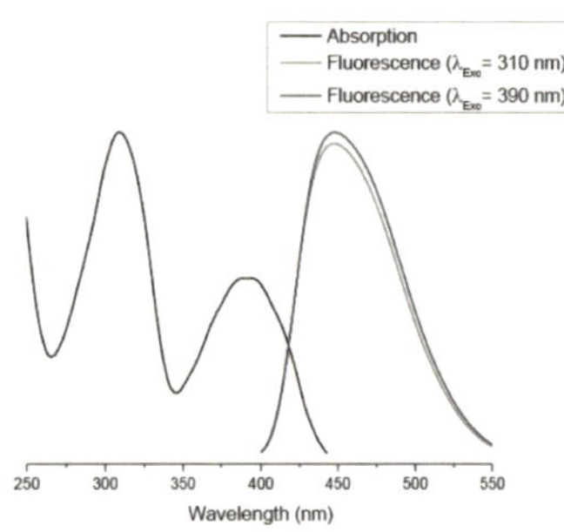  |
| 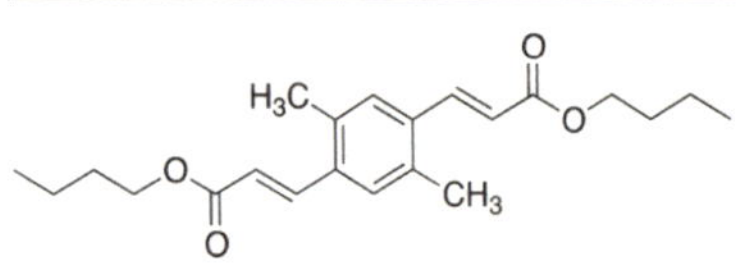  **16** | 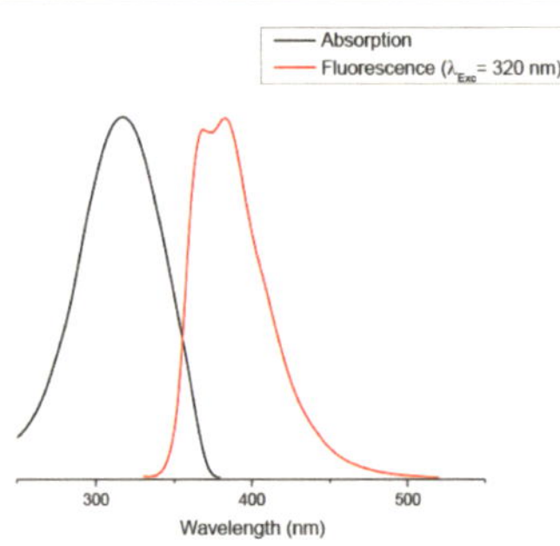  |
| 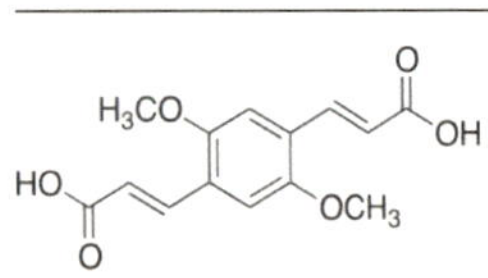  **17** | 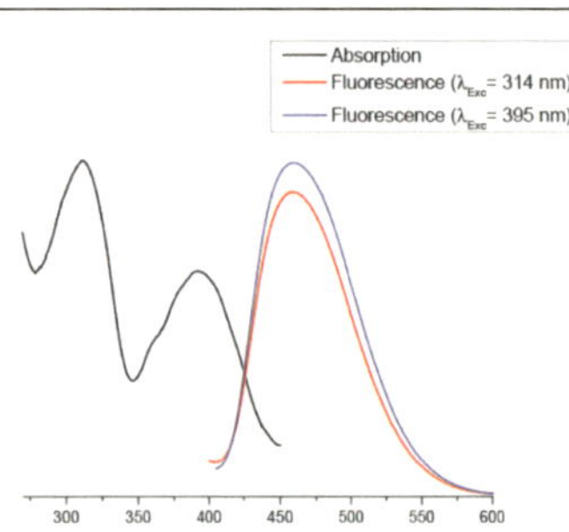  |

http://dx.doi.org/10.5772/63394

| Structure | UV and FL spectra |
|---|---|
| 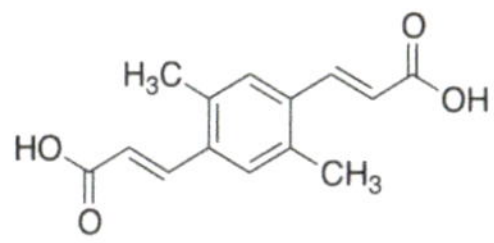  **18** | 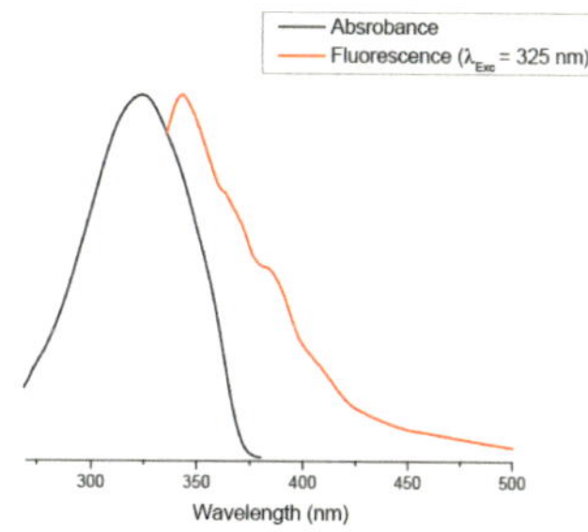  |
| 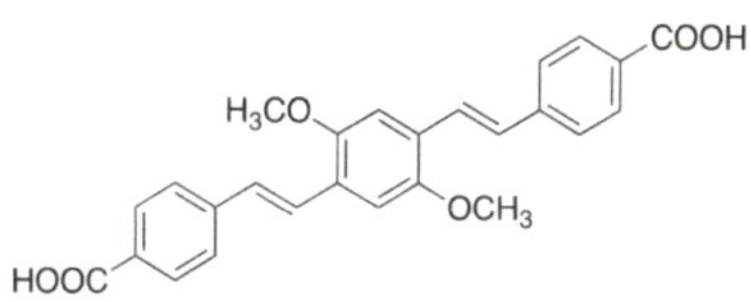  **19** | 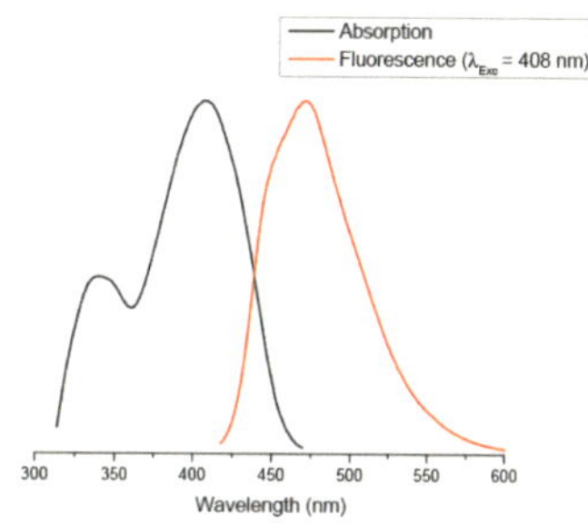  |
| 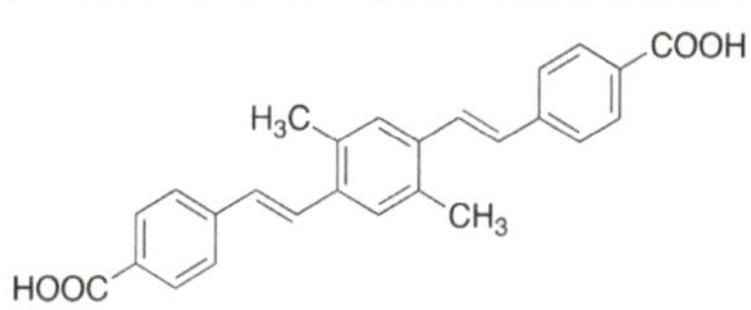  **20** | 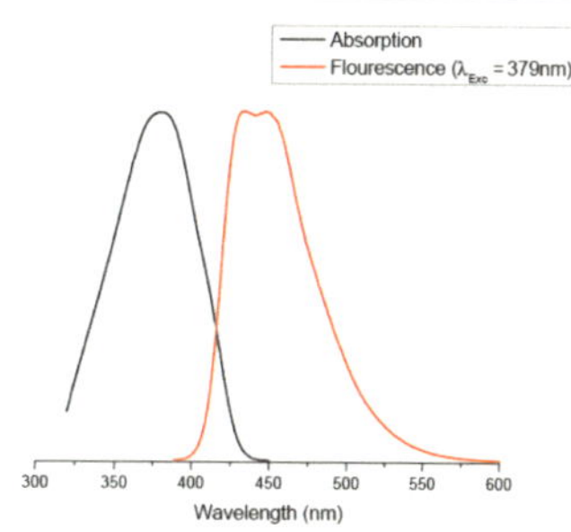  |
| 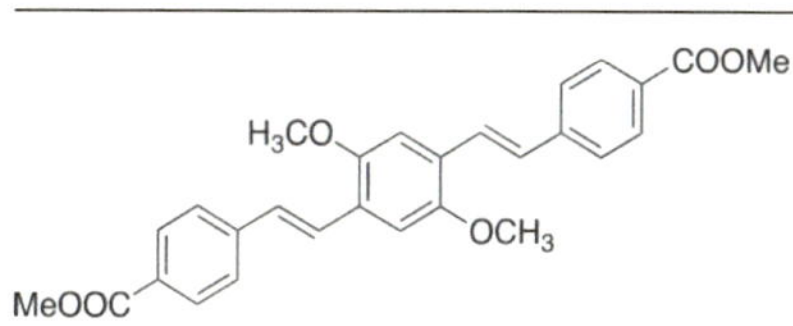  **21** | 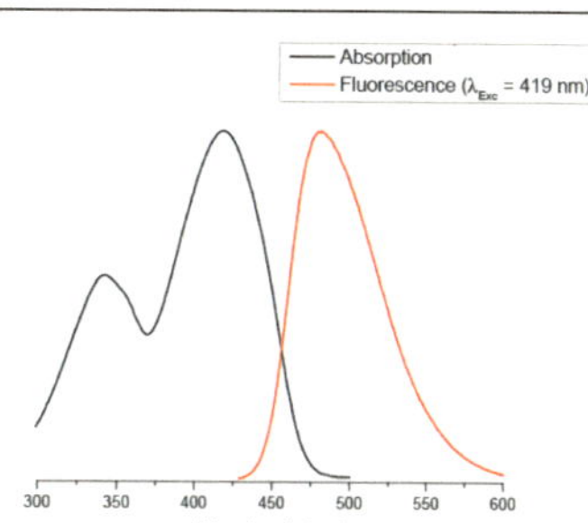  |

| Structure | UV and FL spectra |
|---|---|
| 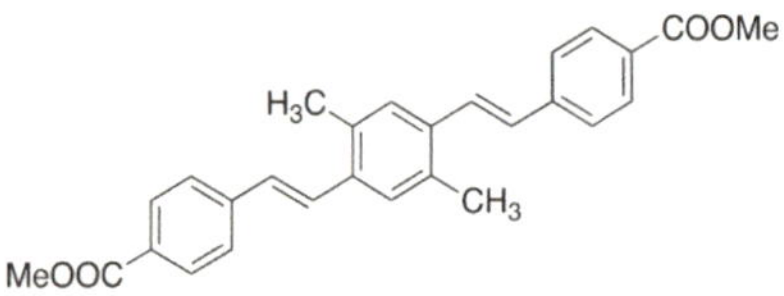 **22** | 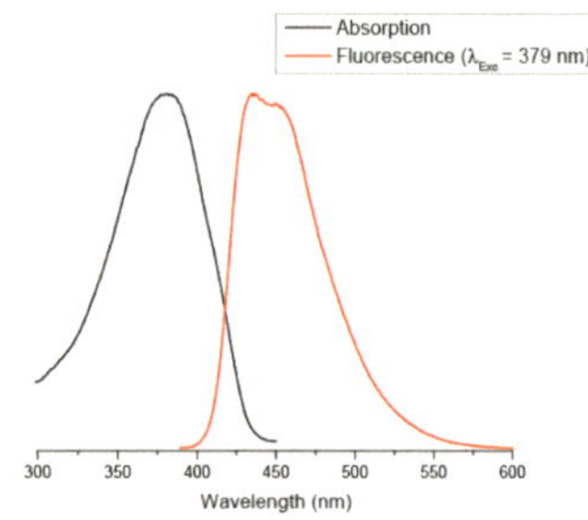 |
| 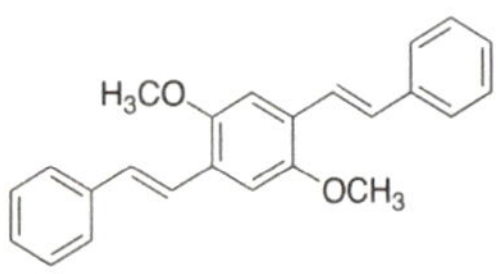 **23** | 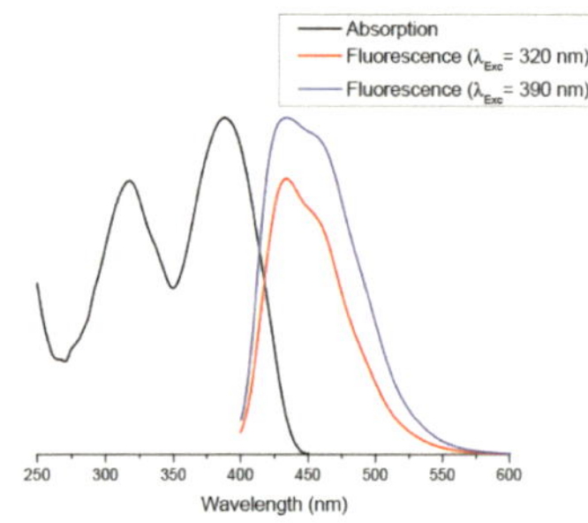 |
| 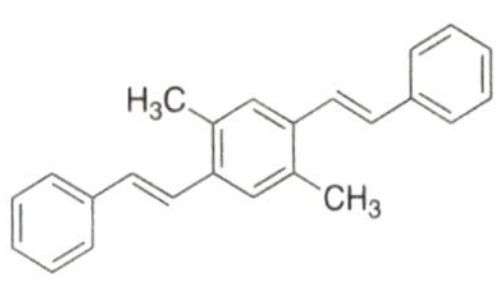 **24** | 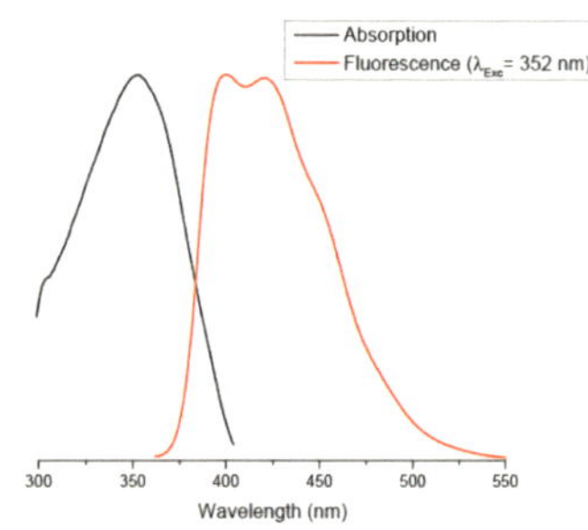 |
| 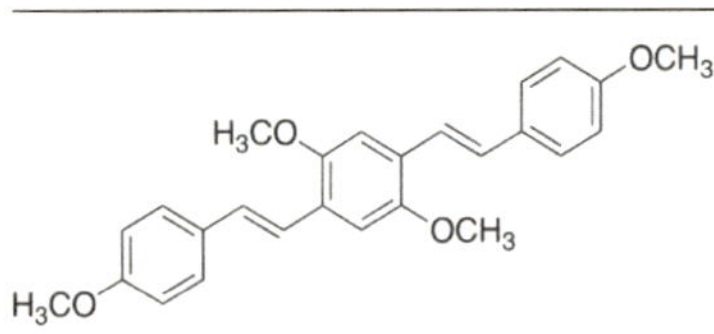 **25** | 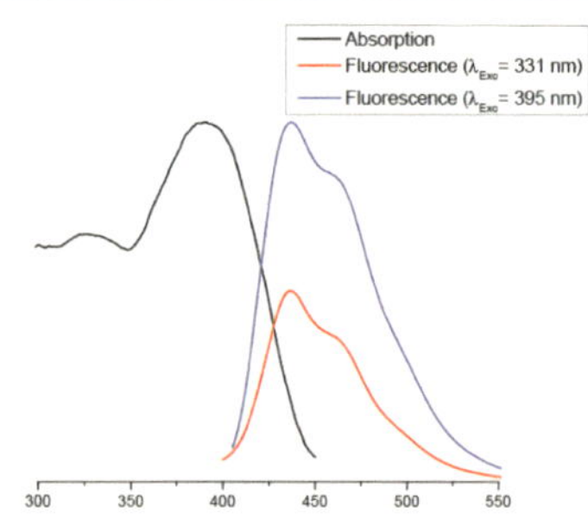 |

http://dx.doi.org/10.5772/63394

| Structure | UV and FL spectra |
| --- | --- |
| 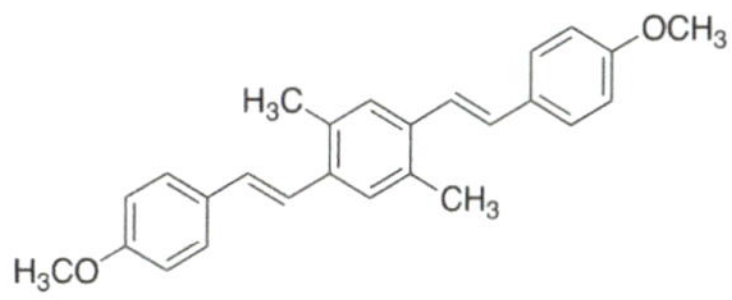 **26** | 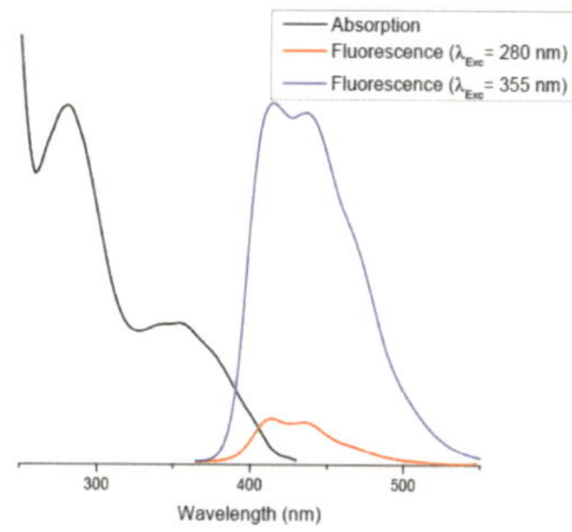  |
| 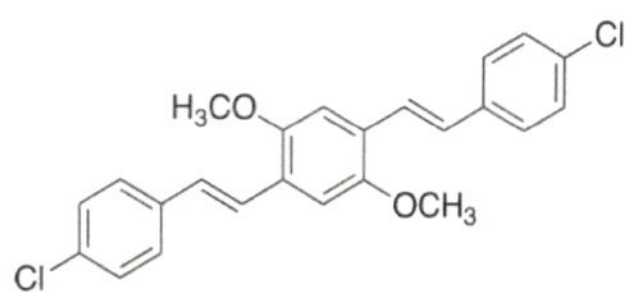 **27** | 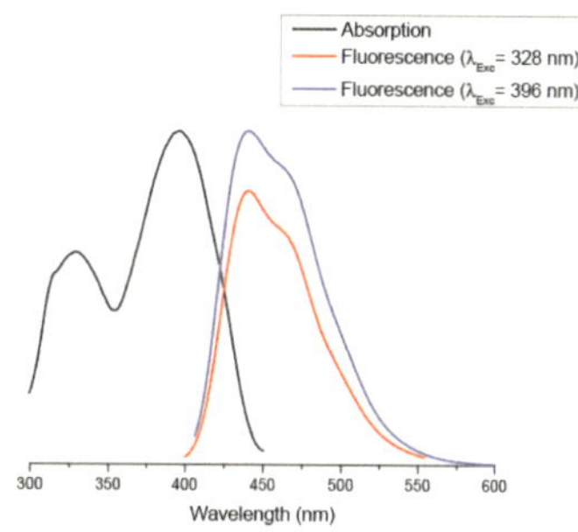  |
| 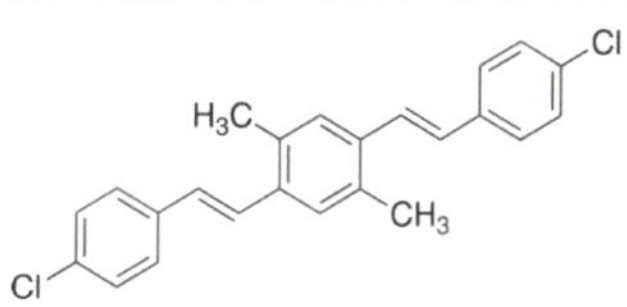 **28** | 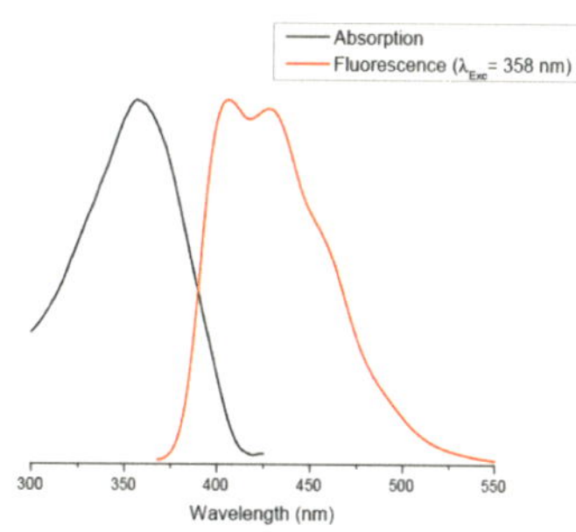  |
| 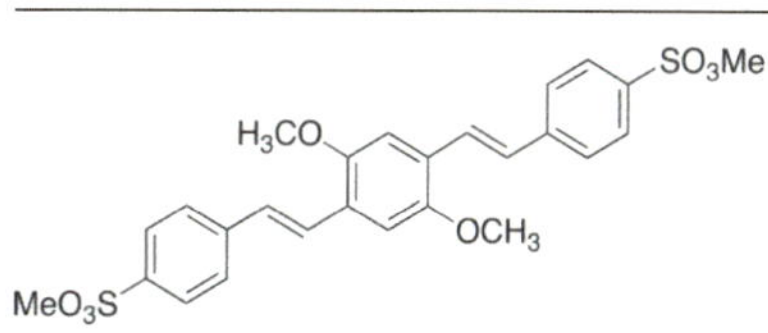 **29** | 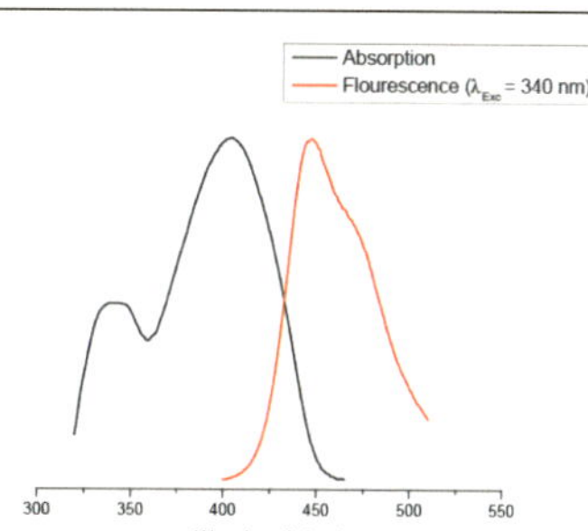  |

| Structure | UV and FL spectra |
|---|---|
| 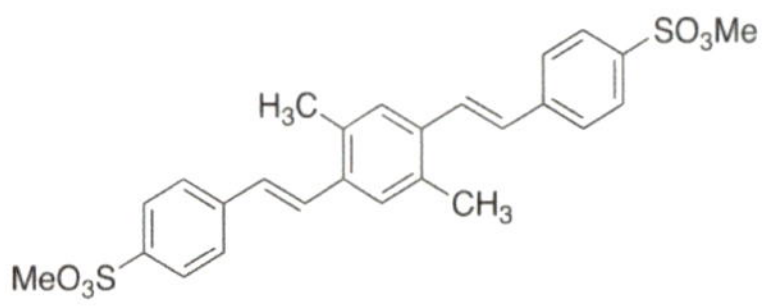  **30** | 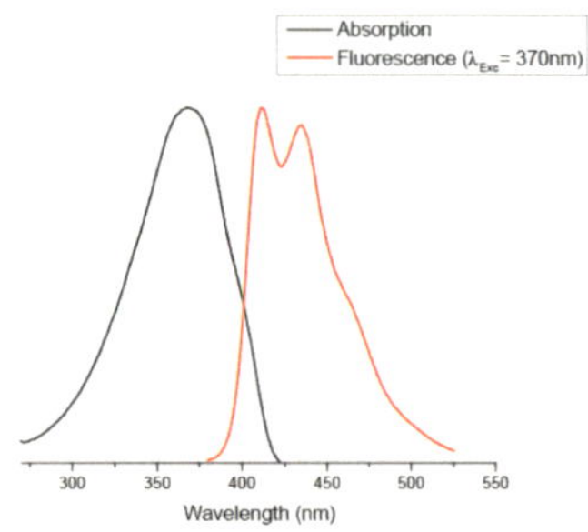  |

**Table 1.** Ultraviolet and fluorescence emission spectra of some oPVs synthesized under solvothermal conditions.

Beyond this, novel protocols for the implementation of the Heck reaction have been developed, reaching the standardization of a very efficient green and economic solvothermal methodology that can be used in Heck polymerizations. In this protocol, minimal amounts of solvents are used, which helps to get very simple purification procedures that as a result gives reaction yields close to 100%, in over 40 oPVs systems synthesized. Thus, it is possible to explore the influence of any substituent over the physical and chemical properties in oPVs in order to obtain, for instance, a more precise description of the relationship between the structure and optoelectronic properties in this class of compounds, simplifying the chemical design of a target PPV with defined properties.

Some oPVs synthesized and their absorption and emission spectra are presented in **Table 1**, where it is seen how the optoelectronic properties of the oligomers change according to their structure and functional groups.

It is very important to highlight that the oligomer approach applied for the oPVs shown in **Table 1** allowed to predict that some of these systems can be used as MALDI matrices, [32–34], UV and visible photocatalyst and organic chromophores for chemosensors [35, 36]. Just as example, compound **27** in **Table 1** due to a very high molar absorptivity at 355 nm (wavelength for Nd:YAG laser in MALDI), low fluorescence quantum yield and crystallographic properties has been studied as matrix for MALDI, showing a very high efficiency at very low laser power, identifying more analytes than other conventional and commercially available matrices. As many authors have established, it is expected that the polymer analogue to compound **27** (currently under analysis) presents a much better behavior as matrix due to the amplification effect related to a greater population of conjugated structures in a close proximity [37], and additionally, a polymer matrix in MALDI might improve the analysis of small analytes, since the polymer matrix will have a very low volatility and fragmentation, leading to a few unwanted overlapping and ghost signals.

Here, employing only one single route, it has been shown that the oligomer approach has a vast scope for various areas like, for instance, chemical synthesis and materials science.

http://dx.doi.org/10.5772/63394

By the optimization of the synthetic conditions to yield small conjugated systems (oligomers) in which it is possible to track in a very easy way changes and effect of the catalyst source, solvent and ligand nature, among other synthetic factors, it is possible to get very efficient catalytic systems that can be used to improve the molecular weight of the analogue polymers. Also, the oligomer approach has shown to be very efficient in predicting exact polymer structures of conjugated systems to applications like matrices in MALDI and chemosensors. All these examples supported by a complete structural and optical characterization make over the oligomers that can be extrapolated to the analogue polymer.

## Author details

Juan C. Cárdenas, Cristian Ochoa-Puentes and Cesar A. Sierra*

*Address all correspondence to: casierraa@unal.edu.co

Universidad Nacional de Colombia, Bogotá, D.C., Colombia

## References

[1] Park YW. Editorial for the conducting polymers for carbon electronics themed issue. Chemical Society Reviews. 2010;39(7):2352-3.

[2] Heeger AJ. Semiconducting and metallic polymers: the fourth generation of polymeric materials. Nobel Lecture. 2000.

[3] Mikroyannidis JA, Tsagkournos DV, Balraju P, Sharma GD. Synthesis and photovoltaic properties of an alternating phenylenevinylene copolymer with substituted-triphenylamine units along the backbone for bulk heterojunction and dye-sensitized solar cells. Journal of Power Sources. 2011;196(4):2364-72.

[4] Chen J-C, Wu H-C, Chiang C-J, Chen T, Xing L. Synthesis and properties of air-stable n-channel semiconductors based on MEH-PPV derivatives containing benzo[c]cinnoline moieties. Journal of Materials Chemistry C. 2014;2(24):4835-46.

[5] Hou J, Fan B, Huo L, He C, Yang C, Li Y. Poly(alkylthio-p-phenylenevinylene): synthesis and electroluminescent and photovoltaic properties. Journal of Polymer Science Part A: Polymer Chemistry. 2006;44(3):1279-90.

[6] Chen L, McBranch DW, Wang H-L, Helgeson R, Wudl F, Whitten DG. Highly sensitive biological and chemical sensors based on reversible fluorescence quenching in a conjugated polymer. Proceedings of the National Academy of Sciences. 1999;96(22):12287-92.

[7] Geens W, Shaheen SE, Wessling B, Brabec CJ, Poortmans J, Serdar Sariciftci N. Dependence of field-effect hole mobility of PPV-based polymer films on the spin-casting solvent. Organic Electronics. 2002;3(3–4):105-10.

[8] Tremblay JF. LG develops bright, long-lasting organic light-emitting diode. Chemical and Engineering News. 2014;92(38):14-6.

[9] Esser B, Swager TM. Detection of ethylene gas by fluorescence turn-on of a conjugated polymer. Angewandte Chemie International Edition. 2010;49(47):8872-5.

[10] Yusuke T, Takashi K, Hiroyuki N, Chun-Young L, Eui-Doo D, Jung-Il J. Poly(2-diphenylamino-1,4-phenylenevinylene): its preparation via chemical vapor deposition polymerization. Science and Technology of Advanced Materials. 2006;7(5):475.

[11] Li P, Wen S, Cheng W, Zhang J, Yao S, Xu B, et al. Oxadiazole containing poly(p-phenylenevinylene)s: synthesis and characterization. New Journal of Chemistry. 2012;36(8):1626-33.

[12] Buchmeiser MR, Unold J, Schneider K, Anderson EB, Hermanutz F, Frank E, et al. A new carbon precursor: synthesis and carbonization of triethylammonium-based poly(p-phenylenevinylene) (PPV) progenitors. Journal of Materials Chemistry A. 2013;1(42):13154-63.

[13] Pu Y-J, Kurata T, Soma M, Kido J, Nishide H. Triphenylamine- and oxadiazole-substituted poly(1,4-phenylenevinylene)s: synthesis, photo-, and electroluminescent properties. Synthetic Metals. 2004;143(2):207-14.

[14] Fan Y-L, Lin K-F. Dependence of the luminescent properties and chain length of poly[2-methoxy-5-(2′-ethyl-hexyloxy)-1,4-phenylene vinylene] on the formation of cis-vinylene bonds during Gilch polymerization. Journal of Polymer Science Part A: Polymer Chemistry. 2005;43(12):2520-6.

[15] Padmanaban G, Ramakrishnan S. Conjugation length control in soluble poly[2-methoxy-5-((2′-ethylhexyl)oxy)-1,4-phenylenevinylene] (MEHPPV): Synthesis, optical properties, and energy transfer. Journal of the American Chemical Society. 2000;122(10):2244-51.

[16] Jaballah N, Chemli M, Hriz K, Fave J-L, Jouini M, Majdoub M. Blue-luminescent poly(p-phenylenevinylene) derivatives: synthesis and effect of side-group size on the optical properties. European Polymer Journal. 2011;47(1):78-87.

[17] Karpagam S, Guhanathan S, Sakthivel P. Applications of wittig reactions in dibenzo 18-crown-6-ether substituted phenylenevinylene oligomer—synthesis, photo luminescent, and dielectric properties. Journal of Applied Polymer Science. 2011;120(2):960-7.

[18] Winkler B, Mau AWH, Dai L. Crown ether substituted phenylenevinylene oligomers: synthesis and electroluminescent properties. Physical Chemistry Chemical Physics. 2000;2(2):291-5.

[19] Cid J-J, Ehli C, Atienza-Castellanos C, Gouloumis A, Maya E-M, Vazquez P, et al. Synthesis, photophysical and electrochemical characterization of phthalocyanine-based poly(p-phenylenevinylene) oligomers. Dalton Transactions. 2009;20:3955-63.

[20] Detert H, Sugiono E. Soluble oligo(phenylenevinylene)s with electron withdrawing substituents for the use in light emitting diodes. Synthetic Metals. 2000;115(1–3):89-92.

[21] Zhang W, Zhu L, Qin J, Yang C. Novel water-soluble red-emitting poly(p-phenylene-vinylene) derivative: synthesis, characterization, and fluorescent acetylcholinesterase assays. The Journal of Physical Chemistry B. 2011;115(42):12059-64.

[22] Beletskaya IP, Cheprakov AV. The Heck reaction as a sharpening stone of palladium catalysis. Chemical Reviews. 2000;100(8):3009-66.

[23] Pasco ST, Lahti PM, Karasz FE. Synthesis of substituted poly(p-phenylenevinylene) copolymers by the Heck method for luminescence studies. Macromolecules. 1999;32(21):6933-7.

[24] Ruan J, Xiao J. From $\alpha$-arylation of olefins to acylation with aldehydes: a journey in regiocontrol of the Heck reaction. Accounts of Chemical Research. 2011;44(8):614-26.

[25] Cárdenas JC, Fadini L, Sierra CA. Triphenylphosphite and ionic liquids: positive effects in the Heck cross-coupling reaction. Tetrahedron Letters. 2010;51(52):6867-70.

[26] Carvajal TR, Kuebler SM, Sierra CA. Synthesis of novel phenylenevinylene linkers with electron-donating substituents by the Heck reaction. Synthetic Metals. 2015;209:183-7.

[27] Díaz C, Alzate D, Rodríguez R, Ochoa C, Sierra CA. High yield and stereospecific synthesis of segmented poly (p-phenylene vinylene) by the Heck reaction. Synthetic Metals. 2013;172:32-6.

[28] Alzate D, Hinestroza JP, Sierra CA. High-yield synthesis of the novel E,E-2,5-dime-thoxy-1,4-bis[2-(4-ethylcarboxylatestyril)] benzene by the Heck reaction. Synthetic Communications. 2013;43(17):2280-5.

[29] Sierra CA, Lahti PM. A photoluminescent, segmented oligo-polyphenylenevinylene copolymer with hydrogen-bonding pendant chains. Chemistry of Materials. 2004;16(1): 55-61.

[30] Cárdenas JC, Ochoa-Puentes C, Gutiérrez-Puebla E, Sierra CA. Synthesis, crystal structure determination and photoluminescence properties of a pure anti trans-trans phenylenevinylene derivative. Synthetic Metals. 2016 5//;215:194-9.

[31] Estrada-Florez S, Ochoa-Puentes C, Sierra CA. Synthesis, structural characterization and evaluation of the optical properties of phenylenevinylene oligomers obtained by Mizoroki-Heck cross coupling reaction. Journal of Luminscence. 2016_407, submitted.

[32] Castro B. Synthesis of phenylenevinylene systems with potential to donate or accept protons and their possible application as MALDI matrices [M.Sc. thesis]. Bogotá D.C.:

Universidad Nacional de Colombia; 2014. Paper in preparation. M.Sc. thesis available in spanish at www.unal.edu.co

[33] Cely M. Synthesis of phenylenevinylene compounds as possible electronic transfer MALDI matrices [M.Sc. thesis]. Bogotá D.C.: Universidad Nacional de Colombia; 2015. Paper in preparation.

[34] Castellanos L. Development of a MALDI matrix for mass spectrometry based on phenylenenvinilene systems and acrylic acids [M.Sc. thesis]. Bucaramanga: Universidad Industrial de Santander; 2015. Paper in preparation.

[35] Rios Carvajal T. Synthesis and characterization of metal-organic frameworks (MOF) with phenylenevinylene linkers modified with electrodonating groups [M.Sc. thesis]. Bogotá D.C.: Universidad Nacional de Colombia; 2014. Paper in preparation.

[36] Cárdenas JC. Synthesys, characterization and evaluation of metal-organic frameworks (MOFs) as possible methane sensors [Ph.D. thesis]. Bogotá D.C.: Universidad Nacional de Colombia; 2016. Paper in preparation.

[37] Van Hutten PF, Krasnikov VV, Hadziioannou G. A model oligomer approach to light-emitting semiconducting polymers. Accounts of Chemical Research. 1999;32(3):257-65.

Chapter 11

# Intercalation of Poly(bis-(methoxyethoxyethoxy)phosphazene) into Lithium Hectorite

Iskandar Saada, Rabin Bissessur,
Douglas C. Dahn and Matthieu Hughes

Additional information is available at the end of the chapter

http://dx.doi.org/10.5772/64580

**Abstract**

Poly(bis-(methoxyethoxyethoxy)phosphazene) (MEEP) intercalated into lithium hectorite was investigated for its potential application as a solid polymer electrolyte in lithium-ion polymer batteries. Varying amounts of MEEP were intercalated into lithium hectorite, and the physical properties of the nanocomposites were monitored using powder X-ray diffraction, thermogravimetric analysis, differential scanning calorimetry, and attenuated total reflectance spectroscopy. Alternating current (AC) impedance spectroscopy was used to determine the ionic conductivity of the nanocomposites when complexed with lithium triflate salt.

**Keywords:** lithium hectorite, poly(bis-(methoxyethoxyethoxy)phosphazene), nanocomposites, solid polymer electrolytes

## 1. Introduction

The electrolyte in a lithium-ion battery is the medium through which lithium ions flow between anode and cathode. It is electrically insulating, and thus prohibits the passage of electrons. Much work has been done to develop solid-state electrolyte materials such as solid polymer electrolytes (SPEs) which have the advantage of enhanced safety compared to conventional liquid organic electrolytes. Some of the most recent research on electrolyte materials has focused on utilizing polymers such as poly(ethylene oxide) (PEO) [1] and polyphosphazenes [2, 3]. Current research on polyphosphazenes focuses on exploiting their fire-resistant properties [4], ionic conductivity [5], and as phosphazene-based dye-sensitized solar cells [6]. Polyphosphazenes are amorphous polymers with low glass transition temperatures ($T_g$), which dictates their polymer-chain flexibility and aids in lithium-ion mobility [7].

Since the first report on poly(bis-(methoxyethoxyethoxy)phosphazene) (MEEP), **Figure 1**, an array of phosphazene-based polymers have been synthesized with diverse alkyl ether and alkoxy side groups, ultimately yielding ionically conductive polymers that are flexible. When polyphosphazenes such as MEEP are complexed with lithium salts (e.g., lithium triflate), they have been shown to possess enhanced ionic conductivity compared to PEO-based SPEs [8, 9]. Previous studies have reported that $(MEEP)_4LiCF_3SO_3$ yields ionic conductivity in the range of $2 \times 10^{-5}$ to $1 \times 10^{-4}$ S/cm at ambient temperatures [10, 11]. However, the dimensional stability of MEEP is low, so it leaks out of cells at ambient temperatures. Researchers have attempted to enhance the dimensional stability of polyphosphazenes by investigating polymer blends with PEO [12], inducing cross-linking via $^{60}$Co-gamma irradiation [8] and preparing of polyphosphazene-silicate networks [13]. One other promising approach is the intercalation of polyphosphazenes into layered structures, which act as hosts for the polymers. Due to the weak electrostatic interactions holding the layers of two-dimensional structures, the layers may be exfoliated allowing for the intercalation of ionically conductive polymers. This yields nanocomposite materials with ionically conductive properties, along with enhanced mechanical and thermal durability provided by the layered structure. Intercalation of MEEP into two-dimensional layered structures has been investigated using layered structures such as graphite oxide [10], molybdenum disulfide [14], sodium montmorillonite [15], and sodium hectorite [16], and has typically yielded ionically conductive nanocomposite materials with enhanced physical properties. In this chapter, we report on the intercalation of various amounts of MEEP into lithium hectorite.

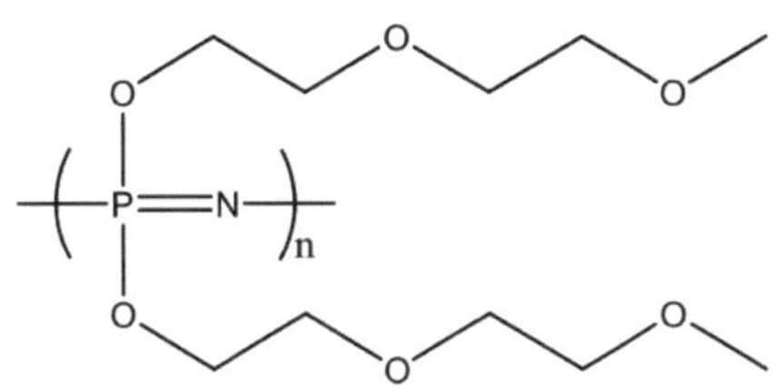

**Figure 1.** Structure of MEEP.

Hectorite belongs to the family of smectite clays, and has a 2:1 structural arrangement composed of two tetrahedral silicate layers ($T_d$) encompassing an octahedral layer ($O_h$), as shown in **Figure 2** [17]. Their industrial applications range from pharmaceutical drug additives [18] to automobile parts [19], and potentially as SPEs in batteries [20, 21]. Hectorite has negatively charged layers, which are compensated with cations such as $Na^+$ and $Li^+$ to balance the overall charge. Hectorite is an appealing layered structure for the intercalation of MEEP due to its high thermal stability, high surface area, exfoliating/restacking capability, and high cation exchange capacity [22]. In this chapter, we exploit the cation exchange capability of hectorite by driving out the naturally lying sodium ions with lithium ions, and working with the lithiated form of hectorite [23]. The intercalation of varying stoichiometric molar ratios of MEEP was conducted with the lithiated form of hectorite.

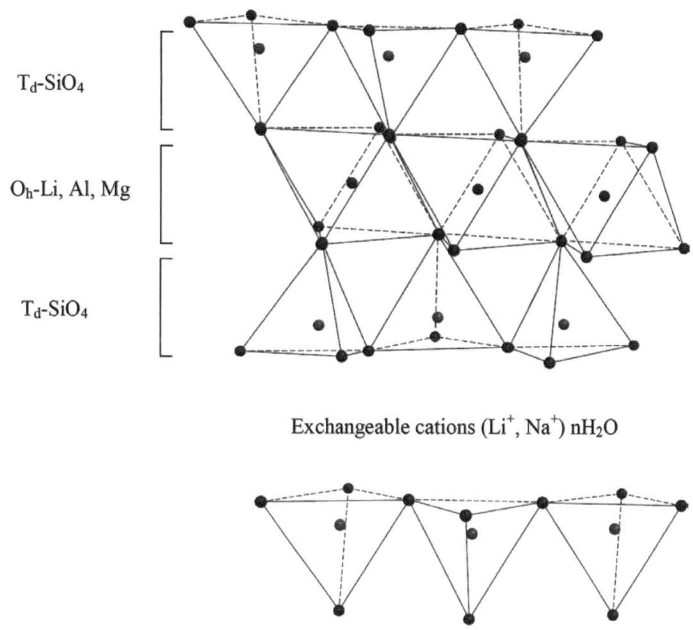


**Figure 2.** Structure of hectorite.

The aim of this chapter is to present the effects that varying polymeric molar ratios to Li-hectorite have on the physical properties of the synthesized nanocomposites, and to investigate the ionic conductivity of the salt-complexed nanocomposites. The synthesized nanocomposites were characterized using thermogravimetric analysis (TGA), differential scanning calorimetry (DSC), powder X-ray diffraction (XRD), and attenuated total reflectance (ATR). The ionic resistance of the materials was determined using AC impedance spectroscopy.

## 2. Experimental

### 2.1. Purification and lithiation of hectorite

Sodium hectorite (SHCa-1) was purchased from Source Clays Repository. Since the fine powder has calcium carbonate and other impurities, a purification process was performed as outlined in the literature [16]. When the purification process was complete, a cation exchange was done in order to replace the sodium ions with lithium ions [23]. This process was carried out twice to ensure maximum substitution of lithium ions. Elemental analysis was used to monitor the sodium- and lithium-ion content at Guelph Chemical Laboratories Ltd. Ontario, Canada. The data indicated an increase in lithium-ion proportion from $Li_{0.5}$ $Na_{0.8}$ $Si_1$ (sodium hectorite) to $Li_3$ $Na_{0.4}$ $Si_1$ (lithium hectorite).

### 2.2. Synthesis of MEEP and MEEP-salt complex

The synthesis of MEEP was performed as described in the literature [24]. $(MEEP)_4\ LiCF_3SO_3$ has been found to exhibit high ionic conductivity compared to other polymer:salt ratios [13, 25], so this polymer:salt ratio was used in this work. The synthesized polymer was stored in a vacuum desiccator, and salt-polymer complexes were used as soon as they were prepared in order to minimize exposure to humidity. Hereafter, the pure (uncomplexed) polymer will be referred to as MEEP, while the salt-complexed polymer will be referred to as Li-MEEP.

### 2.3. Preparation of nanocomposites

A general procedure was employed for the intercalation of MEEP into Li-hectorite. Li-hectorite (0.10 g, $2.6 \times 10^{-4}$ mol) was suspended in deionized water and left to stir until fully suspended in water (typically 30 min). The polymer with molar ratio of 0.5, 1, 2, or 4 to Li-hectorite was dissolved in 5 mL of deionized water. A pipette was used to transfer the polymer solution to the Li-hectorite suspension at a rate of one drop per second. The progress of the reactions was monitored via XRD. The products were then isolated via freeze drying, and stored in a vacuum desiccator.

### 2.4. Materials characterization

Powder X-ray diffraction was conducted on a Bruker AXS D8 Advance diffractometer. The instrument is equipped with a graphite monochromator, variable divergence slit, variable antiscatter slit, and a scintillation detector. Cu ($k\alpha$) radiation ($\lambda = 1.542$ Å) was utilized and the data were collected at room temperature on glass substrates.

Thermogravimetric analysis (TGA) was performed on a TA Q500 using a heating rate of 10°C/min, with the use of platinum pans under dry-compressed air. Samples were freeze dried prior to TGA analysis in order to minimize their moisture content.

Differential scanning calorimetry (DSC) was performed on a TA Q100 using heat/cool/heat cycles. Samples were crimped in aluminum pans, and ran under nitrogen flow at a rate of 50 mL/min.

Attenuated total reflectance spectroscopy (ATR) data were collected using a Bruker Alpha A-T (resolution 0.9, 128 scans).

AC impedance spectroscopy (IS) was conducted to determine the ionic conductivity of the salt-complexed materials [26]. The samples were tested using rectangular glass substrates with two rectangular stainless steel electrodes on the opposite ends of the substrates. For the intercalated nanocomposites, samples were cast onto the substrates, between the electrodes, after 3 days of reaction time in order to ensure complete intercalation. Typical samples had a width of about 9 mm, were 20–90-μm thick, and had a length between the electrodes in the direction of current of about 6 mm. In most cases, the films were not uniform in thickness, and this was the main source of uncertainty in the ionic conductivity values that were obtained. The frequency range used was 10 kHz to 0.01 Hz. In order to remove moisture, the samples were placed under vacuum for at least 24 h at room temperature prior to the data collection. The temperature of the samples was controlled using a Cryodyne 350CP refrigerator and a Lakeshore 321 temperature controller. The data collection was performed using a Solartron 1250 frequency response analyzer and a home-built accessory circuit for high-impedance samples. The conductivity was then determined by fitting the IS data to an equivalent circuit model using LEVMW [27].

# 3. Results and discussion

## 3.1. MEEP/Lithium hectorite

### *3.1.1. Powder X-ray diffraction*

Powder X-ray diffraction was used to monitor the intercalation process and the effect of the molar ratio of MEEP to Li-hectorite on the polymer loading in the layered structure.

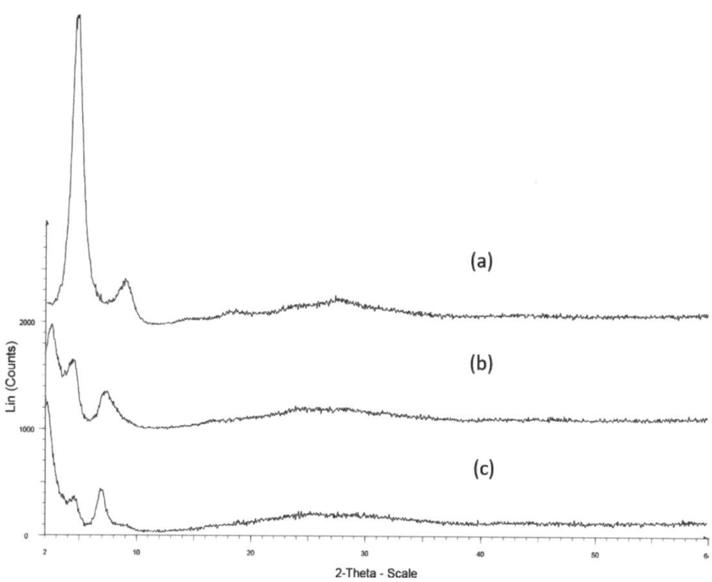


**Figure 3.** XRD data for (a) MEEP:Li-hectorite (0.5:1), (b) MEEP:Li-hectorite (2:1), and (c) MEEP:Li-hectorite (4:1).

The diffractograms for MEEP:Li-hectorite (0.5:1), MEEP:Li-hectorite (2:1), and MEEP:Li-hectorite (4:1) nanocomposites are displayed in **Figure 3** to illustrate the enhancement in basal spacing (d-spacing) of the layered host upon the intercalation of MEEP. The net interlayer expansion is obtained by subtracting the basal spacing of dry Li-hectorite heated to 650°C (d-spacing = 9.5 Å) from the basal spacing of the synthesized nanocomposite. For example, MEEP:Li-hectorite (0.5:1) nanocomposite has a basal spacing of 18.9 Å, which corresponds to an interlayer expansion of 9.40 Å. The XRD data for all the nanocomposites are summarized in **Table 1**.

| Material | Basal spacing (Å) | Net expansion (Å) | Average crystallite size (Å) |
|---|---|---|---|
| MEEP:Li-hectorite (0.5:1) | 18.9 | 9.40 | 73 |
| MEEP:Li-hectorite (1:1) | 21.7 | 12.2 | 74 |
| MEEP:Li-hectorite (2:1) | 36.4 | 26.9 | 126 |
| MEEP:Li-hectorite (4:1) | 41.5 | 32.0 | 135 |
| Na-hectorite | 10.0 | – | 140 |
| Dry Li- hectorite | 9.50 | – | 182 |

**Table 1.** Summary XRD data of hectorites and MEEP:Li-hectorite nanocomposites.

The synthesized nanocomposites are crystalline as indicated by XRD, and a significant increase in basal spacing is observed as the ratio of MEEP to Li-hectorite is increased. From the XRD diffractograms of the nanocomposites, the average crystallite size was determined using the Scherrer formula [28]. The crystallite size appeared to increase upon increasing the MEEP molar ratio to Li-hectorite, which is possibly due to the significant enhancement in basal spacing upon loading of the polymer into the layered structure.

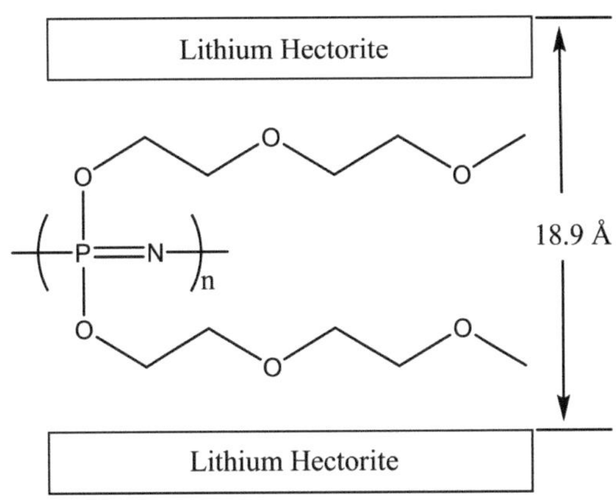


**Figure 4.** Schematic arrangement of MEEP in lithium hectorite (0.5:1).

The dimensions of MEEP were estimated using Spartan '08. The average dimension was determined for the largest possible distance between the ether oxygens on the R-groups in MEEP, and was found to be approximately 7.9 Å for one unit of MEEP [29]. This indicates that a single layer of MEEP is inserted between the Li-hectorite sheets for the MEEP:Li-hectorite (0.5:1) nanocomposite, which had a net layer expansion of 9.40 Å. The difference of 1.5 Å between the observed basal spacing of the 0.5:1 nanocomposite and the calculated dimensions of MEEP could be due to the manner in which MEEP is oriented within the layers. MEEP is a highly flexible polymer and may not necessarily be oriented within the layers as depicted in **Figure 4**. For example, a helical conformation may also be possible.

### *3.1.2. Thermogravimetric analysis*

Thermogravimetric analysis was used to compare the thermal stability of MEEP, MEEP:Li-hectorite, Li-MEEP, and Li-MEEP:Li-hectorite. This was completed in order to monitor the thermal stability of nanocomposites with different polymer concentrations and to compare the stability of salt-complexed materials with their uncomplexed counterparts. The MEEP:Li-hectorite thermograms were also used to calculate the stoichiometry of the synthesized nanocomposites.

The thermogram of pristine MEEP (**Figure 5(c)**) shows that it undergoes a major decomposition between 230 and 350°C, followed by an onward gradual weight loss. Once MEEP is complexed with lithium triflate ($LiCF_3SO_3$), its decomposition is slightly compromised due to the presence of the inorganic salt, which appeared to decrease the polymer onset decomposition temperature by approximately 30°C; thereafter, complete decomposition of the triflate salt is observed at 420°C (**Figure 5(d)**).

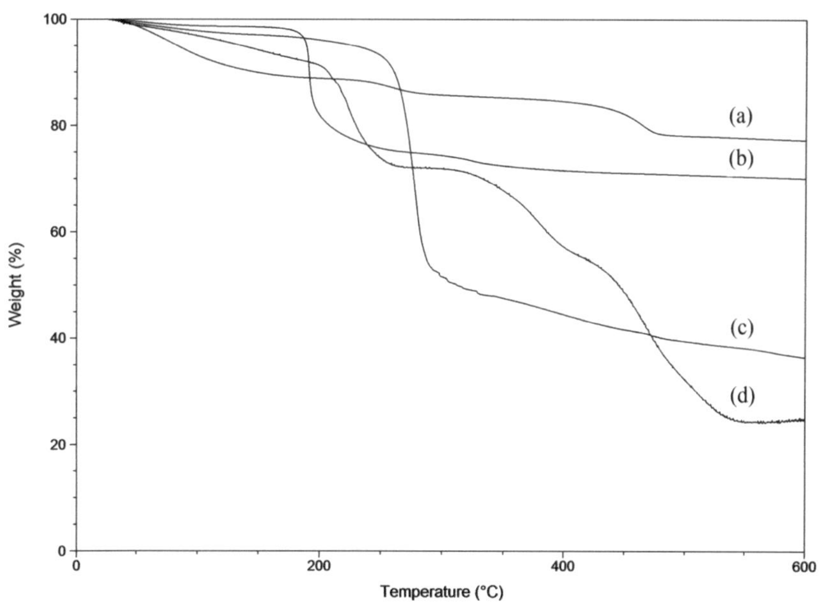


**Figure 5.** TGA data for (a) Li-MEEP:Li-hectorite (1:1), (b) MEEP:Li-hectorite(1:1), (c) MEEP, and (d) Li-MEEP.

Upon polymer intercalation, the nanocomposite thermograms appeared to have three weight loss steps, where the thermogram of, for example, MEEP:Li-hectorite (1:1) (**Figure 5(b)**) illustrates a small loss of water near 100°C, followed by the decomposition at 178°C corresponding to the presence of externally lying MEEP. The final decomposition is the gradual decomposition of the intercalated MEEP occurring at 305°C. This is quite similar to the thermal behavior of Li-MEEP:Li-hectorite (1:1) (**Figure 5(a)**), except for the decomposition of the salt which occurs at around 420°C. The TGA data are summarized in **Table 2**.

| Material | $T_a$ (°C) | $T_b$ (°C) | $\Delta T_c$ (°C) |
|---|---|---|---|
| MEEP | 240 | N/A | N/A |
| Li-MEEP | 204 | N/A | N/A |
| MEEP:Li-hectorite (0.5:1) | 205 | 312 | 72 |
| MEEP:Li-hectorite (1:1) | 178 | 305 | 65 |
| MEEP:Li-hectorite (2:1) | 176 | 315 | 75 |
| MEEP:Li-hectorite (4:1) | 170 | 324 | 84 |

$T_a$, onset decomposition temperature of pure MEEP or externally lying MEEP in nanocomposite; $T_b$, onset decomposition temperature of intercalated MEEP in nanocomposite; and $\Delta T_c$, difference in decomposition temperature of the intercalated MEEP and pristine MEEP.

**Table 2.** Thermogravimetric data.

As shown in **Table 2**, the onset decomposition temperatures of the externally lying MEEP ($T_a$) in the nanocomposites were lower than that of the bulk polymer, which occurs at 240°C. However, the onset decomposition temperatures of the intercalated MEEP ($T_b$) in the nanocomposites are significantly higher than that of the pure polymer, indicating enhancement in the thermal stability of polymer when sandwiched between the layers of hectorite.

| MEEP:Li-hectorite (mol ratio) | Stoichiometry |
|---|---|
| 0.5:1 | $(H_2O)_{0.61}$ $(MEEP_{Ext})_{0.17}$ $(MEEP_{In})_{0.023}$(Li-hectorite) |
| 1:1 | $(H_2O)_{0.42}$ $(MEEP_{Ext})_{0.44}$ $(MEEP_{In})_{0.073}$(Li-hectorite) |
| 2:1 | $(H_2O)_{0.48}$ $(MEEP_{Ext})_{0.71}$$(MEEP_{In})_{0.12}$(Li-hectorite) |
| 4:1 | $(H_2O)_{0.34}$ $(MEEP_{Ext})_{0.84}$ $(MEEP_{In})_{0.14}$(Li-hectorite) |

**Table 3.** Stoichiometry of MEEP:Li-hectorite nanocomposites.

Since the nanocomposites displayed three decomposition steps, the stoichiometry was calculated in order to compare the spread between externally lying and intercalated polymer. It is important to note that the stoichiometry was calculated only for the MEEP:Li-hectorite nanocomposites, and not for the Li-MEEP:Li-hectorite nanocomposites due to the presence of lithium salt. The stoichiometry data of the nanocomposites are displayed in **Table 3**.

As shown in **Table 3**, when the molar ratio of MEEP is increased with respect to Li-hectorite, there is an increase in the amount of the externally lying polymer and intercalated polymer. These observations are in agreement with the XRD data (**Table 1**), which indicate that the basal spacing of the intercalated nanocomposite increases when the amount of MEEP is increased. In fact, MEEP:Li-hectorite (4:1) has almost twice as much of the intercalated polymer compared to MEEP:Li-hectorite (1:1).

http://dx.doi.org/10.5772/64580

### 3.1.3. *Differential scanning calorimetry*

Differential scanning calorimetry was used to monitor the glass transition temperature ($T_g$) of pristine MEEP, Li-MEEP, and their corresponding synthesized nanocomposites (**Figure 6**).

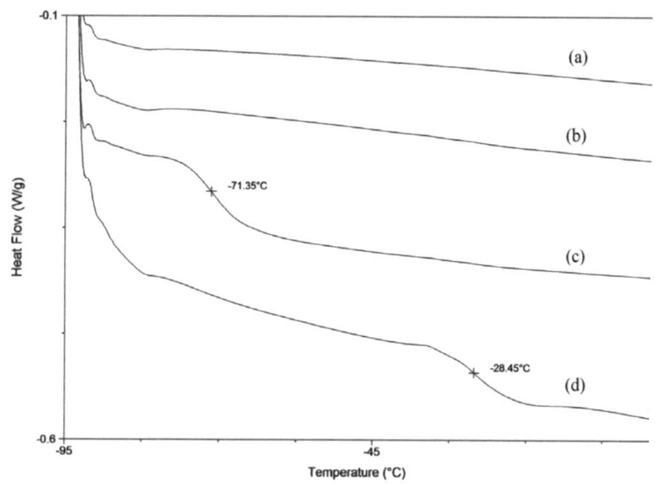


**Figure 6.** DSC data for (a) MEEP:Li-hectorite (1:1), (b) Li-MEEP:Li-hectorite (1:1), (c) MEEP, and (d) Li-MEEP.

| Material | ($T_g$, °C) |
|---|---|
| MEEP | –71 |
| Li-MEEP | –28 |
| MEEP:Li-hectorite (0.5:1) | – |
| MEEP:Li-hectorite (1:1) | – |
| MEEP:Li-hectorite (2:1) | – |
| MEEP:Li-hectorite (4:1) | – |

**Table 4.** Summary of DSC data.

The DSC of pure MEEP indicates a $T_g$ of –71°C, which is in fairly good agreement with the previously reported literature value of –83°C [30]. Upon complexing MEEP with lithium triflate ($LiCF_3SO_3$), the glass transition temperature ($T_g$) significantly increases to –28°C due to the crystalline nature of lithium triflate. However, upon intercalation of pristine MEEP or Li-MEEP into Li-hectorite, a $T_g$ is not observed for the ratios used. Due to the lack of glass transition temperatures in both Li-MEEP:Li-hectorite and MEEP:Li-hectorite nanocomposites, it is believed that (1) the oxygen atoms of the polymer are potentially interacting with the tetrahedrally coordinated silicon atoms in the hectorite sheets and restricting chain mobility or (2) the polymer is no longer flexible when it is intercalated in the layers of hectorite. The glass transition temperatures are displayed in **Table 4**.

### *3.1.4. Attenuated total reflectance*

ATR spectroscopy was used to monitor the bond vibrations in MEEP, MEEP:Li-hectorite, Li-MEEP, and Li-MEEP:Li-hectorite. More specifically, it was important to determine whether intercalating MEEP or Li-MEEP into Li-hectorite hinders the flexibility of the polymer, and ultimately its ionic conductivity (**Figure 7**).

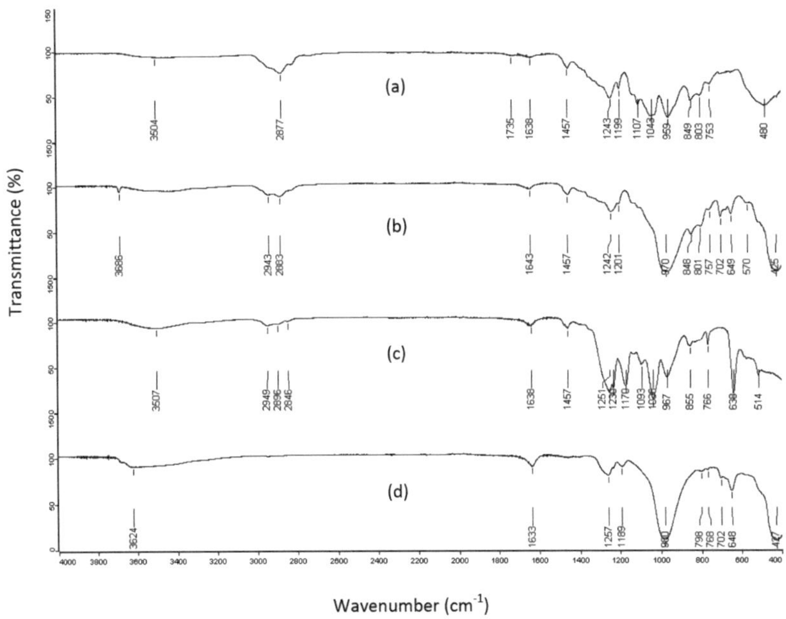


**Figure 7.** ATR results for (a) MEEP, (b) MEEP:Li-hectorite (1:1), (c) Li-MEEP, and (d) Li-MEEP:Li-hectorite(1:1).

| **Major vibrations ($cm^{-1}$)** | **MEEP** | **MEEP:Li-hectorite(1:1)** | **Li-MEEP** | **Li-MEEP:Li-hectorite(1:1)** |
|---|---|---|---|---|
| HOH stretch | N/A | 3686 | N/A | 3624 |
| $Sp^3$ C-H stretch/bend | 2877/1457 | 2883/1457 | 2896/1457 | 2887/1457 |
| P=N | 1243 | 1242 | 1251 | 1257 |
| P-O-C | 959 | 967 | 980 | 972 |
| C-O-C ether | 1199–1043 | 1201 | 1170–1036 | 1189 |
| PNP skeletal | 849/803/753 | 848/801/757 | 855/766 | 798/768 |
| $CF_3$ | N/A | N/A | 1249 | 1253 |
| C-F deformation | N/A | N/A | 638 | 638 |
| $SO_3$ asymmetric bend | N/A | N/A | 573 | 573 |
| $SO_3$ symmetric bend | N/A | N/A | 518 | 518 |

**Table 5.** Summary of IR data.

From the IR data, it is observed that the P = N vibration in the pure polymer occurs at 1243 $cm^{-1}$. Upon complexation with lithium triflate, the P = N vibration of the polymer shifts to 1251 $cm^{-1}$, and to 1257 $cm^{-1}$ in the synthesized nanocomposites. This increase in vibrational energy of the P = N bond is indicative of the increased rigidity of the polymer backbone. The P-O-C vibration in MEEP shifts from 959 to 967 $cm^{-1}$ upon intercalation into Li-hectorite, indicating increased rigidity of the polymer side chains. Thus, the IR data support lack of polymer flexibility upon intercalation, and are in very good agreement with the DSC results. The IR data are summarized in **Table 5** [8].

### *3.1.5. AC impedance spectroscopy*

Impedance measurements were conducted on Li-MEEP and the Li-MEEP:Li-hectorite nanocomposites. Since the DSC and IR data indicated that the polymer-chain flexibility was restricted in the nanocomposites, it was necessary to investigate the ionic conductivity properties of Li-MEEP prior to intercalation, and post-intercalation. A complex plane plot of the impedance of an Li-MEEP sample is given in **Figure 8**. High-frequency data (10 kHz) is near the origin, and low-frequency (0.01 Hz) at the upper right.

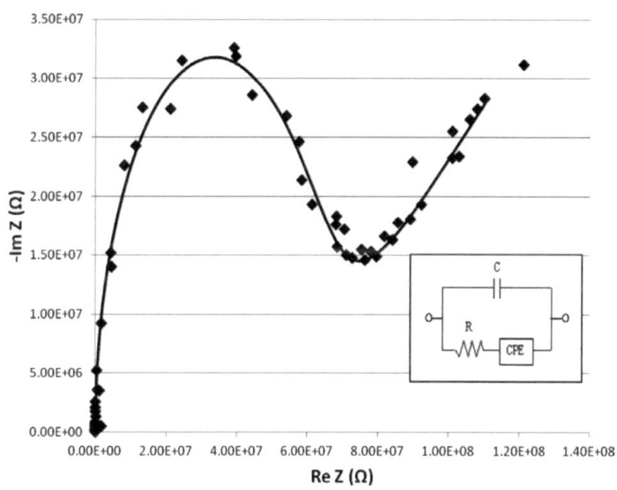


**Figure 8.** Complex-plane plot of Li-MEEP impedance at 300 K.

As shown in **Figure 8**, Li-MEEP demonstrates a curve typical for an ionic conductor. The value of Re(Z) at which Im(Z) goes through a minimum (about $7 \times 10^7$ Ω in **Figure 8**) corresponds approximately to the resistance (*R*) of the sample. The value of *R* and the dimensions of the polymeric film were used to calculate the ionic conductivity of the sample. In order to obtain more accurate values of *R* from the impedance data, a complex nonlinear least-squares fit was done with an equivalent circuit model using the program LEVMW [27]. A three-component equivalent circuit consisting of a resistor (R), a constant-phase element (CPE) that models the effects of the blocking electrodes, and a parallel capacitor (C) was used and is shown in the inset of **Figure 8**. Using the resistance value *R* and the dimensions of the polymeric film, the ionic conductivity of Li-MEEP was calculated.

Ionic conductivity measurements were performed within the temperature range of 220–310 K, and the results for a typical Li-MEEP sample are displayed in **Figure 9**. As is typical for ionically conducting polymers, the conductivity drops rapidly as the temperature is reduced, and for this sample it was too small to be measured with our system below 260 K.

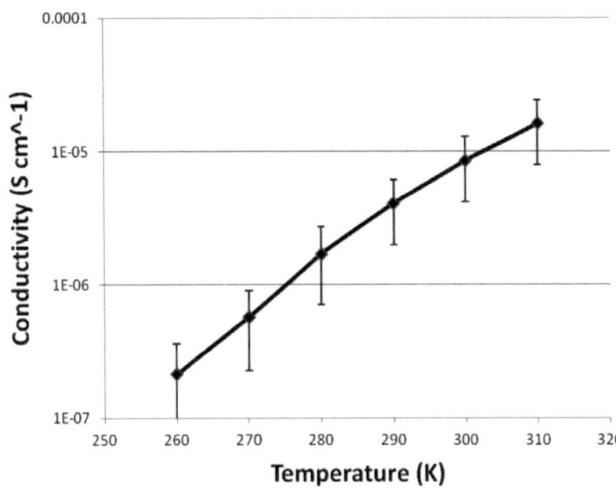


**Figure 9.** Variable-temperature ionic conductivity of Li-MEEP.

The room temperature ionic conductivity of Li-MEEP samples ranged from about $4 \times 10^{-6}$ to $1.3 \times 10^{-5}$ S/cm, which is in agreement with the literature [11, 12]. However, upon intercalation of Li-MEEP into Li-hectorite, the conductivity of the resulting nanocomposite material was below our detection limit, which in the case of these cast film samples is about $10^{-7}$ S/cm. These observations further indicate that there is an interaction occurring between the polymer and the layered structure that is ultimately inhibiting the nanocomposites from conducting lithium ions.

## 4. Conclusions

MEEP has been intercalated into Li-hectorite. A series of nanocomposites have been created by varying the mole ratio of the polymer with respect to the clay. The nanocomposites were characterized by TGA, DSC, FTIR, XRD, and AC impedance spectroscopy. XRD data confirm the successful and complete intercalation of MEEP into the layered silicate. TGA data confirm that increasing the molar ratio of MEEP with respect to the Li-hectorite results in a larger amount of the polymer within the layers. In fact, MEEP:Li-hectorite (4:1) has approximately twice the amount of intercalated polymer versus MEEP:Li-hectorite (1:1). TGA also indicates an enhancement in thermal stability of the intercalated polymer versus the pristine polymer for all nanocomposites. For instance, the intercalated polymer in MEEP:Li-hectorite (1:1) is thermally stable up to 305°C, while the pure polymer decomposes at 240°C. The room temperature ionic conductivity of Li-MEEP was determined to be about 1 × $10^{-5}$ S/cm; however, the polymer-salt complex displayed high ionic resistance when intercalated into Li-hectorite, and hence the ionic conductivity of the nanocomposites was too small to be determined (below about $10^{-7}$ S/cm). The impedance data are in good agreement with the DSC results, where no glass transition temperature has been detected for all the synthesized nanocomposites. ATR spectroscopy also confirms that the rigidity of the polymer backbone increases upon complexation with lithium triflate, and when the lithium complexed MEEP is intercalated into Li-hectorite.

## Acknowledgements

The authors are grateful for the financial support from the Natural Sciences and Engineering Research Council (NSERC) of Canada, Canada Foundation for Innovation (CFI), Atlantic Innovation Fund of Canada (AIF), and UPEI.

## Author details

Iskandar Saada[1], Rabin Bissessur[1*], Douglas C. Dahn[2] and Matthieu Hughes[2]

*Address all correspondence to: rabissessur@upei.ca

1 Chemistry Department, University of Prince Edward Island, Charlottetown, PE, Canada

2 Physics Department, University of Prince Edward Island, Charlottetown, PE, Canada

## References

[1] Johan MR, Shy OH, Ibrahim S, Yassin SSM, Hui TY. Effects of $Al_2O_3$ nanofiller and EC plasticizer on the ionic conductivity enhancement of solid PEO-$LiCF_3SO_3$ solid polymer electrolyte. Solid State Ionics. 2011;196:41–47. DOI:10.1016/j.ssi.2011.06.001

[2] Burjanadze M, Karatas Y, Kaskhedikar N, Kogel LM, Kloss S, Gentschev A-C, Hiller MM, Muller RA, Stolina R, Vettikuzha P, Weimhöfer H-D. Salt-in-polymer electrolytes for lithium ion batteries based on organo-functionalized polyphosphazenes and polysiloxanes. Z. Phys. Chem. 2010;224:1439–1473. DOI:10.1524/zpch.2010.0046

[3] Allcock HR, Kellam III EC. The synthesis and applications of novel aryloxy/oligoethyleneoxy substituted polyphosphazenes as solid polymer electrolytes. Solid State Ionics. 2003;156:401–414.

[4] Allcock HR, Taylor JP. Phosphorylation of phosphazenes and its effects on thermal properties and fire retardant behaviour. Polymer Eng. Sci. 2010;40:1177–1189.

[5] Fei S-T, Allcock HR. Methoxyethoxyethoxyphosphazenes as ionic conductive fire retardant additives for lithium battery systems. J. Power Sources. 2010;195:2082–2088. DOI:10.1016/j.jpowsour.2009.09.043

[6] Fei S-T, Lee S-HA, Pursel SM, Basham J, Hess A, Grimes CA, Horn MW, Mallouk TE, Allcock HR. Electrolyte infiltration in phosphazene-based dye-sensitized solar cells. J. Power Sources. 2011;196:5223–5230.

[7] Allcock HR, O'Connor SJM, Olmeijer DL, Napierala ME, Cameron CG. Polyphosphazenes bearing branched and linear oligoethyleneoxy side groups as solid solvents for ionic conduction. Macromolecules. 1996;29:7544–7552.

[8] Nazri G, MacArthur DM, Ogara JF. Polyphosphazene electrolytes for lithium batteries. Chem. Mater. 1989;1:370–374. DOI:10.1021/cm00003a019

[9] Abraham KM, Alamgir M. Dimensionally stable MEEP-based polymer electrolytes and solid-state lithium batteries. Chem. Mater. 1991;3:339–348. DOI:10.1021/cm00014a027

[10] Scully SF, Bissessur R, MacLean KW, Dahn DC. Inclusion of poly[bis(methoxyethoxyethoxy)phosphazene] into layered graphite oxide. Solid State Ionics. 2009;180:216–221.

[11] Blonsky PM, Shriver DF, Austin P, Allcock HR. Polyphosphazene solid electrolytes. J. Am. Chem. Soc. 1984;106:6854–6855. DOI:10.1021/ja00334a071

[12] Silva VPR, Silva GG, Caliman V, Rieumont J, de Miranda-Pinto COB, Archanjo BS, Neves BRA. Morphology, crystalline structure and thermal properties of PEO/MEEP blends. Eur. Polymer J. 2007;43:3283–3291.

[13] Allcock HR, Chang Y, Welna DT. Ionic conductivity of covalently interconnected polyphosphazene–silicate hybrid networks. Solid State Ionics. 2006;177:569–572. DOI: 10.1016/j.ssi.2005.11.017

[14] Bissessur R, Gallant D, Brüning R. New poly[bis-(methoxyethoxyethoxy)phosphazene]–$MoS_2$ nanocomposite. Solid State Ionics. 2003;158:205–209. DOI:10.1016/S0167-2738(02)00774-9

[15] Hutchison JC, Bissessur R, Shriver DF. Enhancement of ion mobility in aluminosilicate-polyphosphazene nanocomposites. Mat. Res. Soc. Symp. Proc. 1996;457:489–494.

[16] Scully SF, Bissessur R. Encapsulation of polymer electrolytes into hectorite. Appl. Clay. Sci. 2010;47:444–447. DOI:10.1016/j.clay.2009.12.023

[17] Madejová J, Bujdák J, Janek M, Komadel P. Comparative FT-IR study of structural modifications during acid treatment of dioctahedral smectites and hectorite. Spectrochim. Acta. Part A. 1998;54:1397–1406.

[18] Carretero MI, Pozo M. Clay and non-clay minerals in the pharmaceutical industry Part I. Excipients and medical applications. Appl. Clay Sci. 2009;46:73–80. DOI:10.1016/j.clay.2009.07.017

[19] Okada A, Usuki A. The chemistry of polymer-clay hybrids. Mater. Sci. Eng. C. 1995;3:109–115.

[20] Sandí G, Carrado KA, Joachin H, Lu W, Prakash J. Polymer nanocomposites for lithium battery applications. J. Power Sources. 2003;119–121:492–496. DOI:10.1016/S0378-7753(03)00272-6

[21] Riley M, Fedkiw PS, Khan SA. Transport properties of lithium hectorite-based composite electrolytes. J. Electrom. Soc. 2002;149:A667–A674. DOI:10.1149/1.1470652

[22] Carrado KA, Forman JE, Botto RE, Winans RE. Incorporation of phthalocyanines by cationic and anionic clays via ion exchange and direct synthesis. Chem. Mater. 1993;5:472–478.

[23] Singhal RG, Capracotta MD, Martin JD, Khan SA, Fedkiw PS. Transport properties of hectorite based nanocomposite single ion conductors. J. Power Sources. 2004;128:247–255.

[24] Allcock HR, Austin PE, Neenan TX, Sisko JT, Blonsky PM, Shriver DF. Polyphosphazenes with etheric side groups: prospective biomedical and solid electrolyte polymers. Macromolecules. 1986;19:1508–1512.

[25] Blonsky PM, Shriver DF, Austin P, Allcock HR. Complex formation and ionic conductivity of polyphosphazene solid electrolytes. Solid State Ionics. 1986;18–19:258–264.

[26] Barsoukov E, Macdondald JR, editors. Impedance Spectroscopy Theory, Experiments, and Applications, 2nd ed. Hoboken, New Jersey, Wiley; 2005.

[27] Macdonald JR. 2015. Available at: jrossmacdonald.com/levmlevmw [accessed 25 June 2015].

[28] Monshi A, Foroughi MR, Monshi MR. Modified Scherrer equation to estimate more accurately nano-crystalline size using XRD. World J. Nano Sci. Eng. 2012;2:154–160.

[29] Spartan '08. Wavefunction Inc., Irvine, CA; 2008.

[30] Allcock HR, Napierala ME, Olmeijer DL, Cameron CG, Kuharcik SE, Reed CS, O'Connor SJM. New macromolecules for solid polymeric electrolytes. Electrochim. Acta. 1998;43:1145–1150.